Diabetes and the Kidney

Contributions to Nephrology

Vol. 170

Series Editor

Claudio Ronco Vicenza

Diabetes and the Kidney

Volume Editors

Kar Neng Lai Hong Kong
Sydney C.W. Tang Hong Kong

35 figures, 5 in color, and 13 tables, 2011

KARGER Basel · Freiburg · Paris · London · New York · New Delhi · Bangkok · Beijing · Tokyo · Kuala Lumpur · Singapore · Sydney

Contributions to Nephrology

(Founded 1975 by Geoffrey M. Berlyne)

Kar Neng Lai
Department of Medicine
University of Hong Kong
Queen Mary Hospital
Hong Kong SAR (China)

Sydney C.W. Tang
Department of Medicine
University of Hong Kong
Queen Mary Hospital
Hong Kong SAR (China)

Library of Congress Cataloging-in-Publication Data

Diabetes and the kidney / volume editors, Kar Neng Lai, Sydney C.W. Tang.
    p. ; cm. -- (Contributions to nephrology, ISSN 0302-5144 ; v. 170)
  Includes bibliographical references and indexes.
  ISBN 978-3-8055-9742-5 (hard cover : alk. paper) -- ISBN 978-3-8055-9743-2
(e-ISBN)
  1. Diabetic nephropathies. I. Lai, Kar Neng. II. Tang, Sydney C. W. III.
Series: Contributions to nephrology ; v. 170. 0302-5144
  [DNLM: 1. Diabetic Nephropathies. W1 CO778UN v.170 2011 / WK 835]
  RC918.D53D54 2011
  616.6'1--dc22

                                                    2011010486

© Copyright 2011 by S. Karger AG, P.O. Box, CH–4009 Basel (Switzerland)
www.karger.com
Printed in Switzerland on acid-free and non-aging paper (ISO 9706) by Reinhardt Druck, Basel
ISSN 0302–5144
ISBN 978–3–8055–9742–5
e-ISBN 978–3–8055–9743–2

Contents

List of Contributors

Charles E. Alpers
Department of Pathology
University of Washington
Seattle, WA 98195-6100 (USA)

Robert C. Atkins
School of Public Health and Preventive
Medicine
Monash University
Melbourne, VIC 3004 (Australia)

Hnin Hnin Aung
Division of Cardiovascular Medicine
University of California Davis
School of Medicine
451 E. Health Sciences Drive
Davis, CA 95616 (USA)

Yolanda E. Bogaert
Department of Medicine
University of Colorado Denver
School of Medicine
Aurora, CO 80045 (USA)

Matthew D. Breyer
Biotechnology Discovery Research
Lilly Research Laboratories
355 E. Merrill St.
Indianapolis, IN 46285 (USA)

Caroline Brorsson
Glostrup Research Institute
Glostrup University Hospital
2600 Glostrup (Denmark)

Alan Cass
Renal & Metabolic Division
The George Institute for Global Health
Sydney Medical School
University of Sydney
Camperdown, NSW 2050 (Australia)

Juliana C.N. Chan
Department of Medicine and Therapeutics
and Hong Kong Institute of Diabetes and
Obesity
The Chinese University of Hong Kong
The Prince of Wales Hospital
Hong Kong SAR (China)

Devasmita Choudhury
Division of Nephrology
Department of Medicine
University of Texas Southwestern
Medical Center Dallas
Dallas, TX 75216 (USA)

Arthur C.K. Chung
Li Ka Shing Institute of Health Sciences
the Chinese University of Hong Kong
Shatin
Hong Kong SAR (China)

Robyn Cunard
Center for Renal Translational Medicine
Division of Nephrology-Hypertension
Department of Medicine
University of California San Diego
San Diego, CA 92093 (USA)

Paola Fioretto
Department of Medical and Surgical
Sciences
University of Padova
Medical School
via Giustiniani 2
35128 Padova (Italy)

Agnes B. Fogo
Department of Pathology
Vanderbilt University Medical Center
Nashville, TN 37232 (USA)

Hunjoo Ha
Department of Bioinspired Science
Division of Life and Pharmaceutical
Sciences , College of Pharmacy
Ewha Womans University
11-1 Daehyun Dong, Seodaemun Ku
Seoul 120-750 (Korea)

Joelle Guitard
Department of Nephrology, Dialysis
and Organ Transplantation
CHU Rangueil
31059 Toulouse (France)

John E. Hall
Department of Physiology & Biophysics
University of Mississippi Medical Center
2500 North State Street
Jackson, MS 39216-4505 (USA)

Richard J. Johnson
Division of Renal Diseases and
Hypertension
University of Colorado Denver
Aurora, CO 80045 (USA)

Min Jun
Renal & Metabolic Division
The George Institute for Global Health
Sydney Medical School
University of Sydney
Camperdown, NSW 2050 (Australia)

Nassim Kamar
Department of Nephrology, Dialysis,
and Organ Transplantation
CHU Rangueil
31059 Toulouse Cedex 9 (France)

Charbel C. Khoury
Feinberg School of Medicine
Northwestern University
Chicago, IL 60611 (USA)

Kar Neng Lai
Department of Medicine
University of Hong Kong
Queen Mary Hospital
Hong Kong SAR (China)

Hui Yao Lan
Department of Medicine & Therapeutics
and Li Ka Shing Institute of Health
Sciences
601 LiHS, Prince of Wales Hospital
the Chinese University of Hong Kong
Shatin
Hong Kong SAR (China)

Joseph C.K. Leung
Department of Medicine
University of Hong Kong
Queen Mary Hospital
Hong Kong SAR (China)

Moshe Levi
Division of Nephrology and Hypertension
Department of Medicine
University of Colorado Denver
12700 East 19th Avenue
Aurora, CO 80045 (USA)

Andrea O.Y. Luk
Department of Medicine and Therapeutics
The Chinese University of Hong Kong
The Prince of Wales Hospital
Hong Kong SAR (China)

Christine Maric
Department of Physiology & Biophysics
University of Mississippi Medical Center
2500 North State Street
Jackson, MS 39216-4505 (USA)

Anna Mathew
Center for Renal Translational Medicine
Division of Nephrology-Hypertension
Department of Medicine
University of California San Diego
San Diego, CA 92093 (USA)

Michael Mauer
Departments of Pediatrics and Medicine
University of Minnesota
Medical School
Minneapolis, MN 55455 (USA)

Carmen Mora-Fernández
Research Unit
University Hospital Nuestra Señora de
Candelaria
38010 Santa Cruz de Tenerife (Spain)

Behzad Najafian
Department of Pathology
University of Washington
Seattle, WA 98195-6100 (USA)

Takahiko Nakagawa
Division of Renal Diseases and
Hypertension
University of Colorado Denver
Aurora, CO 80045 (USA)

Rama Natarajan
Division of Cellular and Molecular
Diabetes Research, Department of
Diabetes
Beckman Research Institute of City of
Hope
Duarte, CA 91010 (USA)

Juan F. Navarro-González
Nephrology Service and Research Unit,
University Hospital Nuestra Señora de
Candelaria
38010 Santa Cruz de Tenerife (Spain)

Kit Fai Ng
Division of Cardiovascular Medicine
University of California Davis School of
Medicine
451 E. Health Sciences Drive
Davis, CA 95616 (USA)

Hyunjin Noh
Hyonam Kidney Laboratory, Division of
Nephrology, Department of Internal
Medicine, Soon Chun Hyang University
Seoul 140–743 (Korea)

Vlado Perkovic
Renal & Metabolic Division
The George Institute for Global Health
Sydney Medical School
University of Sydney
Camperdown, NSW 2050 (Australia)

Flemming Pociot
Glostrup Research Institute
Glostrup University Hospital
2600 Glostrup (Denmark)

Anne T. Reutens
School of Public Health and Preventive
Medicine
Monash University
Melbourne, VIC 3004 (Australia)

Eberhard Ritz
Department of Nephrology
Heidelberg University
69120 Heidelberg (Germany)

John C. Rutledge
Division of Cardiovascular Medicine
University of California Davis
School of Medicine
451 E. Health Sciences Drive
Davis, CA 95616 (USA)

Paola Romagnani
Department of Clinical Pathophysiology
Nephrology Section
University of Florence
Viale Pieraccini 6
50139 Florence (Italy)

Lionel Rostaing
Department of Nephrology, Dialysis, and
Organ Transplantation
CHU Rangueil, 31059 Toulouse (France)

Ivan Rychlík
2nd Department of Medicine
3rd Faculty of Medicine
Charles University
Šrobárova 50
10034 Prague (Czech Republic)

Robert W. Schrier
Department of Medicine
University of Colorado Denver
School of Medicine
Aurora, CO 80045 (USA)

Kumar Sharma
Center for Renal Translational Medicine
Division of Nephrology-Hypertension
Department of Medicine
University of California San Diego
San Diego, CA 92093 (USA)

Sydney C.W. Tang
Department of Medicine
University of Hong Kong
Queen Mary Hospital
Hong Kong SAR (China)

Merlin C. Thomas
Baker IDI Heart and Diabetes Institute
St Kilda Rd Central
Melbourne, VIC 8008 (Australia)

Visith Thongboonkerd
Medical Proteomics Unit
Office for Research and Development
Faculty of Medicine Siriraj Hospital
Mahidol University
2 Prannok Road, Bangkoknoi
Bangkok 10700 (Thailand)

Rose Z.W. Ting
Department of Medicine and Therapeutics
The Chinese University of Hong Kong
The Prince of Wales Hospital
Hong Kong SAR (China)

Louisa M. Villeneuve
Division of Cellular and Molecular
Diabetes Research
Department of Diabetes
Beckman Research Institute of City of
Hope
Duarte, CA 91010 (USA)

Xiao-Xi Zeng
Department of Medicine
West China Hospital of Sichuan University
610041 Chengdu (China)

Xiaoxin Wang
Division of Renal Diseases and
Hypertension
Department of Medicine
University of Colorado Denver
12700 East 19th Avenue
Aurora, CO 80045 (USA)

Fuad N. Ziyadeh
Departments of Internal Medicine and
Biochemistry, Faculty of Medicine,
American University of Beirut
Bliss Street, Beirut (Lebanon)

Preface

Diabetic nephropathy is a leading cause of end-stage renal disease in many developed countries where type 2 diabetes mellitus has already reached an epidemic tsunami by the beginning of this millennium. In these countries, diabetic nephropathy is the primary renal diagnosis in 25–50% of people starting chronic renal replacement therapy. This global and growing threat of diabetic nephropathy due mainly to type 2 diabetes has long caught the attention of nephrologists and diabetologists. Yet, over the past decade, the magnitude of the problem has continued to grow without any sign of reaching its plateau. There are a number of astounding reasons. First, there is the relentless increase in the prevalence of type 2 diabetes as a result of the metabolic syndrome. Despite an ongoing emphasis on the need for lifestyle changes, such as weight loss, diet control, and increased physical exercise, the obesity and diabetes epidemic continues to escalate and the number of people with type 2 diabetes, currently estimated at 150 million, is predicted to double by 2025. This problem is compounded by the fact that the risk of diabetes increases with age and that there is an aging trend in the global population.

Surely, the risk factors and pathophysiology of diabetic nephropathy are not entirely understood, which implies that there is clearly a long way to go before effective renoprotective or preventive strategies for diabetic nephropathy can be implemented for our patients. On the other hand, significant progress has been made in recent years in understanding the pathologic, genetic, pathogenetic, and epigenetic mechanisms of this condition. Furthermore, many large-scale randomized controlled trials have emerged to attest some of these hypotheses. It will therefore be timely for the publication of a textbook to provide an updated overview of all the available clinical and basic scientific data for clinicians and researchers in the field. It is with such intention in mind that we gathered leading world experts actively engaged in diabetic nephropathy research from bench to bedside to compile this exciting monograph on diabetes and the kidney that covers every aspect of transcriptional and translational research in diabetic nephropathy. We wish to thank all the 54 contributors in making this comprehensive volume a success. We hope that while it will provide an update to everyone, this book will also serve as a reference to stimulate further novel investigations to advance our knowledge about all the unknowns in diabetic nephropathy.

Kar Neng Lai, Hong Kong
Sydney C.W. Tang, Hong Kong

Lai KN, Tang SCW (eds): Diabetes and the Kidney.
Contrib Nephrol. Basel, Karger, 2011, vol 170, pp 1–7

Epidemiology of Diabetic Nephropathy

Anne T. Reutens · Robert C. Atkins

School of Public Health and Preventive Medicine, Monash University, and Baker IDI Heart
and Diabetes Institute, Melbourne, Vic., Australia

Abstract

Diabetic nephropathy affects approximately one third of people with type 1 or type 2 diabetes mellitus. Risk factors affecting progression of kidney disease include baseline albumin excretion, age, glycemic control, blood pressure, serum cholesterol and use of renin-angiotensin system blockers. As the total number of people with diabetes is projected to increase substantially to 2050, the prevalence of diabetic nephropathy will rise dramatically, with concomitant increase in associated cardiovascular mortality and end-stage renal disease. This will produce significant social and economic ramifications, particularly in the developing world.

Definition of Diabetic Nephropathy

Early diabetic nephropathy (DN) may be identified by persistent microalbuminuria [1, 2], defined as an albumin excretion rate of 20–200 µg/min or 30–300 mg/24 h, or a spot urine albumin to creatinine ratio of 30–300 mg/g (3.5–35 mg/mmol) in males and 20–200 mg/g (2.5–25 mg/mmol) in females. Overt DN is marked by proteinuria >500 mg/24 h or albuminuria >300 mg/24 h. Decreased estimated glomerular filtration rate (eGFR) <60 ml/min/1.73 m^2 may be another manifestation of overt DN.

Prevalence of Diabetic Nephropathy

The global prevalence of diabetes in adults is estimated to rise from 6.4% (285 million) in 2010 to 7.7% (439 million) in 2030 [3]. The increase will be sharpest in developing countries (69%) compared to developed countries (20%). Over 90% of these people with diabetes will have type 2 diabetes [4]. 36% of the extra

154 million people with diabetes will come from India and China. Such a large worldwide rise in prevalence predominantly of type 2 diabetes over the next 20 years will greatly increase the prevalence of diabetic kidney disease, which develops in about one third of people with type 2 diabetes. The global prevalence of microalbuminuria in diabetes is 39% [5]. Data from AusDiab, The Australian Diabetes, Obesity and Lifestyle Study, a population-based study of Australian adults, showed that incidence of albuminuria in the general population was 0.83% per year. People with diabetes had an incidence of albuminuria of 3.1% per year [6]. DN is already the major cause of end-stage renal disease (ESRD) worldwide, accounting for approximately one third of cases. This is illustrated by the experience in Australia, where people with type 2 diabetes starting dialysis increased 5-fold from 1993 to 2007 [7].

Diabetic kidney disease will impose heavy economic burdens, not only for treatment of ESRD but also because of the associated macrovascular complications of DN [8]. Already, the annual cost of treatment of ESRD in patients with type 2 diabetes in the US is estimated to be USD 39.35 billion in 2010, increased from USD 16.74 billion in 1998 [9]. Within developed countries, certain ethnic groups including Native Americans, African-Americans and indigenous Australians have a higher prevalence of ESRD due to diabetes [10]. The concern is that those countries and minorities that will be most affected by the epidemic of DN will be the least able to afford the costs.

Classic Albuminuric Pathway

This pathway refers to progression to nephropathy from the stage of normoalbuminuria, through microalbuminuria, to macroalbuminuria/proteinuria and reduced GFR. In patients with type 1 diabetes, the 25-year cumulative incidence of DN in cohorts diagnosed prior to 1960 was reported by the Steno Memorial Hospital as 41% [11], and 35% after 40 years of disease from Joslin Clinic cohorts [12]. Some latest studies show a lower incidence of kidney disease. In the DCCT-EDIC study of type 1 diabetes, which followed 1,439 newly diagnosed people for a mean duration of 19 years, 11.4% developed an eGFR <60 ml/min/1.73m^2 [13]. Of these, 16% had developed microalbuminuria and 61% developed macroalbuminuria before reaching this stage. eGFR declined at a greater rate in those with macroalbuminuria (5.7% per year and 5.1% per year with current or past history of macroalbuminuria respectively) compared to those with current or previous microalbuminuria (1.8 and 1.4% per year, respectively) or those with a normal albumin excretion rate (1.2% per year). However, in the Pittsburgh Epidemiology of Childhood-Onset Diabetes Complications Study, a prospective observational study which stratified participants into cohorts according to year of diagnosis from 1950 to 1980, there was no significant difference in incidence of overt nephropathy at 25 years between cohorts, and the pooled incidence was 32% [14].

A 2006 Austrian report of the Lainz cohort which followed 648 people with type 1 diabetes for 20 years showed that 13.0% had died and 5.6% required renal replacement therapy (RRT) [15]. Long-term follow up of 1,075 people with type 1 diabetes from the Allegheny county, Pennsylvania, gave a cumulative incidence of RRT after 25 years of diabetes of 11.3% [16]. The incidence of ESRD in type 1 diabetes is declining; cumulative incidence rates at 20 years of ESRD for patients diagnosed between 1965 and 1969, 1970 and 1974, and 1975 and 1979 were 9.1, 4.7, and 3.6%. The same trend was seen in a Finnish cohort of 20,005 people with type 1 diabetes diagnosed between 1965 and 1999 [17]. The 20-year incidence of ESRD was 2.2% at 20 years and 7.8% at 30 years after diagnosis, with those diagnosed between 1965 and 1969 having the worst prognosis. Similar improvement was seen in the Pittsburgh study, in which 20-year rate of ESRD was 16% in those diagnosed in the 1950s, compared with 4% in those diagnosed after 1964 [14]. In people aged <45 years, the incidence of diabetes-related ESRD decreased by 4.3% per year from 1990 to 2006 according to analysis of US data done by the US Renal Data System (USRDS) [18]. This age group would be mainly those with type 1 diabetes.

For type 2 diabetes, the United Kingdom Prospective Diabetes Study (UKPDS) followed 5,097 people who were newly diagnosed with type 2 diabetes from 1977 to 1997 to determine the rates of progression of kidney disease. 2.0% per year progressed from normo- to microalbuminuria, 2.8% per year from micro- to macroalbuminuria, and 2.3% per year from macroalbuminuria to elevated plasma creatinine or RRT [19]. By 10 years after diagnosis, 27% of patients had some evidence of renal disease (24.9% with microalbuminuria, 5.3% macroalbuminuria. 0.8% with high creatinine level or needing RRT). In the Casale Monferrato study, in which 1,103 patients with type 2 diabetes were followed for a median of 5.33 years, 3.7% of patients per year progressed to overt nephropathy, and those with baseline microalbuminuria had a 42% increased risk of progression compared to those with normoalbuminuria [20]. In a prospective follow-up of 191 outpatients with type 2 diabetes, the 5-year cumulative incidence of DN was 23%. The risk factors for progression were increased baseline albumin excretion, HbA1c, age, increased serum cholesterol, male gender and presence of retinopathy [21]. Certain ethnic groups have particular susceptibility to DN. For example, Pima Indians (from Arizona) with type 2 diabetes had a higher incidence of ESRD 15 years after onset of proteinuria (61%) compared to US Caucasians (17%) [22].

From the USRDS analysis, the incidence of ESRD due to diabetes started to decline from 1996 onwards for people aged ≥65 years, by 3.9% per year in the 45–64 year age group, 3.4% per year for those aged 65–74 years and by 2.1% in those ≥75 years [18]. The ethnic groups with the highest rates of diabetes-related ESRD were African-Americans, Native Americans and Hispanics (425, 333 and 310 per million population, respectively, compared to 117.8 in white people) [23].

The pathway from microalbuminuria to ESRD can be reversed. Studies of the earliest stage of diabetic kidney disease show that a high proportion of people with microalbuminuria remit to normoalbuminuria. The Joslin Study of the Natural History of Microalbuminuria followed 386 people with type 1 diabetes and persistent microalbuminuria [24]. Regression of microalbuminuria was defined as a 50% reduction in urinary albumin excretion. Over 6 years, the cumulative incidence of regression was 58%. Similar results were obtained in a study of 216 Japanese patients with microalbuminuria and type 2 diabetes, in which the 6-year cumulative incidence of regression was 54% and remission (to normoalbuminuria) was 51% [25]. 28% progressed to overt nephropathy. Factors that predicted regression/remission were lower systolic blood pressure, lower HbA1c, use of renin-angiotensin system blockers and short duration of microalbuminuria. Factors predicting progression of DN were higher HbA1c and baseline albumin excretion rate, current smoking, retinopathy, higher waist-hip ratio and body mass index [26, 27].

The Nonalbuminuric Pathway

People who develop renal impairment associated with diabetes may proceed via a nonproteinuric pathway, particularly in older age groups. The Joslin study of 79 patients with type 1 diabetes who were initially normoalbuminuric and then developed microalbuminuria gave a 12-year cumulative incidence of stage 3–5 chronic kidney disease of 29% [28]. Of these, about half did not have proteinuria but had persistent microalbuminuria or regressed to normoalbuminuria. The occurrence of an isolated reduction in GFR without antecedent microalbuminuria in patients with biopsy-proven diabetic kidney disease was first documented in 8 females with type 1 diabetes [29]. A similar finding of diminished creatinine clearance with normoalbuminuria was confirmed in type 2 diabetes, with both males and females involved (with a female preponderance) [30–32]. 30% of people with type 2 diabetes and eGFR <60 ml/min/1.73m^2 had no albuminuria and no diabetic retinopathy in the nationally representative US NHANES study [32], and 55% of people with eGFR <60 ml/min/1.73 m^2 had a persistently normal urinary albumin excretion in a nationally representative Australian cohort of people with type 2 diabetes [33]. The mechanisms for this alternate pathway of kidney disease in diabetes are under investigation.

Association of Diabetic Nephropathy with Cardiovascular Mortality

A recent collaborative meta-analysis of 105,872 participants in studies of participants derived from the general population showed that albuminuria ≥1.1 mg/mmol (10 mg/g) and low eGFR <60 ml/min/1.73 m^2 were independently predictive of all-cause and cardiovascular mortality [34]. The HOPE study demonstrated that microalbuminuria at baseline predicted cardiovascular events (myocardial infarction, stroke, cardiovascular death), all cause mortality and hospitalization for congestive cardiac failure in people with or without diabetes [35]. In the UKPDS, there was a trend towards increased cardiovascular mortality with worse renal function [19]. The annual cardiovascular death rate was 0.7% in those with no kidney disease, 2.0% with microalbuminuria, 3.5% with macroalbuminuria and 12.1% in those with an elevated plasma creatinine or on RRT. People with macroalbuminuria were more likely to die than to progress to elevated plasma creatinine or RRT (4.6% annual rate of death from any cause compared to 2.3% annual rate of progression).

Conclusion

DN affects approximately one third of those with diabetes. Improved management has led to reduction in incidence of progression to ESRD in this group. However, cardiovascular mortality is greatly increased in people with overt DN. The number of people affected by chronic kidney disease due to diabetes will escalate in the coming decades, and this will impose great strains on health delivery and costs, particularly in developing countries.

References

1   Mogensen C, Christensen C: Predicting diabetic nephropathy in insulin-dependent patients. N Eng J Med 1984;311:89–93.

2   Mogensen C: Microalbuminuria predicts clinical proteinuria and early mortality in maturity onset diabetes. N Eng J Med 1986; 310:356–360.

3   Shaw J, Sicree R, Zimmet P: Global estimates of the prevalence of diabetes for 2010 and 2030. Diabetes Res Clin Pract 2010;87:4–14.

4   Zimmet P, Alberti K, Shaw J: Global and societal implications of the diabetes epidemic. Nature 2001;414:782–787.

5   Rossing P: Diabetic nephropathy: worldwide epidemic and effects of current treatment on natural history. Curr Diab Rep 2006;6: 479–483.

6   Barr E, Magliano D, Zimmet P, et al: AusDiab 2005. The Australian Diabetes, Obesity and Lifestyle Study. Tracking the accelerating epidemic: its causes and outcomes. Melbourne, International Diabetes Institute, 2006.

7   McDonald S, Excell L, Livingston B: Appendix II, in ANZDATA Registry Report 2008. Adelaide, 2008.

8   Postma M, de Zeeuw D: The economic benefits of preventing end-stage renal disease in patients with type 2 diabetes mellitus. Nephrol Dial Transplant 2009;24:2975–2983.

9   Trivedi H, Pang M, Campbell A, Saab P: Slowing the progression of chronic renal failure: economic benefits and patients' perspectives. Am J Kidney Dis 2002;39: 721–729.

10 Lopes A: End-stage renal disease due to diabetes in racial/ethnic minorities and disadvantaged populations. Ethn Dis 2009; 19(suppl 1):S1–47–51.

11 Andersen A, Christiansen J, Andersen J, Kreiner S, Deckert T: Diabetic nephropathy in type 1 (insulin-dependent) diabetes: an epidemiological study. Diabetologia 1983;25: 496–501.

12 Krolewski A, Warram J, Christlieb A, Busick
    E, Kahn C: The changing natural history of
    nephropathy in type 1 diabetes. Am J Med
    1985;78:785–794.
13 Molitch M, Steffes M, Sun W, et al:
    Development and progression of renal
    insufficiency with and without albuminuria
    in adults with type 1 diabetes in the Diabetes
    Control and Complications Trial and the
    Epidemiology of Diabetes Interventions and
    Complications Study. Diabetes Care 2010;
    33:1536–1543.
14 Pambianco G, Costacou T, Ellis D, Becker
    D, Klein R, Orchard T: The 30-year natural
    history of type 1 diabetes complications.
    The Pittsburgh Epidemiology of Diabetes
    Complications Study experience. Diabetes
    2006;55:1463–1469.
15 Stadler M, Auinger M, Anderwald C, et
    al: Long-term mortality and incidence of
    renal dialysis and transplantation in type 1
    diabetes mellitus. J Clin Endocrinol Metab
    2006;91:3814–3820.
16 Nishimura R, Dorman J, Bosnyak Z, Tajima
    N, Becker D, Orchard T: Incidence of
    ESRD and survival after renal replacement
    therapy in patients with type 1 diabetes: a
    report from the Allegheny County Registry.
    Am J Kidney Dis 2003;42:117–124.
17 Finne P, Reunanen A, Stenman S, Groop P,
    Gronhagen-Riska C: Incidence of end-stage
    renal disease in patients with type 1 diabetes.
    JAMA 2005 294:1782–1787.
18 Burrows N, Li Y, Geiss L: Incidence of treat-
    ment for end-stage renal disease among
    individuals with diabetes in the U.S. contin-
    ues to decline. Diabetes Care 2010;33:73–77.
19 Adler A, Stevens R, Manley S, Bilous R, Cull
    C, Holman R, UKPDS group: Development
    and progression of nephropathy in type 2
    diabetes: the United Kingdom Prospective
    Diabetes Study (UKPDS 64). Kidney Int
    2003;63:225–232.
20 Bruno G, Merletti F, Biggeri A, et al:
    Progression to overt nephropathy in type
    2 diabetes: the Casale Monferrato Study.
    Diabetes Care 2003;26:2150–2155.
21 Gall M, Hougaard P, Borch-Johnsen K,
    Parving H: Risk factors for development of
    incipient and overt diabetic nephropathy in
    patients with non-insulin dependent diabetes
    mellitus: prospective, observational study.
    BMJ 1997;314:783–788.

22 Nelson R, Knowler W, McCance D, et al:
    Determinants of end-stage renal disease
    in Pima Indians with type 2 (non-insulin-
    dependent) diabetes mellitus and proteinu-
    ria. Diabetologia 1993;36:1087–1093.
23 US Renal Data System: USRDS 2010 Annual
    Report: Atlas of Chronic Kidney Disease and
    End-Stage Renal Disease in the United States.
    Bethesda, National Institutes of Health,
    National Institute of Diabetes, Digestive and
    Kidney Diseases, 2010.
24 Perkins B, Ficociello L, Silva K, Finkelstein
    D, Warram J, Krolewski A: Regression of
    microalbuminuria in type 1 diabetes. N Engl
    J Med 2003;348:2285–2293.
25 Araki S, Haneda M, Sugimoto T, Isono M,
    Isshiki K, Kashiwagi A, Koya D: Factors
    associated with frequent remission of
    microalbuminuria in patients with type 2
    diabetes. Diabetes 2005;54:2983–2987.
26 Rossing P, Hougaard P, Parving H: Risk fac-
    tors for development of incipient and overt
    diabetic nephropathy in type 1 diabetic
    patients: a 10-year prospective observational
    study. Diabetes Care 2002;25:859–864.
27 Vergouwe Y, Soedamah-Muthu S, Zgibor J,
    et al: Progression to microalbuminuria in
    type 1 diabetes: development and validation
    of a prediction rule. Diabetologia 2010;53:
    254–262.
28 Perkins B, Ficociello L, Roshan B, Warram J,
    Krolewski A: In patients with type 1 diabetes
    and new-onset microalbuminuria the devel-
    opment of advanced chronic kidney disease
    may not require progression to proteinuria.
    Kidney Int 2010;77:57–64.
29 Lane P, Steffes M, Mauer S: Glomerular
    structure in IDDM women with low glomer-
    ular filtration rate and normal urinary albu-
    min excretion. Diabetes 1992;41:581–586.
30 Tsalamandris C, Allen T, Gilbert R, Sinha
    A, Panagiotopoulos S, Cooper M, Jerums
    G: Progressive decline in renal function in
    diabetic patients with and without albuminu-
    ria. Diabetes 1994;43:649–655.
31 MacIsaac R, Tsalamandris C,
    Panagiotopoulos S, Smith T, McNeil K,
    Jerums G: Nonalbuminuric renal insuf-
    ficiency in type 2 diabetes. Diabetes Care
    2004;27:195–200.

32 Kramer H, Nguyen Q, Curhan G, Hsu C: Renal insufficiency in the absence of albuminuria and retinopathy among adults with type 2 diabetes mellitus. JAMA 2003; 289:3273–3277.

33 Thomas M, Macisaac R, Jerums G, Weekes A, Moran J, Shaw J, Atkins R: Nonalbuminuric renal impairment in type 2 diabetic patients and in the general population (national evaluation of the frequency of renal impairment cO-existing with NIDDM [NEFRON] 11). Diabetes Care 2009;32:1497–1502.

34 Chronic Kidney Disease Prognosis Consortium, Matsushita K, van der Velde M, et al: Association of estimated glomerular filtration rate and albuminuria with all-cause and cardiovascular mortality in general population cohorts: a collaborative meta-analysis. Lancet 2010;375:2073–2081.

35 Gerstein H, Mann J, Yi Q, et al: Albuminuria and risk of cardiovascular events, death, and heart failure in diabetic and nondiabetic individuals. JAMA 2001;286:421–426.

Anne T. Reutens, MBBS, PhD, FRACP
School of Public Health and Preventive Medicine,
Monash University, and Baker IDI Heart and Diabetes Institute
Melbourne, Vic. 3004 (Australia)
Tel. +61 3 8532 1800, Fax +61 3 8532 1100, E-Mail anne.reutens@monash.edu

Lai KN, Tang SCW (eds): Diabetes and the Kidney.
Contrib Nephrol. Basel, Karger, 2011, vol 170, pp 8–18

Genetics of Diabetic Nephropathy in Diverse Ethnic Groups

Caroline Brorsson · Flemming Pociot

Glostrup Research Institute, Glostrup University Hospital, Glostrup, Denmark

Abstract

Genetic susceptibility is considered an important factor for the development and progression to diabetic nephropathy (DN), and for more than 20 years researchers have tried to unravel the genetic determinants of the disease. It is now clear that the pathogenesis of DN is most likely multifactorial and attributed to several genetic and environmental risk factors. Several candidate genes have been shown to be associated with the disease, but the results have not been consistent and most of the genes conferring risk to DN remain to be identified. In addition, studies have suggested that there might be differences in susceptibility loci and/or alleles between diverse populations. Recent developments in genotyping technology and increased information on the human genome have facilitated genome-wide association scans (GWAS) for investigating novel disease susceptibility across the entire human genome. The few GWAS performed for DN so far in combination with improved understanding of the human genome have identified novel risk loci and emphasized the importance of performing detailed genetic studies across diverse ethnic populations to fully unravel the genetic susceptibility to DN.

Diabetic nephropathy (DN) is a complex phenotype likely caused by the interaction between susceptibility alleles and environmental factors. There is evidence for a genetic susceptibility to diabetic kidney disease, but despite intensive research efforts through primarily candidate gene searches, it has proved difficult to identify the causative genes.

The first evidence for genetic factors underlying DN was gathered in epidemiological studies that displayed familial aggregation of DN in both type 1 and type 2 diabetes. Studies in type 1 diabetes mellitus (T1DM) demonstrated that siblings of diabetic subjects with kidney disease had a 3-fold increase in the risk of DN compared to siblings of diabetic subjects without nephropathy [1–3]. Familial clustering of nephropathy was also observed in type 2 diabetes mellitus (T2DM) [4, 5] and the risk was found to differ between different ethnic groups [6, 7]. These early studies were the starting point for the search for the genetic determinants of DN. However, despite large efforts aimed at identifying the susceptibility genes, the studies have not provided conclusive evidence for a causative role of any specific candidate gene. A plausible reason for the lack of replication among the reports includes limited sample sizes of the individual studies rendering them largely underpowered to detect genetic variants with small individual effects on disease risk. In the last few years, genome-wide association scans (GWAS) have produced an unprecedented number of robustly associated genetic risk factors for common complex diseases such as T1DM and T2DM [8–11]. With the completion of the first GWAS in DN in T1DM published and more on their way, we can now begin to unravel the genes and molecular pathways that underlie the development of DN that hopefully in the near future will lead to the development of new treatment strategies.

Approaches to Study the Genetic Causes of Diabetic Nephropathy

The two main approaches to study the molecular genetics of common disease are linkage and association studies, both of which have been extensively used in DN. Linkage scans are family based and evaluate polymorphic genetic markers that are uniformly distributed across the genome for cosegregation with a trait in affected families. Classical linkage scans were performed using microsatellite markers, which are highly polymorphic tandem repeats of DNA sequence; however, more recent studies use the more common single nucleotide polymorphisms (SNPs), which are mostly biallelic and more informative of the surrounding region. Genetic association studies commonly investigate SNPs for differences in allele frequency between unrelated diseased (case) and healthy (control) subjects as well as in family trios using family-based association tests, e.g. the transmission disequilibrium test. Association studies are generally more powerful for the identification of genetic variants with modest effect on the risk of disease compared to linkage studies. However, linkage scans are hypothesis free, whereas association studies of candidate genes require prior biological knowledge of the candidate gene as well as the disease pathogenesis. In recent years, GWAS have contributed immensely to the understanding of the genetic basis of a wide range of common diseases. GWAS studies are conducted using a dense set of SNPs, usually in the range of a hundred thousand to a million, which are distributed across the genome. A common feature of the GWAS published to date is that they still have relatively low power to detect variants with small effect sizes, and the main limitation to detect these is the need for very large sample sizes in order to gain statistical power. This problem is very relevant to the field of DN, in which it has proven difficult to gain access to large, well-characterized sample cohorts for large-scale genetic studies. As for other complex traits, we expect the contribution of individual DN risk genes to have odds ratios in the range of 1.05–1.5. One emerging solution to this problem is to combine sample sets from multiple GWAS studies of similar traits into a meta-analysis, which has successfully been done for T1DM and T2DM [9–11]. It should be stressed that exact phenotyping in studies of DN is crucial, since it is not straightforward to combine data derived from studies focusing on albuminuria, overt proteinuria or end-stage renal disease (ESRD). Furthermore, it is currently not clear whether DN in type 1 and type 2 diabetes has a shared genetic component. Phenotypic definition of DN has proven simpler in patients with type 1 diabetes; this is why efforts to discover genetic risk variants of DN might be more successful in these patients.

Genome-Wide Linkage Scans in Diabetic Nephropathy

Several genome-wide linkage scans have been conducted for DN in different populations [12–19]. Most of the analyses have been limited to small numbers of families from different ethnic groups. Nevertheless, a few consistent results have emerged. One of the most promising findings in type 1 diabetes nephropathy (T1DN) is a linkage peak near the gene encoding angiotensin II receptor type 1 *(AGTR1)* located on chromosome 3q25 first identified in Caucasian subjects [13]. An association study evaluating 14 candidate genes in the 3q region in three independent European populations demonstrated that the susceptibility variant might be located in the promoter region of the adiponectin gene (*ADIPOQ*), rather than *AGTR1* [20]; however, the results have been inconclusive. In type 2 diabetes nephropathy (T2DN) a linkage peak in a region on chromosome 18q22–23 has consistently been identified in Turkish and Pima Indian [14], African-American [15], Caucasian and American Indian [18] family cohorts. A fine-mapping effort of this region identified the carnosinase 1 gene *(CNDP1)* as the most likely candidate gene for DN [21]. Further support for the *CNDP1* gene was reported in a larger study of T2D ESRD in a European-American study population [22]. Although many linkage scans have been published for DN, few of the findings have been replicated, probably due to the relatively small sample sizes and the limited power of linkage scans to identify risk variants with minor effects on the disease susceptibility.

Candidate Genes Associated with Diabetic Nephropathy

A large number of genetic association studies have been conducted for DN generating several candidate genes. However, similar to the linkage scans, replication of the findings in independent cohorts has proven difficult, and none of the genes has so far been robustly replicated (table 1).

Table 1. Genetic association of candidate genes in different subgroups

Gene (minor allele)	Group	Odds ratio (95% CI)	Significance
ACE (rs179975 D)	all	1.24 (1.12–1.37)	+
ACE (rs179975 D)	ESRD	1.39 (1.21–1.6)	+
ACE (rs179975 D)	T2DM	1.33 (1.16–1.52)	+
ACE (rs179975 D)	Asian	1.28 (1.10–1.56)	+
ACE (rs179975 D)	proteinuria	1.2 (1.07–1.36)	+
ACE (rs179975 D)	T1DM	1.13 (1.04–1.23)	+
ACE (rs179975 D)	European	1.1 (0.99–1.22)	–
ADIPOQ (rs17300539 A)	all	1.37 (0.93–2.03)	–
AGTR1 (rs5186 C)	all	0.99 (0.91–1.08)	–
AGTR1 (rs5186 C)	European	1.06 (0.94–1.19)	–
AGTR1 (rs5186 C)	T1DM	1.04 (0.92–1.17)	–
APOE (E2)	all	1.7 (1.12–2.58)	+
APOE (E2)	Asian	2.35 (1.29–4.30)	+
APOE (E2)	T2DM	2.21 (1.22–4.00)	+
APOE (E2)	T1DM European	1.48 (0.84–2.58)	–
APOE (E4)	all	0.78 (0.62–0.98)	+
CARS (rs451041 A)	all	1.37 (1.21–1.54)	+
CARS (rs739401 C)	all	1.32 (1.15–1.51)	+
CNDP1 (D18S880 5L)	all	0.92 (0.82–1.01)	–
CNDP1 (D18S880 5L)	T2DM	0.77 (0.61–0.97)	+
CPVL/CHN2 (rs39059 G)	all	0.74 (0.64–0.85)	+
CPVL/CHN2 (rs39075 A)	all	0.84 (0.62–1.14)	–
ELMO1 (rs741301 G)	all	1.31 (0.82–2.08)	–
ELMO1 (rs741301 G)	Asian	1.58 (1.28–1.94)	+
FRMD3 (rs1888747 C)	all	0.74 (0.65–0.83)	+
FRMD3 (rs10868025 A)	all	0.72 (0.64–0.81)	+

Genetic heterogeneity between populations and DN phenotypes is apparent for many of the loci. The lack of association for the *ELMO1* gene can probably be explained by the lack of meta-analysis for the genetic markers that were associated in the African-American and European populations, that differed compared to the Japanese population. Data were collected from table 1, figure 2 and figure 3 of a recent meta-analysis by Mooyaart et al. [45]. + = Positive association in the meta-analysis; – = no association in the meta-analysis.

Among the candidate genes for DN, the genes involved in the renin-angiotensin system have been studied most extensively [see chapter by Lai et al., this vol., pp. 135–144]. In addition to *AGTR1*, an association between DN and variation in the angiotensin 1-converting enzyme gene *(ACE)* on chromosome 17q23 was first reported in T1DM subjects [23, 24]. A 287-bp insertion/deletion polymorphism in intron 16 of *ACE* (ACE I/D) was originally found to be predisposing to DN for carriers of the deletion allele, which was also associated with a higher expression of ACE in plasma/serum in nondiabetic and diabetic subjects [23, 25]. The association between polymorphisms in *ACE* and various traits of DN has since been extensively investigated by a number of studies in both T1DN and T2DN; however, with conflicting results [26].

Genetic variation in the gene encoding apolipoprotein E (*APOE*) on chromosome 19q has been associated with DN. ApoE consists of the three major isoforms E2, E3 and E4 encoded by the common alleles ε2, ε3, ε4, of which the ε2 allele has been associated with an increased risk of mainly T1DN in Caucasians [27–29] but also in Japanese subjects with T2DN [30]. The E2 isoform has also been associated with a reduced affinity for its receptor [31]. The ε4 allele on the other hand has been associated with protection from T2DN in Japanese individuals [32] and the E4 isoform with an increased affinity for its receptor [31]. Although conflicting results have been found in other studies, a recent meta-analysis of 23 published studies reported that the ε2 allele was significantly associated with DN [33]. However, no association was demonstrated for the ε4 allele, and significant heterogeneity was detected between the studies which most likely contributed to the conflicting results.

The gene engulfment and cell motility 1 *(ELMO1)* located on chromosome 7p14 was first found to be associated with DN in a gene-focused GWAS of Japanese subjects with T2DN [34]. The association has subsequently been replicated in African-American [35] and Caucasian populations [36]. Interestingly, the genetic variants in *ELMO1* predisposing to DN were different in the three studies indicating that the diverse ancestral background gives rise to genetic heterogeneity between different ethnic populations.

Genome-Wide Association Scans in Different Ethnic Groups/Diverse Populations

The last few years have witnessed the feasibility and rapid expansion of GWAS, also for DN. GWAS data are available from two studies (The Genetics of Kidneys in Diabetes, GoKinD, study and The Diabetes Control and Complications Trial and Epidemiology of Diabetes Interventions and Complications Study, DCCT and EDIC), but larger sample sizes and samples from diverse populations would increase the chances for gene discovery. Importantly, new data are being generated in both US, UK and Finland [37, 38].

GWAS have, with few exceptions, been focused on populations of European descent. However, the degree to which observation gained from these studies is transferable to other populations has not been extensively explored. Nevertheless, a number of population-genetic factors have facilitated the success of GWAS in European populations. These include a less pronounced variation of allele frequency across groups and a more moderate linkage disequilibrium (LD) compared with other populations [39–41]. It has also facilitated the use of shared controls in large studies of multiple phenotypes [8].

GWAS in non-European populations pose several challenges. Will the same results be detected in different populations? Will causal variants have comparable allele frequencies and disease risk in diverse populations? What factors will be the sources of different results across groups?

We argue that it is crucial to carry out genetic studies in diverse populations not only for the ultimate goal of bringing medical advances from genetic studies to populations worldwide but also for the considerable scientific benefits in characterizing risk variants beyond what can be achieved with populations of European descent alone.

Several observations suggest that no single population is sufficient for fully uncovering the variants underlying disease in all populations. These various reasons: differences in disease allele frequency and LD patterns, phenotypic prevalence differences, differences in effect size and differences in rare variants provide the motivation for genetic studies in diverse populations.

The case for using diverse populations in genetic studies is emphasized by the observation that the proportion of phenotypic variation explained by gene variants identified through GWAS is typically small [42]. GWAS have focused on common variants (minor allele frequency, MAF: >5%), and rare variants (MAF: 1–5%) have not been examined to the same extent. Rare variants are one possible genetic source for the unexplained phenotypic variation of DN. Rare variants are usually more recent in origin, and therefore more likely to be geographically confined. Hence, separate populations are more likely to differ in rare alleles than in common alleles.

Genome-Wide Association Scan in Type 1 Diabetes Nephropathy

The first GWAS in DN was completed under the auspices of GoKinD, which has collected a large cohort of T1DM patients with or without DN. The initial study included 1,700 subjects genotyped for over 360,000 SNPs on the Affymetrix 5.0 500K array [43]. Although none of the SNPs achieved genome-wide significance after correcting for the number of tests conducted, 11 SNPs in four chromosomal regions achieved p values $<1 \times 10^{-5}$ and were considered for replication in the DCCT/EDIC study. The association signals were located on chromosome 9q near the gene ezrin, radixin, moesin (FERM) domain-containing

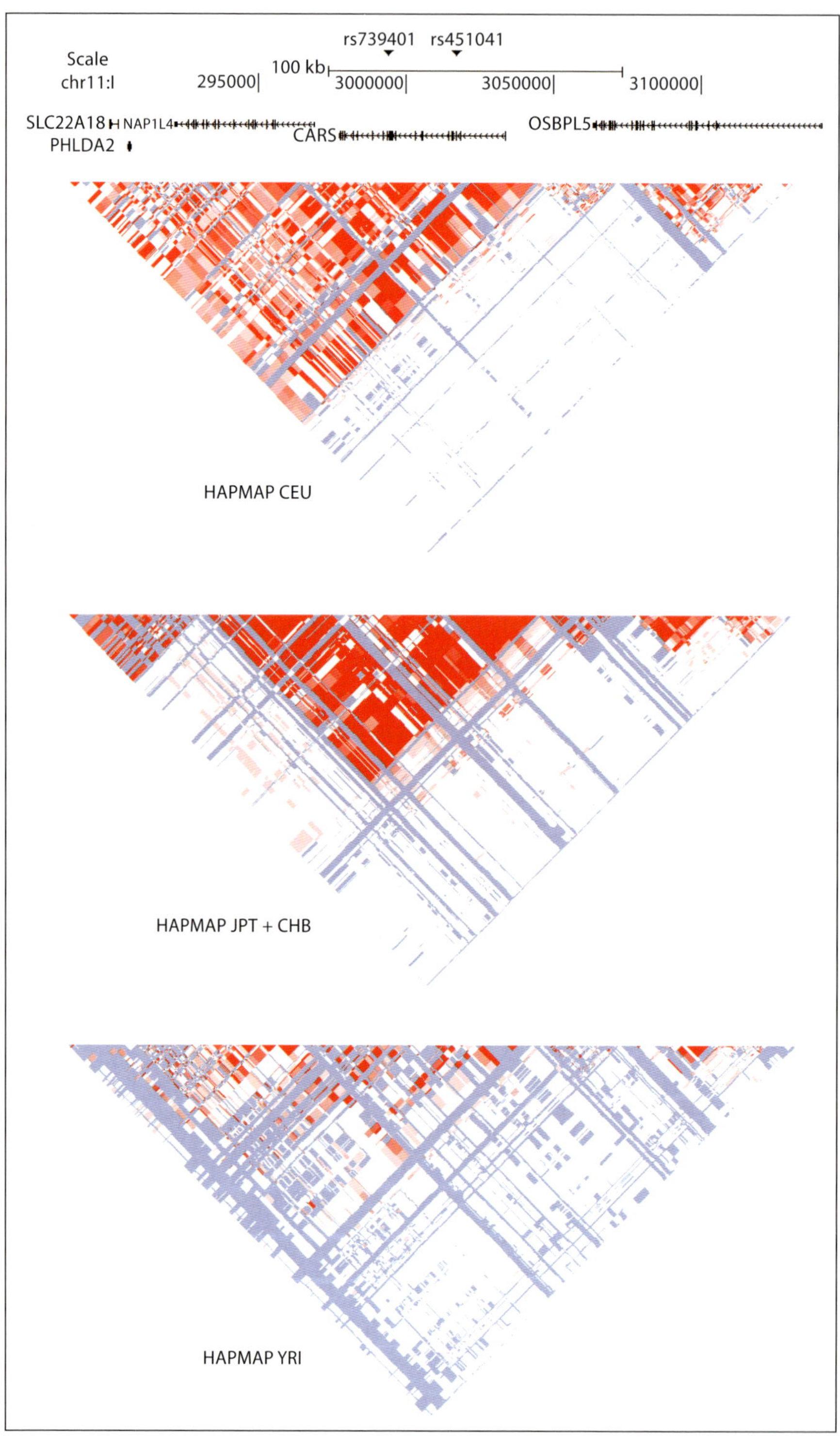

rs739401
rs451041
Scale
chr11:I
100 kb
295000|
3000000|
3050000|
3100000|
SLC22A18
NAP1L4
PHLDA2
CARS
OSBPL5
HAPMAP CEU
HAPMAP JPT + CHB
HAPMAP YRI

3 *(FRMD3)*, on chromosome 7p near the genes β-chimerin *(CHN2)* and serine carboxypeptidase vitellogenic-like *(CPVL)*, on chromosome 11p in an intron of the gene cysteinyl-tRNA synthetase *(CARS)* and in an intergenic region on chromosome 13q. The association signals on chromosome 9q and 11p were nominally significant in the replication study. The association on chromosome 13q has since been replicated in a study of T2DN in Japanese subjects [44], while the other regions are still awaiting confirmation in independent cohorts. Genetic heterogeneity between populations could be one contributing factor for the lack of replication in the Japanese population. Figure 1 illustrates the diverse pattern of LD across the associated region on chromosome 11p for the three HapMap populations.

Conclusions

Genetic susceptibility plays an important role in the pathogenesis of DN. Candidate gene and genome-wide linkage studies have suggested several susceptibility genes across diverse ethnic groups, but most of these remain to be robustly replicated. Most recently, GWAS have been performed in DN cohorts, and a better understanding of the genetic architecture of DN risk is emerging. Several observations suggest that no single population is sufficient for fully uncovering the risk genes underlying DN, and studies in diverse human populations should improve the potential of GWAS.

Fig. 1. LD plot of the chromosome 11p region associated with DN in the GWAS by Pezzolesi et al. [43]. The differences in LD patterns between the three HapMap populations are clearly visible in the genomic region surrounding the candidate gene *CARS*. The location of the two risk variants rs739401 and rs451041 is indicated by arrows. Plots of LD scores were produced in the UCSC genome browser (http://genome.ucsc.edu). CEU = European ancestry; JPT + CHB = Asian ancestry (Japan and China); YRI = African ancestry (Yoruba).

References

1 Seaquist ER, Goetz FC, Rich S, Barbosa J: Familial clustering of diabetic kidney disease. Evidence for genetic susceptibility to diabetic nephropathy. N Engl J Med 1989;320: 1161–1165.
2 Borch-Johnsen K, Norgaard K, Hommel E, Mathiesen ER, Jensen JS, et al: Is diabetic nephropathy an inherited complication? Kidney Int 1992;41:719–722.
3 Quinn M, Angelico MC, Warram JH, Krolewski AS: Familial factors determine the development of diabetic nephropathy in patients with IDDM. Diabetologia 1996;39: 940–945.
4 Pettitt DJ, Saad MF, Bennett PH, Nelson RG, Knowler WC: Familial predisposition to renal disease in two generations of Pima Indians with type 2 (non-insulin-dependent) diabetes mellitus. Diabetologia 1990;33: 438–443.
5 Fava S, Azzopardi J, Hattersley AT, Watkins PJ: Increased prevalence of proteinuria in diabetic sibs of proteinuric type 2 diabetic subjects. Am J Kidney Dis 2000;35:708–712.
6 Nelson RG, Newman JM, Knowler WC, Sievers ML, Kunzelman CL, et al: Incidence of end-stage renal disease in type 2 (non-insulin-dependent) diabetes mellitus in Pima Indians. Diabetologia 1988;31:730–736.
7 Chandie Shaw PK, Baboe F, van Es LA, van der Vijver JC, van de Ree MA, et al: South-Asian type 2 diabetic patients have higher incidence and faster progression of renal disease compared with Dutch-European diabetic patients. Diabetes Care 2006;29: 1383–1385.
8 Wellcome Trust Case Control Consortium: Genome-wide association study of 14,000 cases of seven common diseases and 3,000 shared controls. Nature 2007;447:661–678.
9 Barrett JC, Clayton DG, Concannon P, Akolkar B, Cooper JD, et al: Genome-wide association study and meta-analysis find that over 40 loci affect risk of type 1 diabetes. Nat Genet 2009;41:703–707.
10 Cooper JD, Smyth DJ, Smiles AM, Plagnol V, Walker NM, et al: Meta-analysis of genome-wide association study data identifies additional type 1 diabetes risk loci. Nat Genet 2008;40:1399–1401.
11 Zeggini E, Scott LJ, Saxena R, Voight BF, Marchini JL, et al: Meta-analysis of genome-wide association data and large-scale replication identifies additional susceptibility loci for type 2 diabetes. Nat Genet 2008;40: 638–645.
12 Imperatore G, Hanson RL, Pettitt DJ, Kobes S, Bennett PH, et al: Sib-pair linkage analysis for susceptibility genes for microvascular complications among Pima Indians with type 2 diabetes. Pima Diabetes Genes Group. Diabetes 1998;47:821–830.
13 Moczulski DK, Rogus JJ, Antonellis A, Warram JH, Krolewski AS: Major susceptibility locus for nephropathy in type 1 diabetes on chromosome 3q: results of novel discordant sib-pair analysis. Diabetes 1998; 47:1164–1169.
14 Vardarli I, Baier LJ, Hanson RL, Akkoyun I, Fischer C, et al: Gene for susceptibility to diabetic nephropathy in type 2 diabetes maps to 18q22.3–23. Kidney Int 2002;62: 2176–2183.
15 Bowden DW, Colicigno CJ, Langefeld CD, Sale MM, Williams A, et al: A genome scan for diabetic nephropathy in African Americans. Kidney Int 2004;66:1517–1526.
16 Placha G, Poznik GD, Dunn J, Smiles A, Krolewski B, et al: A genome-wide linkage scan for genes controlling variation in renal function estimated by serum cystatin C levels in extended families with type 2 diabetes. Diabetes 2006;55:3358–3365.
17 Chen G, Adeyemo AA, Zhou J, Chen Y, Doumatey A, et al: A genome-wide search for linkage to renal function phenotypes in West Africans with type 2 diabetes. Am J Kidney Dis 2007;49:394–400.
18 Iyengar SK, Abboud HE, Goddard KA, Saad MF, Adler SG, et al: Genome-wide scans for diabetic nephropathy and albuminuria in multiethnic populations: the family investigation of nephropathy and diabetes (FIND). Diabetes 2007;56:1577–1585.
19 Osterholm AM, He B, Pitkaniemi J, Albinsson L, Berg T, et al: Genome-wide scan for type 1 diabetic nephropathy in the Finnish population reveals suggestive linkage to a single locus on chromosome 3q. Kidney Int 2007;71:140–145.

20 Vionnet N, Tregouet D, Kazeem G, Gut I, Groop PH, et al: Analysis of 14 candidate genes for diabetic nephropathy on chromosome 3q in European populations: strongest evidence for association with a variant in the promoter region of the adiponectin gene. Diabetes 2006;55:3166–3174.

21 Janssen B, Hohenadel D, Brinkkoetter P, Peters V, Rind N, et al: Carnosine as a protective factor in diabetic nephropathy: association with a leucine repeat of the carnosinase gene CNDP1. Diabetes 2005;54:2320–2327.

22 Freedman BI, Hicks PJ, Sale MM, Pierson ED, Langefeld CD, et al: A leucine repeat in the carnosinase gene CNDP1 is associated with diabetic end-stage renal disease in European Americans. Nephrol Dial Transplant 2007;22:1131–1135.

23 Marre M, Bernadet P, Gallois Y, Savagner F, Guyene TT, et al: Relationships between angiotensin I converting enzyme gene polymorphism, plasma levels, and diabetic retinal and renal complications. Diabetes 1994;43:384–388.

24 Doria A, Warram JH, Krolewski AS: Genetic predisposition to diabetic nephropathy. Evidence for a role of the angiotensin I-converting enzyme gene. Diabetes 1994;43:690–695.

25 Rigat B, Hubert C, Alhencgelas F, Cambien F, Corvol P, et al: An insertion deletion polymorphism in the angiotensin I-converting enzyme gene accounting for half the variance of serum enzyme levels. J Clin Invest 1990;86:1343–1346.

26 Ng DP, Tai BC, Koh D, Tan KW, Chia KS: Angiotensin-I converting enzyme insertion/deletion polymorphism and its association with diabetic nephropathy: a meta-analysis of studies reported between 1994 and 2004 and comprising 14,727 subjects. Diabetologia 2005;48:1008–1016.

27 Chowdhury TA, Dyer PH, Kumar S, Gibson SP, Rowe BR, et al: Association of apolipoprotein epsilon2 allele with diabetic nephropathy in Caucasian subjects with IDDM. Diabetes 1998;47:278–280.

28 Araki S, Moczulski DK, Hanna L, Scott LJ, Warram JH, et al: APOE polymorphisms and the development of diabetic nephropathy in type 1 diabetes: results of case-control and family-based studies. Diabetes 2000;49:2190–2195.

29 Werle E, Fiehn W, Hasslacher C: Apolipoprotein E polymorphism and renal function in German type 1 and type 2 diabetic patients. Diabetes Care 1998;21:994–998.

30 Eto M, Horita K, Morikawa A, Nakata H, Okada M, et al: Increased frequency of apolipoprotein epsilon 2 allele in non-insulin dependent diabetic (NIDDM) patients with nephropathy. Clin Genet 1995;48:288–292.

31 Mahley RW: Apolipoprotein-E – cholesterol transport protein with expanding role in cell biology. Science 1988;240:622–630.

32 Kimura H, Suzuki Y, Gejyo F, Karasawa R, Miyazaki R, et al: Apolipoprotein E4 reduces risk of diabetic nephropathy in patients with NIDDM. Am J Kidney Dis 1998;31:666–673.

33 Li Y, Tang K, Zhang Z, Zhang M, Zeng Z, et al: Genetic diversity of the apolipoprotein E gene and diabetic nephropathy: a meta-analysis. Mol Biol Rep 2010, Epub ahead of print.

34 Shimazaki A, Kawamura Y, Kanazawa A, Sekine A, Saito S, et al: Genetic variations in the gene encoding ELMO1 are associated with susceptibility to diabetic nephropathy. Diabetes 2005;54:1171–1178.

35 Leak TS, Perlegas PS, Smith SG, Keene KL, Hicks PJ, et al: Variants in intron 13 of the ELMO1 gene are associated with diabetic nephropathy in African Americans. Ann Hum Genet 2009;73:152–159.

36 Pezzolesi MG, Katavetin P, Kure M, Poznik GD, Skupien J, et al: Confirmation of genetic associations at ELMO1 in the GoKinD collection supports its role as a susceptibility gene in diabetic nephropathy. Diabetes 2009;58:2698–2702.

37 Pezzolesi MG, Poznik GD, Skupien J, Smiles A, Mychaleckyj JC, et al: Confirmation of genetic associations from a genome-wide association scan of the GoKinD collection identifies common type 1 and type 2 diabetic nephropathy loci on chromosome 11p and 13q (program 1898); in 60th Annu Meet Am Soc Hum Genet, Washington, November 2010.

38 Salem RM, Sandholm N, McKnight AJ, Isakova T, Brennan E, et al: Meta-analysis of genome-wide association studies of diabetic nephropathy (program 1039); in 60th Annu Meet Am Soc Hum Genet, Washington, November 2010.

39 Cardon LR, Palmer LJ: Population stratification and spurious allelic association. Lancet 2003;361:598–604.

40 Tishkoff SA, Kidd KK: Implications of biogeography of human populations for 'race' and medicine. Nat Genet 2004;36:S21–S27.

41 Jakobsson M, Scholz SW, Scheet P, Gibbs JR, VanLiere JM, et al: Genotype, haplotype and copy-number variation in worldwide human populations. Nature 2008;451:998–1003.

42 Manolio TA, Collins FS, Cox NJ, Goldstein DB, Hindorff LA, et al: Finding the missing heritability of complex diseases. Nature 2009; 461:747–753.

43 Pezzolesi MG, Poznik GD, Mychaleckyj JC, Paterson AD, Barati MT, et al: Genome-wide association scan for diabetic nephropathy susceptibility genes in type 1 diabetes. Diabetes 2009;58:1403–1410.

44 Maeda S, Araki S, Babazono T, Toyoda M, Umezono T, et al: Replication study for the association between four loci identified by a genome-wide association study on European American subjects with type 1 diabetes and susceptibility to diabetic nephropathy in Japanese subjects with type 2 diabetes. Diabetes 2010;59:2075–2079.

45 Mooyaart AL, Valk EJ, van Es LA, Bruijn JA, de Heer E, et al: Genetic associations in diabetic nephropathy: a meta-analysis. Diabetologia 2011;54:544–553.

Caroline Brorsson, MSc, PhD
Glostrup Research Institute, Glostrup University Hospital
DK–2600 Glostrup (Denmark)
Tel. +45 38633560, E-Mail Cbro0026@glo.regionh.dk

Lai KN, Tang SCW (eds): Diabetes and the Kidney.
Contrib Nephrol. Basel, Karger, 2011, vol 170, pp 19–27

Clinical Manifestation and Natural History of Diabetic Nephropathy

Eberhard Ritz[a] · Xiao-Xi Zeng[b] · Ivan Rychlík[c]

[a]Department of Nephrology, Heidelberg University, Heidelberg, Germany; [b]Department of Medicine, West China Hospital of Sichuan University, Chengdu, China; [c]2nd Department of Medicine, 3rd Faculty of Medicine, Charles University, Prague, Czech Republic

Abstract

The prevalence of diabetes, predominantly of type 2, and the incidence of diabetic nephropathy have dramatically increased worldwide. Diabetic patients constitute the largest proportion of patients with end-stage renal disease (ESRD) requiring dialysis or transplantation; in developed countries, this accounts for up to 50% of ESRD patients, but this proportion has stabilized and possibly somewhat decreased in recent years. Chronic kidney disease in diabetic patients is more heterogeneous than previously thought. The largest proportion suffers from proteinuric diabetic nephropathy with Kimmelstiel-Wilson lesions as the underlying pathology, but reduced glomerular filtration rate in the absence of albuminuria/proteinuria is recognized in an increasing proportion of type 2 diabetic patients. Of particular interest is the recent recognition of vascular lesions in the brain and retina as predictors of nonproteinuric nephropathy with reduced GFR; although currently unproven, such lesions may also be of potential relevance for target blood pressure. Because of the high prevalence of type 2 diabetes in the population, coexisting primary kidney disease and diabetic nephropathy occur in a sizable proportion of type 2 diabetic patients with ESRD. The optimal point to start treatment differs according to target organs. There is no doubt that in proteinuric diabetic patients the earlier the treatment (blood pressure lowering, renin-angiotensin system blockade) is started, the greater is the benefit – at least in patients with proteinuric disease and no major cardiovascular damage. In our opinion, there is no one target blood pressure that fits all patients. Survival of patients with diabetic nephropathy is to a large extent determined by cardiovascular comorbidity. It is currently a matter of debate whether the current definition of type 2 diabetes is appropriate. Some recent findings suggest that minor renal hemodynamic and morphological changes are seen even in (prediabetic) patients who fail to meet the current definition of type 2 diabetes.

Copyright © 2011 S. Karger AG, Basel

History

The glycosuria of type 2 diabetes had been known for millennia [1], and pro-teinuria as well as glomerular lesions of type 2 diabetes for one and one half centuries, respectively; what has been new in the past 2–3 decades, however, is the dramatic increase in end-stage kidney disease in type 2 diabetes – 'a medical catastrophe of worldwide dimension' [2], presumably the result of higher preva-lence of type 2 diabetes secondary to obesity, ageing, physical inactivity, etc. on the one hand and better survival because of treatment of diabetic complications, particularly cardiovascular ones [3].

Epidemiology of End-Stage Renal Disease in Diabetes

Currently, diabetic nephropathy has reached epidemic proportions in many parts of the world. It has become the primary diagnosis in almost 50% of people starting renal replacement therapy in certain Asian countries [4], in Europe and in the US. As one example, in Hong Kong the incidence of dialysis-dependent diabetics increased from 25% in 1996 to 46% in 2004 [4], and similar increases were seen in the US [5] Europe [6] and other parts of the world. Of particu-lar concern is the observation that nephropathy of type 2 diabetes is no lon-ger exclusively a disease of the elderly [7]. These considerations prompted the International Society Nephrology to introduce for the World Kidney Day 2010 the motto: 'Diabetic Kidney Disease: Act Now or Pay Later' [8].

More recently, the relentless increase has flattened out both in Europe [9] and the US; the 2009 data of USRDS (www.usrds.org) showed that today dia-betes accounts for 44% of new admissions for renal replacement therapy. Since one decade, a decrease has been noted of type 2 diabetic patients developing end-stage renal disease (ESRD) as a proportion of all diagnosed diabetics (fig. 1) [10], potentially reflecting better management of type 2 diabetes and of chronic kidney disease (CKD), specifically blood pressure control and renin-angiotensin system (RAS) blockade [11]. The epidemiology is confronted, however, with uncertainties concerning numerator and denominator. Potential lead time bias has to be considered, i.e. a false decrease because of a greater denominator as a result of earlier diagnosis of type 2 diabetes. On the other hand, a high propor-tion of undiagnosed prediabetic and diabetic individuals is found in the gen-eral population [12]. As a further confounder, we observed that the diagnosis of diabetic nephropathy had not been made by the referring physician in 10–20% of diabetic patients admitted for ESRD because of the absence of hyperglyce-mia secondary to anorexia and weight loss in advanced uremia [13]; this may also explain the observation that 'burnt-out' diabetes is seen in ESRD [14] and apparent de novo diabetes in a substantial proportion of patients after starting dialysis.

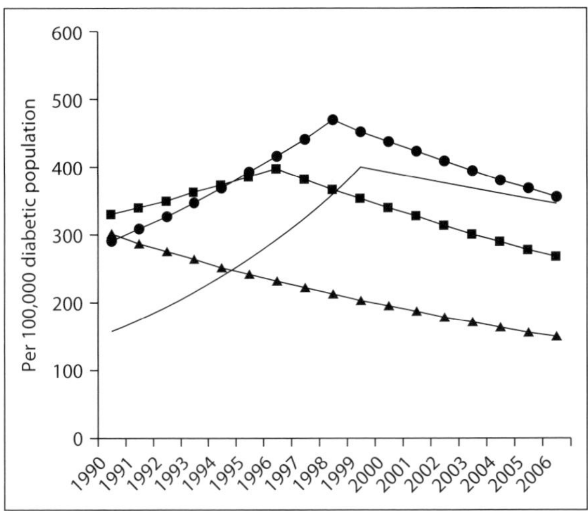

Fig. 1. Modeled age-specific incidence of diabetes-related ESRD in the population with diabetes (US, 1990–2006). Predicted values: ▲ = <45 years; ■ = 45–64 years; ● = 65–74 years; — = ≥75 years. Predicted values were modeled using joinpoint regression analysis [10].

There are interesting differences between ethnicities in the incidence of diabetes and diabetic nephropathy on the one hand and survival before and after renal replacement therapy on the other [15, 16]: ESRD as a consequence of type 2 diabetes is somewhat less common in European populations and is more prevalent in migrant ethnic minority populations. In contrast to the higher incidence of ESRD in diabetes, survival of diabetic patients on dialysis tends to be higher in Asia compared to Europe, as shown in the Taiwan registry [17]. Such interethnic differences of survival rates on dialysis have also been observed in Israel where ethnical groups have a higher incidence of ESRD, but higher rates of survival than Jewish Israelis [18].

Beyond Kimmelstiel-Wilson Nephropathy

While in the past the view prevailed that in type 2 diabetes nephropathy was due to Kimmelstiel-Wilson lesions of the glomerulus and characterized by progression from microalbuminuria to proteinuria and ESRD, it has recently been established beyond doubt that a nonproteinuric type of diabetic nephropathy is quite prevalent indeed [19, 20] (fig. 2), and this may be the result of microvascular disease as suggested by the observation that brain microinfarcts on MRI predict subsequent loss of renal function [21]. This novel form raises obvious issues of the safety of target blood pressures and the relevance (or its absence) of RAS blockade in this novel form.

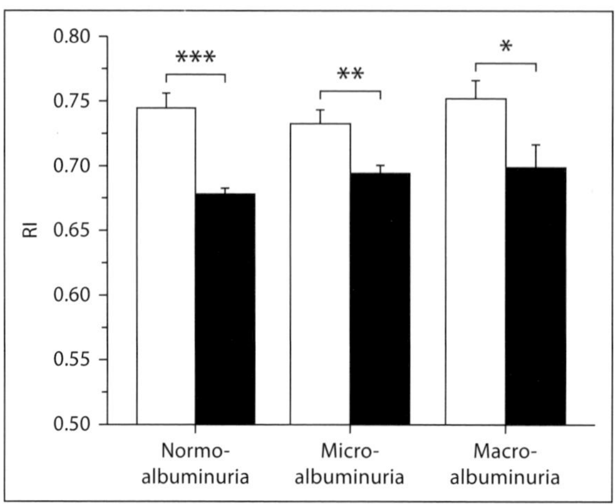

Fig. 2. Intrarenal arterial resistance index (RI) in 325 patients with type 2 diabetes stratified according to eGFR, i.e. < or ≥60 ml/min per 1.73 m^2, and albuminuric status, i.e. normo- (AER <20), micro- (20–200), or macroalbuminuria (>200). ■ = GFR <60 ml/min per 1.73 m^2 (n = 93); □ = eGFR ≥60 ml/min per 1.73 m^2 (n = 232). * $p < 0.05$, ** $p < 0.01$, *** $p < 0.001$ [19].

Renal artery stenosis may present as uncontrolled hypertension and progressive renal failure; it is definitely more frequent than in nondiabetic patients, but its frequency in diabetic patients with advanced renal failure is low [22].

Another addition to the spectrum of presentations is acute renal failure or acute kidney injury (AKI). It is known that episodes of AKI increase the risk of progression to ESRD [23]; even in nondiabetic individuals, glycemia is a predictor of AKI [24]. Preexisting renal damage, as obviously present in diabetes with CKD, predisposes strongly to AKI [25] and subsequently higher risk of progression to ESRD.

Special Issues of Nephropathy in Type 2 Diabetes

A brief comment on two accessory findings in nephropathy of type 2 diabetes is appropriate. In type 1 diabetes, diabetic nephropathy is almost always accompanied by extrarenal signs of diabetic microvascular disease, e.g. retinopathy and neuropathy, but this is much less consistent in type 2 diabetes. If retinopathy is absent in type 1 diabetes, other causes of CKD must be considered [26], but in type 2 diabetes absence of retinopathy does by no means exclude the diagnosis of diabetic nephropathy [27].

The frequent presence of microhematuria may pose a diagnostic problem. There had been suggestions that microhematuria pointed to the presence of nondiabetic glomerulopathy, e.g. thin basement membrane disease or IgA nephropathy [28]. However, the hallmark of glomerulonephritis, i.e. acanthocyturia (dysmorphic erythrocytes), is uncommon in diabetic nephropathy. Hematuria was found in 62% of patients with the clinical diagnosis of diabetic nephropathy, but acanthocyturia was found in only 4%, in contrast to the high prevalence of 40% in patients with glomerulonephritis [29].

The coexistence of diabetic nephropathy with primary kidney disease is not unanticipated given the frequencies of these two conditions, depending on ethnicity, geographic location, biopsy policy, etc. Renal biopsy studies in type 2 diabetes documented how misleading clinical data can be [30].

Prevention of Progression to ESRD in Diabetic Nephropathy with Impaired Renal Function

In both type 1 and type 2 diabetes, controlled trials documented that progression to CKD can be attenuated, specifically by interventions, glycemic control, blood pressure control, RAS blockade and possibly statins.

A crucially important point is early treatment. This is illustrated by comparing the degree to which progression was attenuated in early diabetic nephropathy (DETAIL study) [31] in contrast to the IDNT [32, 33] and RENAAL trials [33]. The argument of early intervention gains further momentum by the observation that albuminuria even within the normoalbuminuric range is a predictor of ESRD [34] and that in normoalbuminuric type 2 diabetic patients, ACE inhibitors [35] or ARB [Haller, in press] reduce progression. That not only blood pressure, but also proteinuria is an independent treatment target is illustrated by the analysis of Atkins et al. [36] of the data of the IDNT study. Because RAS blockade attenuates progression only in proteinuric not in nonproteinuric patients [37], the issue of treatment of nonproteinuric diabetic CKD patients is unresolved; data in nondiabetic individuals suggest no further benefit from aggressively lowering blood pressure below conventional targets [38].

A second issue is blood pressure control. The issue whether dedicated blood pressure lowering to prespecified targets improves outcomes is uncertain. One major problem is that office blood pressure is the worst of all blood pressure parameters, whereas the patient self-measured morning blood pressure (and 24-hour blood pressure) predicted best nephropathy, cardiac and retinal complications [39]. A currently unresolved issue is that in diabetes central blood pressure in the aorta is substantially higher than in the periphery. An important (and often neglected) aspect is that patients with diabetic polyneuropathy frequently experience orthostatic hypotension.

Observational studies clearly showed: the lower the blood pressure, the better the outcome [40]. Furthermore, in all controlled intervention trials in diabetes, the arm aiming for lower values had better outcomes [41].

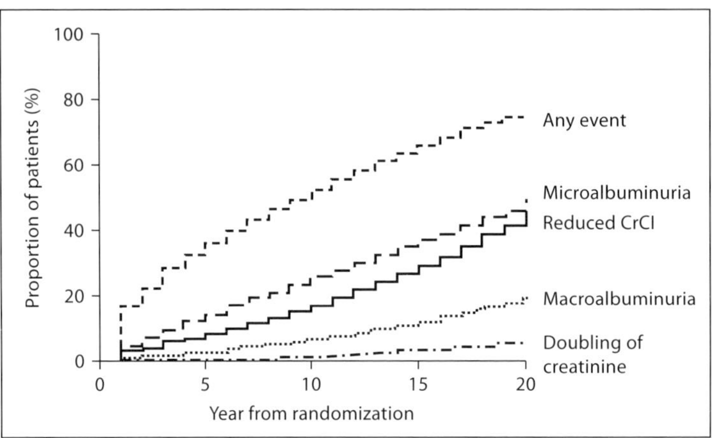

Fig. 3. Kaplan-Meier plots of proportion of patients with microalbuminuria, macroalbuminuria, reduced creatinine clearance (CrCl), doubling of plasma creatinine, or any one of these after diagnosis of diabetes [20].

There is no doubt that in type 2 diabetic patients with moderately advanced target organ damage as shown in the ADVANCE study, endpoints, particularly renal endpoints, are reduced even when blood pressure is lowered to values <120 mm Hg systolic [42]. However, in studies with more advanced diabetic nephropathy, systolic blood pressures <120 mm Hg, and particularly low diastolic blood pressures, were associated with higher mortality [43] and MI, respectively. It is important that blood pressure lowering and reduction in proteinuria are independent and additive [43].

The current controversy surrounding the target is the result of the observation that hypertensive patients (including diabetic patients) of the ON TARGET study (with high cardiovascular comorbidity!) experienced higher mortality if baseline blood pressure was <130 mm Hg or less [44]. It is the opinion of the authors that target blood pressures should be individualized, based on cardiovascular comorbidity and potential orthostatic hypotension.

Beyond Diabetic Nephropathy

At the end, a brief reflection whether treatment of manifest type 2 diabetes, and particularly manifest diabetic nephropathy, does not come to a certain extent too late. One important cause of the current epidemic of type 2 diabetes and diabetic nephropathy is the increasing prevalence of its precursor, the metabolic syndrome, i.e. the prediabetic state the prevalence of which has increased dramatically with the ageing of the population and is particularly frequent in certain ethnicities, e.g. present in no less than 73% of the South Indian Population lead-ing to efforts to introduce preventive programs in South East Asia. It is expected that interventions including weight loss, dietary control, and increased exer-cise reduce the progression to overt diabetes and finally diabetic nephropathy, although some prevention studies with limited observation times so far cannot prove reduction in renal risk, although some benefit was seen with respect to retinopathy [45].

Whether the current definition of diabetes deserves downward revision is currently unclear. At any rate, in the ARIC study with 15 years follow-up, HbA1c values >6–6.5 predicted onset of diabetes during follow-up [46]. In pro-teinuric patients with metabolic syndrome, Mac-Moune et al. [47] observed GBM thickening which preceded overt diabetes by up to 2 years. In patients with metabolic syndrome, Sucurro et al. [48] noted a significant decrease in eGFR prior to the onset of overt diabetes. Finally, there have been reports of diffuse [49] and nodular [50] glomerulosclerosis in patients who later on developed overt diabetes – it is unclear whether the glomerular lesions pre-ceded overt diabetes (prediabetic) or whether they were the remnant of past episodes of manifest diabetes (postdiabetic). At any rate, the true point of onset of diabetic nephropathy, and the validity of our current diagnostic instruments deserve close scrutiny.

References

1 Ardalan MR, Shoja MM, Tubbs RS, Eknoyan G: Diseases of the kidney in medieval Persia – the Hidayat of Al-Akawayni. Nephrol Dial Transplant 2007;22:3413–3121.

2 Ritz E, Rychlík I, Locatelli F, Halimi S: End-stage renal failure in type 2 diabetes: a medi-cal catastrophe of worldwide dimensions. Am J Kidney Dis 1999;34:795–808.

3 van Dieren S, Uiterwaal CS, van der Schouw YT, et al: The global burden of diabetes and its complications: an emerging pandemic. Eur J Cardiovasc Prev Rehabil 2010;17(suppl 1):S3–S8.

4 Tang SC: Diabetic nephropathy: a global and growing threat. Hong Kong Med J 2010;16:244–245.

5 Williams ME: Diabetic CKD/ESRD 2010: a progress report? Semin Dial 2010;23:129–33.

6 Rychlik I, Miltenberger-Miltenyi G, Ritz E: The drama of the continuous increase in end-stage renal failure in patients with type II diabetes mellitus. Nephrol Dial Transpl 1998;13(suppl 8):6–10.

7 Yokoyama H, Okudaira M, Otani T, et al: Higher incidence of diabetic nephropathy in type 2 than in type 1 diabetes in early-onset diabetes in Japan. Kidney Int 2000;58:302–311.

8 Atkins RC, Zimmet P: Diabetic kidney disease: act now or pay later. J Am Soc Hypertens 2010;4:3–6.

9 Zoccali C, Kramer AK, Jager K: The data-bases: renal replacement therapy since 1989 – the European Renal Association and European Dialysis and Transplant Association (ERA-EDTA). Clin J Am Soc Nephrol 2009;4(suppl 1):S18–S22.

10 Burrows NR, Li Y, Geiss LS: Incidence of treatment for end-stage renal disease among individuals with diabetes in the U.S. contin-ues to decline. Diabetes Care 2009;33:73–77.

11 James MT, Hemmelgarn BR, Tonelli M: Early recognition and prevention of chronic kid-ney disease. Lancet 2010;375:1296–1309.

12 Plantinga LC, Crews DC, Coresh J, et al: Prevalence of chronic kidney disease in US adults with undiagnosed diabetes or prediabetes. Clin J Am Soc Nephrol 2010;5: 673–682.

13 Schwenger V, Müssig C, Hergesell O, Zeier M, Ritz E: Incidence and clinical characteristics of renal insufficiency in diabetic patients (in German). Dtsch Med Wochenschr 2001;126:1322–1326.

14 Kovesdy CP, Park JC, Kalantar-Zadeh K: Glycemic control and burnt-out diabetes in ESRD. Semin Dial 2010;23:148–156.

15 Randhawa G: Renal health disparities in the United Kingdom: a focus on ethnicity. Semin Nephrol 2010;30:8–11.

16 Feehally J: Ethnicity and renal replacement therapy. Blood Purif 2010;29:125–129.

17 Kao TW, Huang JW, Hung KY, et al: Life expectancy, expected years of life lost and survival of hemodialysis and peritoneal dialysis patients. J Nephrol 2010;23:677–682.

18 Kalantar-Zadeh K, Golan E, Shohat T, et al: Survival disparities within American and Israeli dialysis populations: learning from similarities and distinctions across race and ethnicity. Semin Dial 2010;23:586–594.

19 MacIsaac RJ, Tsalamandris C, Panagiotopoulos S, et al: Nonalbuminuric renal insufficiency in type 2 diabetes. Diabetes Care 2004;27:195–200.

20 Retnakaran R, Cull CA, Thorne KI, et al: Risk factors for renal dysfunction in type 2 diabetes: U.K. Prospective Diabetes Study 74. Diabetes 2006;55:1832–1839.

21 Uzu T, Kida Y, Shirahashi N, et al: Cerebral microvascular disease predicts renal failure in type 2 diabetes. J Am Soc Nephrol 2010;21:520–526.

22 Myers DI, Poole LJ, Iman K, et al: Renal artery stenosis by three-dimensional magnetic resonance angiography in type 2 diabetics with uncontrolled hypertension and chronic renal insufficiency: prevalence and effect on renal function. Am J Kidney Dis 2003;41:351–359.

23 Ishani A, Xue JL, Himmelfarb J, et al: Acute kidney injury increases risk of ESRD among elderly. J Am Soc Nephrol 2009;20:223–228.

24 Stolker JM, McCullough PA, Rao S, et al: Pre-procedural glucose levels and the risk for contrast-induced acute kidney injury in patients undergoing coronary angiography. J Am Coll Cardiol 2010;55:1433–1440.

25 Singh P, Rifkin DE, Blantz RC: Chronic kidney disease: an inherent risk factor for acute kidney injury? Clin J Am Soc Nephrol 2010;5:1690–1695.

26 Christensen PK, Larsen S, Horn T, et al: Causes of albuminuria in patients with type 2 diabetes without diabetic retinopathy. Kidney Int 2000;58:1719–1731.

27 Chavers BM, Mauer SM, Ramsay RC, Steffes MW: Relationship between retinal and glomerular lesions in IDDM patients. Diabetes 1994;43:441–446.

28 Pham TT, Sim JJ, Kujubu DA, et al: Prevalence of nondiabetic renal disease in diabetic patients. Am J Nephrol 2007;27: 322–328.

29 Heine GH, Sester U, Girndt M, Köhler H: Acanthocytes in the urine: useful tool to differentiate diabetic nephropathy from glomerulonephritis? Diabetes Care 2004;27:190–194.

30 Mazzucco G, Bertani T, Fortunato M, et al: Different patterns of renal damage in type 2 diabetes mellitus: a multicentric study on 393 biopsies. Am J Kidney Dis 2002;39:713–720.

31 Barnett AH, Bain SC, Bouter P, et al: Angiotensin-receptor blockade versus converting-enzyme inhibition in type 2 diabetes and nephropathy. N Engl J Med 2004;351:1952–1961.

32 Lewis EJ, Hunsicker LG, Clarke WR, et al: Renoprotective effect of the angiotensin-receptor antagonist irbesartan in patients with nephropathy due to type 2 diabetes. N Engl J Med 2001;352:851–860.

33 Brenner BM, Cooper ME, de Zeeuw D, et al: Effects of losartan on renal and cardiovascular outcomes in patients with type 2 diabetes and nephropathy. N Engl J Med 2001;345:861–869.

34 Hallan SI, Ritz E, Lydersen S, et al: Combining GFR and albuminuria to classify CKD improves prediction of ESRD. J Am Soc Nephrol 2009;20:1069–1077.

35 Ruggenenti P, Fassi A, Ilieva AP, et al: Preventing microalbuminuria in type 2 diabetes. N Engl J Med 2004;351:1941–1951.

36 Atkins RC, Briganti EM, Lewis JB, et al: Proteinuria reduction and progression to renal failure in patients with type 2 diabetes mellitus and overt nephropathy. Am J Kidney Dis 2005;45:281–287.

37 Jafar TH, Schmid CH, Landa M, et al: Angiotensin-converting enzyme inhibitors and progression of nondiabetic renal disease. A meta-analysis of patient-level data. Ann Intern Med 2001;135:73–87.

38 Appel LJ, Wright JT Jr, Greene T, et al: Intensive blood-pressure control in hypertensive chronic kidney disease. N Engl J Med 2010;363:918–929.

39 Kamoi K, Miyakoshi M, Soda S, et al: Usefulness of home blood pressure measurement in the morning in type 2 diabetic patients. Diabetes Care 2002;25:2218–2223.

40 Adler AI, Stratton IM, Neil HA, et al: Association of systolic blood pressure with macrovascular and microvascular complications of type 2 diabetes (UKPDS 36): prospective observational study. BMJ 2000;321:412–419.

41 Mancia G, Laurent S, Agabiti-Rosei E, et al: Reappraisal of European guidelines on hypertension management: a European Society of Hypertension Task Force document. J Hypertens 2009;27:2121–2158.

42 de Galan BE, Perkovic V, Ninomiya T, et al: Lowering blood pressure reduces renal events in type 2 diabetes. J Am Soc Nephrol 2009;20:883–892.

43 PohlMA, Blumenthal S, Cordonnier DJ, et al: Independent and additive impact of blood pressure control and angiotensin II receptor blockade on renal outcomes in the irbesartan diabetic nephropathy trial: clinical implications and limitations. J Am Soc Nephrol 2005;16:3027–3037.

44 Sleight P Redon J, Verdecchia P, et al: Prognostic value of blood pressure in patients with high vascular risk in the Ongoing Telmisartan Alone and in combination with Ramipril Global Endpoint Trial study. J Hypertens 2009;27:1360–1369.

45 Gong Q, Gregg EW, Wang J, et al: Long-term effects of a randomised trial of a 6-year lifestyle intervention in impaired glucose tolerance on diabetes-related microvascular complications: the China Da Qing Diabetes Prevention Outcome Study. Diabetologia 2011;54:300–307.

46 Selvin E, Steffes MW, Zhu H, et al: Glycated hemoglobin, diabetes, and cardiovascular risk in nondiabetic adults. N Engl J Med 2010;362:800–811.

47 Mac-Moune Lai F, Szeto CC, Choi PC, et al: Isolate diffuse thickening of glomerular capillary basement membrane: a renal lesion in prediabetes? Mod Pathol 2004;17:1506–1512.

48 Succurro E, Arturi F, Lugarà M, et al: One-hour postload plasma glucose levels are associated with kidney dysfunction. Clin J Am Soc Nephrol 2010;5:1922–1927.

49 Altiparmak MR, Pamuk ON, Pamuk GE, et al: Diffuse diabetic glomerulosclerosis in a patient with impaired glucose tolerance: report on a patient who later develops diabetes mellitus. Neth J Med 2002;60:260–262.

50 Souraty P, Nast CC, Mehrotra R, et al: Nodular glomerulosclerosis in a patient with metabolic syndrome without diabetes. Nat Clin Pract Nephrol 2008;4:639–642.

Eberhard Ritz, MD, PhD,
Department of Nephrology, Heidelberg University
Heidelberg (Germany)
Tel. +49 6221 9112 601, Fax +49 6221 603302, E-Mail Prof.E.Ritz@T-online.de

Lai KN, Tang SCW (eds): Diabetes and the Kidney.
Contrib Nephrol. Basel, Karger, 2011, vol 170, pp 28–35

Obesity, Metabolic Syndrome and Diabetic Nephropathy

Christine Maric · John E. Hall

Department of Physiology and Biophysics, University of Mississippi Medical Center,
Jackson, Miss., USA

Abstract

Diabetic nephropathy is becoming an increasingly important cause of morbidity and mortality worldwide owing to the increasing prevalence of type 2 diabetes, largely driven by increasing obesity. There is considerable evidence that obesity, hypertension and other elements of the metabolic syndrome also contribute to the progression of renal disease independent of diabetes. How they interact and contribute to diabetic nephropathy, however, is not completely understood. Clinical diabetic nephropathy is preceded by an increase in glomerular filtration rate, microalbuminuria and glomerular hypertrophy. Poor glycemic control and elevated systolic blood pressure exacerbate proteinuria and renal injury that may culminate in end-stage renal disease. A similar sequence of events may lead to obesity-related renal disease even in the absence of diabetes. This chapter compares and contrasts factors involved in the development of glomerular hemodynamic and kidney pathological processes associated with diabetes and obesity.

Diabetes is the most common cause of end-stage renal disease (ESRD) worldwide and is associated with increased cardiovascular risk, high morbidity and mortality [1, 2]. While the increasing prevalence of ESRD is to some extent due to better survival of diabetic patients with chronic renal disease as a result of more effective treatments, it is mostly due to the dramatic increase in the prevalence of type 2 diabetes. The growing prevalence of obesity is a major driving force for the continued increase in the prevalence of type 2 diabetes [3] as well as the cluster of risk factors that make up the metabolic syndrome, including hypertension, insulin resistance and dyslipidemia. These disorders are intertwined and likely interact to increase the severity of chronic renal disease. Thus, understanding the complex interrelationships between these factors is of vital importance for our understanding of the pathophysiology of diabetic nephropathy and development of effective regimens for its prevention and treatment.

Clinical diabetic nephropathy evolves in a sequence of stages beginning with initial increases in glomerular filtration rate (GFR) and intraglomerular capillary pressure (P_{Gc}), glomerular hypertrophy and microalbuminuria [4, 5]. Poor glycemic control and elevated systolic blood pressure (BP) further exacerbate the disease progression to proteinuria, nodular glomerulosclerosis and tubulointerstitial injury and a decline in GFR which can eventually lead to ESRD [6, 7]. The development of diabetic nephropathy is commonly thought to result from the cumulative interactions among multiple metabolic and hemodynamic factors which activate common intracellular signaling pathways that in turn trigger the production of cytokines and growth factors, leading to renal disease. However, the current consensus on what triggers the development of diabetic nephropathy and how it progresses is evolving in the face of worldwide epidemics of obesity, diabetes and kidney disease. This article reviews and figure 1 summarizes potential mechanisms by which obesity, diabetes, hypertension and elements of the metabolic syndrome may either cause or exacerbate the progression of nephropathy.

Role of Hypertension in Nephropathy Associated with Obesity, Metabolic Syndrome and Diabetes

Basic, clinical and population research indicate that visceral obesity, the main driver for type 2 diabetes and metabolic syndrome, raises BP [8, 9]. Population studies indicate that excess weight gain may account for 78% of primary (essential) hypertension in men and 65% in women [10]. Moreover, the relationship between body mass index and BP is nearly linear in diverse populations throughout the world [11]. Clinical studies also indicate that weight loss reduces BP in most hypertensive subjects and is effective in primary prevention of hypertension [9].

Visceral, but not subcutaneous, obesity appears to induce hypertension initially by increasing renal tubular reabsorption and causing a hypertensive shift of renal pressure natriuresis through multiple mechanisms including activation of the sympathetic nervous system and renin-angiotensin-aldosterone system (RAAS), as well as physical compression of the kidneys [12, 13]. The hypertension that ensues, as well as the increases in P_{GC} and GFR, and the metabolic abnormalities (e.g. dyslipidemia, hyperglycemia) likely interact to cause renal injury. A similar sequence of events may lead to renal injury in diabetes, irrespective of obesity, suggesting that hypertension plays a key role in obesity and diabetes-associated renal injury. Further supporting the central role of

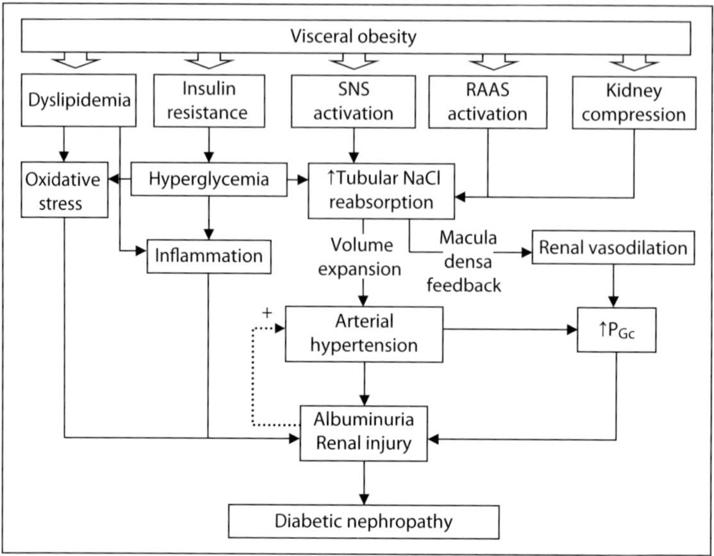

Fig. 1. Interaction between metabolic and hemodynamic pathways in the pathophysiology of diabetic nephropathy. SNS = Sympathetic nervous system.

hypertension in renal injury is the fact that progressive renal injury only occurs when hypertension is superimposed on obesity or diabetes [14]. The importance of rigorous BP control for treating diabetic nephropathy is recognized in current guidelines, with a recommended target of BP less than 130/80 mm Hg [15]. Multiple studies have clearly shown the protective effect on the kidneys of reducing BP in diabetes. In fact, tight BP control in diabetic patients may slow progression of nephropathy to a greater extent than tight control of blood glucose [16].

Effects of Obesity, Metabolic Syndrome and Diabetes on Glomerular Hemodynamics

One of the earliest renal changes in obese humans [17] and in obese dogs fed a high-fat diet for only 5–6 weeks [13] is increased GFR. A likely explanation for increased GFR in obesity is increased salt reabsorption by the proximal tubule or loop of Henle, leading to tubuloglomerular feedback (TGF)-mediated reduction in afferent arteriolar resistance, increased P_{Gc} and increased GFR [18]. The increased GFR initially serves as a compensatory response that permits restoration of salt balance despite continued increases in tubular reabsorption but, over the long-term, contributes to renal injury, especially when combined with elevated BP. The TGF-mediated dilation of afferent arterioles and attendant

Maric · Hall

impairment of renal autoregulation permit increases in BP to be transmitted to the glomerular capillaries causing even greater increases in P_{Gc} and glomerular injury than would occur with comparable increases in BP in kidneys of non-obese, nondiabetic subjects [19].

Renal vasodilation and increases in GFR and P_{Gc} also occur early in the development of diabetes and are associated with a high risk for development of overt diabetic nephropathy [20]. Hyperglycemia is presumed to contribute to the development of glomerular hyperfiltration through mechanisms similar to those occurring in obesity, although the precise mechanisms underlying increased GFR remain inconclusive. Reduced delivery of salt to the macula densa as a consequence of increased proximal reabsorption of glucose and sodium may reduce afferent arteriolar resistance and increase P_{Gc} and GFR via attenuated TGF [21–23]. Also, afferent vasodilation and efferent vasoconstriction in response to circulating or locally formed vasoactive factors (e.g. angiotensin II) produced in response to hyperglycemia or shear stress may promote diabetic glomerular hyperfiltration [24, 25].

Even though the mechanisms explaining the increase in GFR in diabetes and obesity uncomplicated by diabetes may be similar, the factors that trigger TGF-mediated renal vasodilation and glomerular hyperfiltration are different, and some studies suggest that hyperglycemia and obesity may have at least partially additive effects on glomerular hemodynamics [19].

Since the metabolic syndrome is by definition a clustering of several metabolic factors and hypertension, it is often difficult to separate the effects of each element on glomerular hemodynamics and progression of renal injury. Experimental studies, however, provide some clues. For example, mice lacking the gene for the melanocortin-4 receptor are obese, hyperinsulinemic and hyperleptinemic but normotensive at 55 weeks of age [14]. These animals have moderately increased GFR and only modest albuminuria compared with wild-type mice; however, their GFR and albuminuria increased further when rendered hypertensive following treatment with N(G)-nitro-L-arginine methyl ester. These data suggest that elevations in BP exacerbate obesity-related glomerular hyperfiltration and albuminuria, further supporting the concept of an additive, or perhaps synergistic effect of various components of obesity, metabolic syndrome, diabetes and hypertension on glomerular hemodynamics.

It is not clear, however, how these same factors contribute to the decline in GFR characteristic of advanced diabetic nephropathy. Few studies have examined this issue due to the fact that most experimental models of diabetes-related renal injury never really develop overt nephropathy and are in a permanent state of glomerular hyperfiltration. Obesity, metabolic syndrome and diabetes are states of low-grade inflammation and oxidative stress, all of which may lead to kidney damage, progressive loss of nephrons and decline in GFR over time. Another element of the metabolic syndrome, hyperlipidemia, has been linked to reductions in GFR in diabetic nephropathy, especially in the latter stages of the disease. Numerous clinical trials have pointed to the importance of lipid control in preserving GFR in patients with diabetes [26]. However, further studies are needed to determine if the beneficial effects of lipid-lowering agents in diabetic nephropathy are due to improvement in the lipid profile or if there are other renoprotective effects.

Effects of Obesity, Metabolic Syndrome and Diabetes on Albuminuria

Diabetic nephropathy and elements of the metabolic syndrome including insulin resistance and hyperinsulinemia, correlate with the development of microalbuminuria early in the disease process [27, 28]. Microalbuminuria, in turn, signifies increased risk of progression to ESRD and cardiovascular disease [29]. The development of microalbuminuria in diabetic nephropathy was traditionally thought to stem from damage to the glomerular filtration barrier as a consequence of increases in BP which are transmitted to the glomeruli, raising P_G and GFR, and/or hyperglycemia-associated inflammation and oxidative stress [27]. An alternative, or perhaps complementary, explanation is that diabetes also impairs proximal tubular reabsorption of albumin which filters across the glomerular barrier [30].

Hyperlipidemia is known to be a risk factor for the development of albuminuria in patients with diabetes [31]. Further supporting the importance of hyperlipidemia in diabetic nephropathy is the fact that treatment with lipid-lowering agents improves proteinuria. However, the effectiveness of statins on albuminuria in overt diabetic nephropathy has not been established.

Effects of Obesity, Metabolic Syndrome and Diabetes on Glomerular Injury

There are several similarities in the histological appearance of glomeruli from diabetic and obese individuals, and both have early structural changes in the kidney that accompany hyperfiltration and microalbuminuria. The glomeruli of patients with diabetes are characterized by glomerular hypertrophy, widening of the glomerular basement membrane, mesangial expansion and podocytopenia leading to nodular (Kimmelstiel-Wilson) glomerulosclerosis. Similarly, obesity-associated renal injury is characterized by glomerulomegaly, mesangial expansion and podocytopenia leading to focal glomerulosclerosis [19].

Given the similarities in the histological appearance of the glomerulus from diabetic and obese subjects, it is not surprising that the mechanisms leading to these changes are also similar. Diabetes and obesity are both states of low-grade inflammation associated with macrophage infiltration into the adipose tissue and the kidney. The infiltrating macrophages become a source of a whole host of proinflammatory cytokines including tumor necrosis factor-α, interleukin-6, and monocyte chemotactic protein-1 [32]. Furthermore, increased adiposity triggers the release of adipokines into the circulation that in turn may cause renal injury via production of reactive oxygen species (ROS). Persistent hyperglycemia also activates vasoactive hormonal pathways including the RAAS and endothelin. These in turn activate common second messenger signaling pathways such as protein kinase C and MAP kinase and transcription factors such as nuclear factor-κB that lead to the alteration in gene expression of a plethora of growth factors and cytokines such as transforming growth factor-β (TGF-β). TGF-β is a key player in promoting podocyte apoptosis, mesangial cell proliferation and extracellular matrix synthesis, cellular events that are important in the development of diabetes and obesity-associated glomerular injury [33]. Hyperglycemia and associated metabolic disturbances also cause mitochondrial dysfunction and enhanced generation of ROS, which directly alter the expression of key proteins and cytokines causing renal injury.

Kidneys of obese individuals often have glomerular/mesangial lipid deposits (foam cells) present, which supports the concept of lipotoxicity, i.e. lipid-induced renal injury [19]. One of the mechanisms by which hyperlipidemia promotes glomerular injury is through renal upregulation of sterol-regulatory element-binding proteins (SREBP-1 and -2), which in turn promotes podocyte apoptosis and mesangial cell proliferation and cytokine synthesis.

Conclusion

Data from basic and clinical studies suggest that obesity, hypertension, hyperglycemia, hyperlipidemia and other elements of the metabolic syndrome are highly interrelated and contribute to the development and progression of diabetic nephropathy. Given their close relationship, it is often difficult to dissect out their individual effects. It is likely that multiple metabolic abnormalities act in concert to initially cause renal vasodilation, glomerular hyperfiltration and albuminuria that, in turn, effect glomerular pathology, especially when combined with increased BP, and that ultimately progresses to diabetic nephropathy.

Acknowledgement

The authors acknowledge the financial support of NIH/NHLBI (PO1HL51971 to J.E. Hall) and NIH/NIDDK (RO1DK075832 to C. Maric).

References

1 de Zeeuw D, Ramjit D, Zhang Z, Ribeiro AB, Kurokawa K, Lash JP, Chan J, Remuzzi G, Brenner BM, Shahinfar S: Renal risk and renoprotection among ethnic groups with type 2 diabetic nephropathy: a post hoc analysis of RENAAL. Kidney Int 2006;69: 1675–1682.

2 Remuzzi G, Macia M, Ruggenenti P: Prevention and treatment of diabetic renal disease in type 2 diabetes: the BENEDICT study. J Am Soc Nephrol 2006;17:S90–97.

3 Ogden CL, Carroll MD, Curtin LR, McDowell MA, Tabak CJ, Flegal KM: Prevalence of overweight and obesity in the United States, 1999–2004. JAMA 2006;295:1549–1555.

4 Thomson SC, Vallon V, Blantz RC: Kidney function in early diabetes: the tubular hypothesis of glomerular filtration. Am J Physiol Renal Physiol 2004;286: F8–F15.

5 Hostetter TH: Hyperfiltration and glomerulosclerosis. Semin Nephrol 2003;23:194–199.

6 Caramori ML, Mauer M: Diabetes and nephropathy. Curr Opin Nephrol Hypertens 2003;12:273–282.

7 Leon CA, Raij L: Interaction of haemodynamic and metabolic pathways in the genesis of diabetic nephropathy. J Hypertens 2005;23:1931–1937.

8 Hall JE, Jones DW, Kuo JJ, da Silva A, Tallam LS, Liu J: Impact of the obesity epidemic on hypertension and renal disease. Curr Hypertens Rep 2003;5:386–392.

9 Neter JE, Stam BE, Kok FJ, Grobbee DE, Geleijnse JM: Influence of weight reduction on blood pressure: a meta-analysis of randomized controlled trials. Hypertension 2003;42:878–884.

10 Garrison RJ, Kannel WB, Stokes J 3rd, Castelli WP: Incidence and precursors of hypertension in young adults: the Framingham Offspring Study. Prev Med 1987;16:235–251.

11 Kalantar-Zadeh K, Kopple JD: Obesity paradox in patients on maintenance dialysis. Contrib Nephrol 2006;151:57–69.

12 Hall JE, Henegar JR, Dwyer TM, Liu J, Da Silva AA, Kuo JJ, Tallam L: Is obesity a major cause of chronic kidney disease? Adv Ren Replace Ther 2004;11:41–54.

13 Henegar JR, Bigler SA, Henegar LK, Tyagi SC, Hall JE: Functional and structural changes in the kidney in the early stages of obesity. J Am Soc Nephrol 2001;12: 1211–1217.

14 do Carmo JM, Tallam LS, Roberts JV, Brandon EL, Biglane J, da Silva AA, Hall JE: Impact of obesity on renal structure and function in the presence and absence of hypertension: evidence from melanocortin-4 receptor-deficient mice. Am J Physiol Regul Integr Comp Physiol 2009;297:R803–R812.

15 Van Buren PN, Toto R: Hypertension in diabetic nephropathy: epidemiology, mechanisms, and management. Adv Chronic Kidney Dis 2011;18:28–41.

16 Mancia G: Effects of intensive blood pressure control in the management of patients with type 2 diabetes mellitus in the Action to Control Cardiovascular Risk in Diabetes (ACCORD) trial. Circulation 2010;122: 847–849.

17 Chagnac A, Weinstein T, Herman M, Hirsh J, Gafter U, Ori Y: The effects of weight loss on renal function in patients with severe obesity. J Am Soc Nephrol 2003;14:1480–1486.

18 Hall JE: The kidney, hypertension, and obesity. Hypertension 2003;41:625–633.

19 Griffin KA, Kramer H, Bidani AK: Adverse renal consequences of obesity. Am J Physiol Renal Physiol 2008;294:F685–F696.

20 Yip JW, Jones SL, Wiseman MJ, Hill C, Viberti G: Glomerular hyperfiltration in the prediction of nephropathy in IDDM: a 10-year follow-up study. Diabetes 1996;45: 1729–1733.

21 Vallon V, Schroth J, Satriano J, Blantz RC, Thomson SC, Rieg T: Adenosine A(1) receptors determine glomerular hyperfiltration and the salt paradox in early streptozotocin diabetes mellitus. Nephron Physiol 2009;111:30–38.

22 Woods LL, Mizelle HL, Hall JE: Control of renal hemodynamics in hyperglycemia: possible role of tubuloglomerular feedback. Am J Physiol 1987;252:F65–F73.

23 Persson P, Hansell P, Palm F: Tubular reabsorption and diabetes-induced glomerular hyperfiltration. Acta Physiol (Oxf) 2010;200:3–10.

24 Cherney DZ, Scholey JW, Miller JA: Insights into the regulation of renal hemodynamic function in diabetic mellitus. Curr Diabetes Rev 2008;4:280–290.

25 Carmines PK: The renal vascular response to diabetes. Curr Opin Nephrol Hypertens 2010;19:85–90.

26 Fried LF, Orchard TJ, Kasiske BL: Effect of lipid reduction on the progression of renal disease: a meta-analysis. Kidney Int 2001; 59:260–269.

27 Jauregui A, Mintz DH, Mundel P, Fornoni A: Role of altered insulin signaling pathways in the pathogenesis of podocyte malfunction and microalbuminuria. Curr Opin Nephrol Hypertens 2009;18:539–545.

28 de Boer IH, Sibley SD, Kestenbaum B, Sampson JN, Young B, Cleary PA, Steffes MW, Weiss NS, Brunzell JD: Central obesity, incident microalbuminuria, and change in creatinine clearance in the epidemiology of diabetes interventions and complications study. J Am Soc Nephrol 2007;18:235–243.

29 Eijkelkamp WB, Zhang Z, Remuzzi G, Parving HH, Cooper ME, Keane WF, Shahinfar S, Gleim GW, Weir MR, Brenner BM, de Zeeuw D: Albuminuria is a target for renoprotective therapy independent from blood pressure in patients with type 2 diabetic nephropathy: post hoc analysis from the Reduction of Endpoints in NIDDM with the Angiotensin II Antagonist Losartan (RENAAL) trial. J Am Soc Nephrol 2007;18: 1540–1546.

30 Comper WD, Russo LM: The glomerular filter: an imperfect barrier is required for perfect renal function. Curr Opin Nephrol Hypertens 2009;18:336–342.

31 Rutledge JC, Ng KF, Aung HH, Wilson DW: Role of triglyceride-rich lipoproteins in diabetic nephropathy. Nat Rev Nephrol 2010;6: 361–370.

32 King GL: The role of inflammatory cytokines in diabetes and its complications. J Periodontol 2008;79:1527–1534.

33 Ziyadeh FN: Mediators of diabetic renal disease: the case for TGF-beta as the major mediator. J Am Soc Nephrol 2004;15: S55–S57.

Christine Maric, PhD
Department of Physiology & Biophysics, University of Mississippi Medical Center
2500 North State Street
Jackson, MS, 39216-4505 (USA)
Tel. +1 601 984 1818, Fax +1 601 984 1817, E-Mail cmaric@umc.edu

Lai KN, Tang SCW (eds): Diabetes and the Kidney.
Contrib Nephrol. Basel, Karger, 2011, vol 170, pp 36–47

Pathology of Human Diabetic Nephropathy

Behzad Najafian[a] · Charles E. Alpers[a] ·
Agnes B. Fogo[b]

[a]Department of Pathology, University of Washington, Seattle, Wash., and
[b]Department of Pathology, Vanderbilt University Medical Center, Nashville, Tenn., USA

Abstract

Diabetic nephropathy (DN) develops in a subset of diabetic patients, on average about 15 years after onset of metabolic abnormalities. The earliest lesions consist of thickened glomerular basement membranes (GBM), mild mesangial expansion, and arteriolar accumulation of hyaline. Mesangiolysis and exuberant mesangial repair then develop, ultimately resulting in marked increase in mesangial matrix. Established nephropathy is characterized by mesangial expansion which may be nodular, so-called Kimmelstiel-Wilson nodules, hyaline in both afferent and efferent arterioles, and markedly thickened GBM by electron microscopy. Podocyte loss may be a crucial contributor to this progressive sclerosis. There is a distinct predilection of segmental sclerosis to occur at the glomerulotubular junction in type 1 diabetic patients, which can lead to disruption of outflow from the glomerulus, resulting in so-called atubular glomeruli. A recent classification of DN may be useful for categorizing stages of development of the diabetic lesions. Diabetic injury also affects the tubulointerstitium. Tubular basement membranes thicken in parallel to GBM. Early interstitial inflammation with predominantly mononuclear cells is followed by later increased interstitial fibrosis and tubular atrophy. Lesions in type 1 and type 2 diabetic patients share many similarities, with more severe vascular disease and heterogeneous lesions, perhaps reflecting older age and comorbid conditions such as hypertension, in the latter. Diabetic patients typically undergo diagnostic renal biopsies only when the clinical course is not typical for DN, and not surprisingly, a range of other lesions may then be present in addition to DN. DN also recurs or occurs de novo in the transplant, developing more quickly than in the native kidney. Efforts to develop rodent animal models that more completely capture these key features of human DN will allow advances to be made in understanding pathogenesis and targeting novel treatment.

Renal lesions in diabetic patients were described in 1936 in a classic paper by Kimmelstiel and Wilson [1]. They recognized the marked mesangial matrix expansion, described as 'diffuse mesangial (intercapillary) sclerosis'. Since that time, diabetic nephropathy (DN) has emerged as a leading cause of end-stage kidney disease worldwide. We describe the typical pathologic lesions over the course of its development and correlations of structure and function in patients with type 1 versus type 2 diabetes.

Characteristic Lesions of Diabetic Nephropathy

Glomerular lesions constitute the most striking and consistent changes identified in biopsy material from patients with clinical DN (table 1). Established DN is characterized histologically by three lesions, namely mesangial expansion, thickened glomerular basement membranes (GBM) and hyalinosis of afferent and efferent arterioles. A well-defined sequence of glomerular injury has been identified. In the earliest stages of injury, glomeruli may show only hypertrophy or be of normal size without lesions. GBM thickening (fig. 1), diagnosed by electron microscopic measurement, is due to increased accumulation of extracellular matrix and develops within 2 years of onset of diabetes, and progresses with increased duration of diabetes [2].

The earliest detectable light microscopic change is a widening of the mesangium, due to increased accumulation of matrix, best demonstrated with a PAS stain (fig. 2). This process is generally global and diffuse, without predilection for deep glomeruli [3]. Lesions of thickened GBM and mesangial expansion may even be present in adolescent patients after as little as 5 years' duration of diabetes. This process of diffuse mesangial matrix expansion is the hallmark lesion of diabetic glomerulopathy (fig. 2). Increased mesangial cellularity develops early, but in later stages, matrix increase without hypercellularity is typical.

The lesions of diffuse mesangial sclerosis are progressive. As disease advances, nodular accumulations of mesangial matrix, so-called Kimmelstiel-Wilson nodules, develop in about 25% of patients with advanced DN (fig. 2). Nodules typically occur 15 years or more after the onset of type 1 diabetes [4]. The nodules are irregularly distributed amongst glomeruli, and usually only one or two variably sized, round to oval, nodules are found in a particular glomerular tuft. The nodules are acellular to paucicellular, often with a rim of mesangial cells along the periphery, and a distinctive lamellated appearance that is highlighted by silver methenamine stains. Of note, these nodular lesions are not pathognomonic of a diabetic process. Nodular glomerulosclerosis may also be encountered in renal biopsies from patients with monoclonal immunoglobulin or light chain deposition disorders, some cases of membranoproliferative glomerulonephritis, rare cases of resolving postinfectious glomerulonephritis,

Table 1. Summary of pathologic findings in DN

Light microscopy	IF	EM
Glomeruli		
Glomerular hypertrophy	Linear staining of	Thickened basement membranes
Thickened capillary basement	capillary basement	Increased mesangial extracellular
membranes	membranes for IgG	matrix
Diffuse mesangial expansion	and albumin	?Mesangial cell hypertrophy
(sclerosis)		Fibrillar (nonamyloidotic)
Nodular mesangial sclerosis		extracellular matrix
Mesangiolysis		Segmentally denuded GBM
Capillary microaneurysms		(podocyte detachment)
Hyaline deposits ('capsular		
drops' and 'fibrin caps')		
?Mesangial cell proliferation		
Tubules and interstitium		
Atrophy	Linear staining of	Simplification of tubular
Thickened tubular basement	tubular basement	structure in atrophic tubules
membranes	membranes for IgG	Thickened tubular basement
	and albumin	membranes
Interstitial fibrosis		Increased interstitial collagen
Blood vessels		
Hyalinosis of afferent and	No characteristic	Subendothelial and transmural
efferent arterioles	changes	hyaline arterial deposits in small
Intimal sclerosis		arteries and arterioles

amyloidosis, and occasional patients in whom no underlying disease process can be identified (idiopathic nodular glomerulosclerosis) [5]. Correlation with immunofluorescence (IF), and/or electron microscopy (EM) is thus essential to diagnose immune complex or monoclonal protein etiology of such lesions.

Mesangiolysis and capillary microaneurysm formation is central to nodule formation (fig. 2). Mesangiolysis is characterized by the fraying and focal dissolution of the mesangial matrix. Capillary microaneurysms are seen as ballooning of glomerular capillaries with frequent accumulations of hyaline. A sequence of injury of repetitive mesangiolysis, microaneurysm formation, and subsequent capillary collapse and exuberant repair attempts is postulated to lead to the formation of mesangial nodules [6].

Hyalinosis is also typical of DN. This term describes the insudation of plasma protein with a homogeneous, glassy, i.e. 'hyaline' appearance, often with lipid droplets. Hyaline accumulation between the basement membranes of Bowman's capsules and the adjacent parietal epithelium is called a 'capsular drop'. Although capsular drop lesions are not common in DN, they occur even more rarely in other conditions. When hyaline is found within capillary lumina, it has been

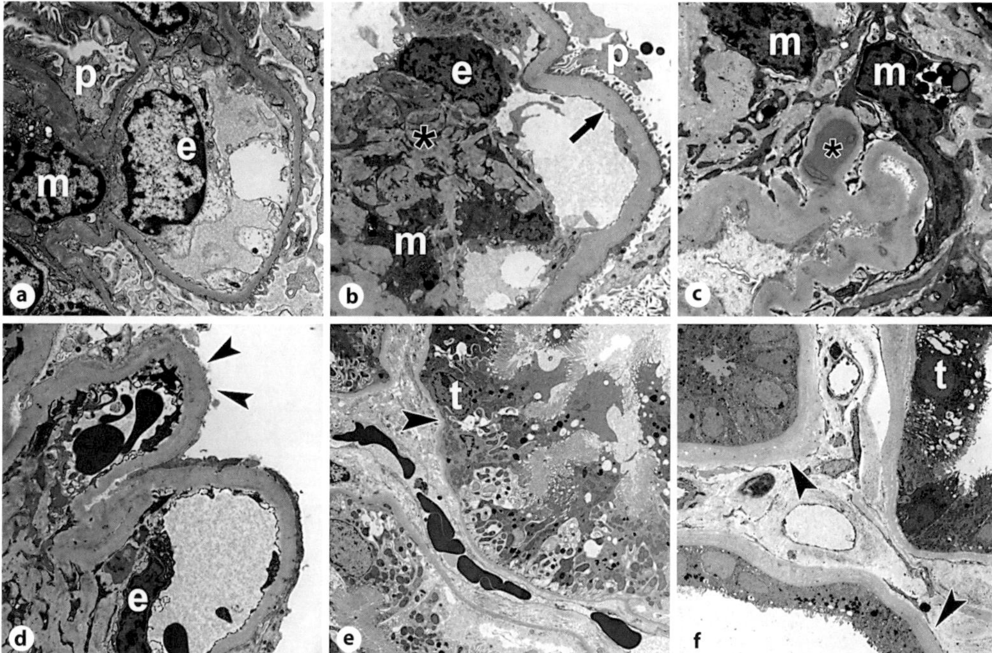

Fig. 1. DN lesions by EM. **a** Normal glomerulus showing normal mesangium and GBM and intact podocytes. ×3,000. **b** Diabetic glomerulopathy characterized by thickening of GBM (arrow) and accumulation of mesangial matrix (asterisk). ×3,000. **c** Accumulation of hyaline (asterisk) in mesangial areas may be confused with immune complex deposition. ×3,000. **d** Segmental denudation of the GBM with loss of overlying podocytes (arrowheads). ×3,000. **e** Normal tubules in a nondiabetic subject with delicate tubular basement membranes (arrowhead). ×3,000. **f** Tubular basement membrane thickening (arrowheads) in a nonatrophic tubule in DN. ×3,000. e = Endothelial cell; m = mesangial cell; p = podocyte; t = tubular epithelial cell.

termed the 'fibrin cap' or 'exudative' lesion. These terms are ill founded, in that there is no fibrin in the lesion, and the basic underlying process is one of insudation. This lesion appears identical to the hyalinosis lesions of focal segmental glomerulosclerosis, and can occur secondarily to any condition that causes sclerosis of glomeruli, and is thought to reflect endothelial injury. 'Hyalinosis' may therefore be a preferable term.

Interestingly, morphometric studies have shown a selective loss of podocytes (see below), whether by detachment or apoptosis. Podocytes have very limited regeneration capacity, and therefore, podocyte loss is proposed to be central for development of progressive sclerosis. Indeed, podocyte loss and sublethal podocyte injury appear to correlate with, and perhaps contribute to, proteinuria [7].

A recent classification schema proposes to stratify the above changes into four levels of severity [4]. Class I describes cases of diabetic injury with only

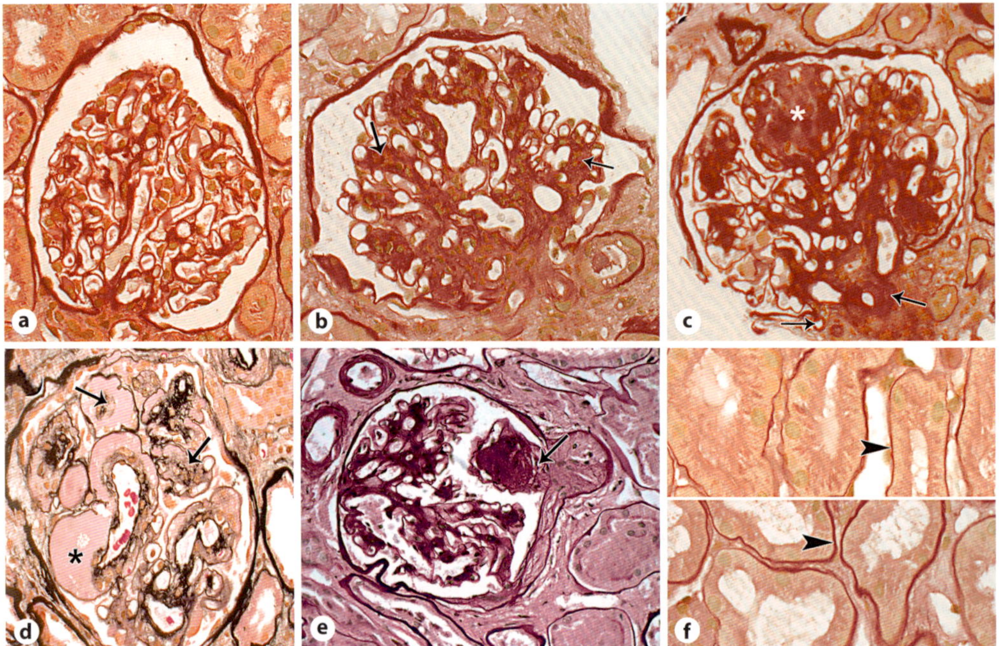

Fig. 2. DN lesions by light microscopy. **a** Normal glomerulus from a patient without diabetes, with normal mesangial area and normal delicate GBMs. PAS. ×40. **b** Glomerulus with diffuse mesangial expansion (arrow) and prominent GBMs, characteristic of diabetic glomerulopathy. PAS. ×40. **c** Glomerulus with a Kimmelstiel-Wilson nodule (asterisk) and hyalinosis of afferent and efferent arterioles (arrows), changes typical of established DN. PAS. ×40. **d** Glomerulus with mesangiolysis (arrow) and microaneurysm formation (asterisk) in diabetic glomerulopathy. Jones Methenamine silver. ×40. **e** Obstruction of the glomerular outflow tract by segmental sclerosis at the glomerulotubular junction (arrow), a lesion encountered in many patients with DN. PAS. ×40. **f** Top: Tubules from a nondiabetic patient. Bottom: Tubular basement membrane thickening (arrowhead) in DN, a lesion that occurs in parallel with GBM thickening in DN. PAS. ×60.

GBM thickening (>2 standard deviations from normal), class II includes mesangial expansion (mild to severe), class III is nodular sclerosis and class IV is advanced DN with all of the preceding changes, and also global sclerosis of >50% of glomeruli.

There are no immune complexes in DN by IF. The kidneys in diabetic patients usually exhibit a characteristic pattern of linear homogeneous staining of IgG, without complement or deposits by EM, along glomerular and tubular basement membranes. This staining is therefore not thought to be indicative of immune injury, but may represent 'stickiness' of the abnormal GBM to antisera used for IF. When superimposed immune complex disease is present in diabetic

Najafian · Alpers · Fogo

patients (see below), granular/chunky disease-specific patterns of IF staining for immunoglobulin and complement are present.

By EM, advanced DN shows diffuse marked, often more than twice normal, thickening of the lamina densa of the GBM (fig. 1). Podocytes show extensive effacement of foot processes and frequent regions where the basement membranes are denuded of overlying foot processes [7] (fig. 1). The mesangium reveals expanded matrix, sometimes with collagen fibrils (fig. 1). Cell debris with remnants of cell organelles is enmeshed in this matrix. Some disorganization and fraying of the mesangium at the interface with the glomerular capillary lumen may be apparent in glomeruli exhibiting mesangiolysis. Hyalinosis lesions have finely granular, electron-dense appearance, occasionally containing lipid droplets, and may be seen in capsular drops, areas of sclerosis with hyalinosis, or distributed within the mesangium and along the capillary walls (fig. 1). These nonimmune hyaline accumulations may be difficult to distinguish from granular, electron-dense immune deposits. The possibility of superimposed immune complex-mediated glomerulonephritis must therefore be made with careful correlation with histologic and IF findings.

Tubular basement membranes are thickened, both in atrophic and intact tubules [8]. Interstitial fibrosis is proportional to tubular atrophy. Hyalinosis of both afferent and efferent arterioles is characteristic of DN, and is distinct from other causes of hyalinosis that affect only the afferent arteriole. In addition, especially in type 2 diabetic patients who previously were typically older than type 1 patients, there often is significant arteriosclerosis of all caliber arteries. Three-dimensional analysis of the arterioles has shown a complex arborization of the arterioles, resulting in the appearance of multiple small arteriole-like vessels at the glomerular vascular pole [9]. Whether this represents abnormal angiogenesis or extrusion of the early branching of the arteriole beyond Bowman's capsule is not known.

Nondiabetic Lesions in Diabetic Patients

The diabetic patient with typical course of DN does not usually undergo renal biopsy. Not surprisingly, diagnostic renal biopsies therefore can show lesions other than or superimposed on DN. All of the common glomerular and interstitial diseases have been documented to occur in diabetic patients. As discussed above, the presence of immune or monoclonal positivity by IF, or deposits by EM, indicate nondiabetic disease. Although DN progression rate is not different in transplanted vs. native kidneys, it has been reported as early as 2 years following transplantation [10]. De novo DN may also develop in patients receiving immunosuppression that induces dysmetabolic syndrome.

Structural-Functional Relationships of Diabetic Nephropathy

Although DN is generally regarded as a one-way path to progressive renal failure, the functional sequence is not always progressive, and many patients can regress from microalbuminuria to normoalbuminuria [11]. Thus, it may be more meaningful to study evolution of lesions in relation to renal function, rather than duration of diabetes. Sequential renal biopsy studies suggest that glomerular lesions are more important determinants of renal dysfunction during most of the natural history of DN, whereas interstitial fibrosis is a stronger determinant of the rate of progression from moderate to severe renal insufficiency [12].

Mesangial expansion, the hallmark of DN, can be detected 4–5 years after the onset of type 1 diabetes [13]. Mesangial matrix is the main component of glomerular mesangial expansion in diabetic glomerulopathy [14]. The rate of mesangial expansion in relation to diabetes duration is nonlinear, and accelerates after 15 years of diabetes [15, 16]. Thus, mesangial expansion is the main lesion correlating with loss of renal function in DN [17]. Mesangial fractional volume closely correlates with a decrease in glomerular capillary filtration surface density. Total glomerular filtration surface is highly correlated with glomerular filtration rate (GFR) across the spectrum from hyperfiltration to renal insufficiency in type 1 diabetes [18]. Mesangial fractional volume and, to a lesser degree, GBM thickening are related to the albumin excretion rate (AER) and high blood pressure [17, 19]. GBM thickening, in contrast to mesangial expansion, tends to increase linearly in relation to diabetes duration [16]. These two parameters together explain 59% of the AER variability in type 1 diabetic patients [19]. Interestingly, even in normoalbuminuric patients with long-standing (20 years) type 1 diabetes there is increase in mesangial fractional volume and GBM width, with some overlap with normal range [19]. Further, those normoalbuminuric patients with more advanced glomerular structural lesions demonstrated greater risk for progression to microalbuminuria and proteinuria [20]. In contrast, GBM thickening and mesangial expansion are universally present in microalbuminuric and proteinuric type 1 diabetic patients [19].

Glomerular number and size may affect progression of DN. Type 1 diabetic patients developing DN after 25 years of diabetes had larger glomeruli compared to patients with DN after only 15 years [21], suggesting that glomerular size or the capacity for glomerular expansion may affect the rate of renal functional decline. Type 1 diabetic patients with severe renal dysfunction had fewer glomeruli compared with patients with mild or no glomerulopathy. This may reflect either complete resorption of sclerotic glomeruli [22] or more susceptibility to glomerulopathy in patients with fewer glomeruli. On the other hand, the rate of development of diabetic glomerulopathy lesions in type 1 diabetic patients with a single transplanted kidney is not different from that in patients with native (two) kidneys [10], suggesting that reduced number of glomeruli does not accelerate the early lesions of diabetic glomerulopathy, although it may affect the progression of the disease in later stages.

Interstitial fibrosis is generally regarded as a structural parameter correlating well with GFR decline; however, this is only true in diabetes when chronic injury is well established. In fact, the fractional volume of interstitium per renal cortex [Vv(Int/cortex)] initially decreases in type 1 diabetic patients [2], perhaps due to tubular hypertrophy. Initial expansion of the cortical interstitium is primarily due to an increase in its cellular component with later expansion due to collagen fibrils at a time when GFR has already declined [23].

Podocyte abnormalities occur early and are postulated to be pivotal for progression of DN. Foot process widening is probably the earliest structural manifestation of podocyte injury in diabetic patients, and was even present in normoalbuminuric type 1 adolescents [24]. Foot process width and AER are highly correlated in type 1 diabetic patients [7]. Podocytes may also detach from the GBM following injury. On average, over 20% of GBM in proteinuric type 1 diabetic patients was completely or partially denuded from podocyte [7]. The extent of denudation directly correlated with AER and paralleled classical changes in diabetic glomerulopathy [7]. Denuded GBM, although present in normoalbuminuric and microalbuminuric patients, was not statistically different from control subjects [5]. This perhaps explains why segmental sclerosis is a common finding in proteinuric type 1 diabetic patients, while rare in normoalbuminuric and microalbuminuric patients [25]. Both number and density of podocytes per glomerulus are decreased in proteinuric or microalbuminuric type 2 diabetic patients [26]. Which of these parameters is a better predictor of renal dysfunction is debatable. These discrepancies could be at least partially explained by different methodologies used in these studies. Podocyte number per glomerulus in microalbuminuric type 2 diabetic Pima Indian patients was the strongest predictor of AER increase and progression to overt nephropathy [26], suggesting that podocyte loss may be an important player in the progression of DN.

Podocyte detachment and loss lead to adhesions and segmental sclerosis, which has a marked predilection for the area of the glomerulotubular junction [Najafian and Mauer, unpubl. data], the location of the so-called tip lesion observed in some nephrotic patients [25, 27]. These glomerulotubular junction abnormalities contribute to GFR loss in type 1 diabetic proteinuric patients through proximal tubular atrophy and progressive closure of the glomerular tubular outlet, ultimately evolving to atubular glomeruli [25]. Serial section studies have shown that only 32% of glomeruli in proteinuric type 1 diabetic patients have normal glomerulotubular junctions and about 17% of glomeruli are atubular [27]. Based on these findings, it has been hypothesized that the accelerated decline in renal function following proteinuria in type 1 DN may be in part due to emergence of novel lesions, such as glomerulotubular junction abnormalities and atubular glomeruli which are tightly associated with proteinuric state [25].

Comparison of Type 1 and Type 2 Diabetes

Type 2 diabetic patients comprise over 80% of diabetes-related end-stage kidney disease cases; however, pathology and structural-functional relationships of DN have been less well studied in type 2 versus type 1 diabetic patients. Although a high incidence of nondiabetic renal lesions has been reported in clinical biopsies from proteinuric type 2 diabetic patients, this was not confirmed in protocol type biopsies [28]. There is more heterogeneity in renal lesions in type 2 diabetic patients when compared with type 1 diabetes [28]. Fioretto et al. [28] proposed a classification system based on renal biopsy lesions in microalbuminuric and proteinuric type 2 diabetic patients: category I with almost normal biopsies (35% of microalbuminuric and 10% of proteinuric patients); category II with classical lesions of DN similar to type I diabetes (30% of microalbuminuric and 55% of proteinuric patients), and category III with disproportionately advanced tubulointerstitial fibrosis, arteriolar hyalinosis, arteriosclerosis or global glomerulosclerosis, despite minor diabetic glomerulopathy. It is not clear whether this heterogeneity is due to other concomitant processes, such as ageing, hypertension, and atherosclerosis or a heterogeneous nature per se of type 2 diabetes lesions. Nevertheless, these categories correlate with some clinical phenotypes. Patients with classical DN lesions (category II) had longer diabetes duration with poorer glycemic control, and faster GFR decline, and always had retinopathy, while retinopathy was absent in category I or III patients [28]. Other studies have also confirmed glomerular structure within the normal range in some proteinuric type 2 diabetic patients in contrast to severe lesions in type 1 diabetic patients with overt nephropathy [29]. Overall, when compared with type 1 diabetic patients with similar renal function, diabetic glomerulopathy is less severe in type 2 diabetic patients, and the relationships between renal function and glomerular structural variables, although still present, are less precise. Cluster analysis showed that about one third of type 2 diabetic patients have greater AER than what was predicted by structural-functional relationship models of type 1 diabetic patients [Fioretto and Najafian, unpubl. data]. These findings highlight a potentially important use of renal biopsy in management of proteinuria in type 2 diabetes, as responsiveness to any specific treatment may be different in patients with different lesions, but with similar levels of proteinuria.

Comparison of Rodent Animal Model Lesions with Human Diabetic Nephropathy

There are multiple murine models, modified by strain and superimposed genetic mutations, of DN, as recently reviewed [6, 30]. Most currently studied mouse models of diabetes that reliably recapitulate early morphological changes in human DN, principally mesangial matrix expansion and in some cases, podocyte loss, include *db/db* and Akita mice. There are few models that recapitulate features of both morphologically early and late DN; of these, mice with deficiency of eNOS (modeling both type 1 and type 2 diabetes, depending on whether the diabetes is induced by the islet cell toxin streptozotocin or by a genetic mutation causing leptin deficiency), OVE26 FVB mice (a type 1 diabetes model), and the recently reported BTBR *ob*/ob mice (modeling type 2 diabetes and obesity) appear to be the most robust. The BTBR *ob*/ob mouse model is particularly noteworthy for the relative rapidity in which lesions develop, making it well suited for studies of new therapeutics. Despite the plethora of diabetic mouse models, all models available to date possess important limitations in their practicality and/or fidelity in recapitulating all of the features of human disease. The tubulointerstitial and vascular lesions of DN have been particularly challenging to model in the mouse. Designing better models of DN that will allow identification of underlying mechanisms remains an important research objective that in turn will facilitate testing of therapeutic interventions that can ameliorate or even reverse the structural alterations of DN.

Conclusion

Diabetes affects all structural compartments of the kidney. Recognition of these lesions and their pathogenesis, natural history and functional correlates, along with development of animal models mirroring these key features, will aid in developing novel strategies for early detection and prevention of progressive DN.

Acknowledgements

This work was supported by grants DK76126 (to C.E.A.) and DK56942 and DK44757 (A.B.F.) from the National Institutes of Health.

References

1   Kimmelstiel P, Wilson C: Intercapillary lesions in glomeruli in kidney. Am J Pathol 1936;12:83–97.

2   Drummond K, Mauer M: The early natural history of nephropathy in type 1 diabetes. II. Early renal structural changes in type 1 diabetes. Diabetes 2002;51:1580–1587.

3   Hørlyck A, Gundersen HJ, Østerby R: The cortical distribution pattern of diabetic glomerulopathy. Diabetologia 1986;29:146–150.

4   Tervaert TW, Mooyaart AL, Amann K, Cohen AH, Cook HT, Drachenberg CB, Ferrario F, Fogo AB, Haas M, de Heer E, Joh K, Noël LH, Radhakrishnan J, Seshan SV, Bajema IM, Bruijn JA, Renal Pathology Society: Pathologic classification of diabetic nephropathy. J Am Soc Nephrol 2010;21: 556–563.

5 Alpers CE, Biava CG: Idiopathic lobular glomerulonephritis (nodular mesangial sclerosis): a distinct diagnostic entity. Clin Nephrol 1989;32:68–74.

6 Alpers CE Hudkins KL: Mouse models of diabetic nephropathy. Curr Opin Nephrol Hypertens 2011, in press.

7 Toyoda M, Najafian B, Kim Y, Caramori ML, Mauer M: Podocyte detachment and reduced glomerular capillary endothelial fenestration in human type 1 diabetic nephropathy. Diabetes 2007;56:2155–2160.

8 Brito PL, Fioretto P, Drummond K, Kim Y, Steffes MW, Basgen JM, Sisson-Ross S, Mauer M: Proximal tubular basement membrane width in insulin-dependent diabetes mellitus. Kidney Int 1998;53:754–761.

9 Min W, Yamanaka N: Three-dimensional analysis of increased vasculature around the glomerular vascular pole in diabetic nephropathy. Virchows Arch A Pathol Anat Histopathol 1993;423:201–207.

10 Chang S, Caramori ML, Moriya R, Mauer M: Having one kidney does not accelerate the rate of development of diabetic nephropathy lesions in type 1 diabetic patients. Diabetes 2008;57:1707–1711.

11 Perkins BA, Ficociello LH, Silva KH, Finkelstein DM, Warram JH, Krolewski AS: Regression of microalbuminuria in type 1 diabetes. N Engl J Med 2003;348:2285–2293.

12 Fioretto P, Steffes MW, Sutherland DE, Mauer M: Sequential renal biopsies in insulin-dependent diabetic patients: structural factors associated with clinical progression. Kidney Int 1995;48:1929–1935.

13 Østerby R: Early phases in the development of diabetic glomerulopathy. Acta Med Scand Suppl 1974;574:3–82.

14 Steffes MW, Bilous RW, Sutherland DE, Mauer SM: Cell and matrix components of the glomerular mesangium in type I diabetes. Diabetes 1992;41:679–684.

15 Mauer SM, Sutherland DE, Steffes MW: Relationship of systemic blood pressure to nephropathology in insulin-dependent diabetes mellitus. Kidney Int 1992;41:736–740.

16 Drummond KN, Kramer MS, Suissa S, Lévy-Marchal C, Dell'Aniello S, Sinaiko A, Mauer M, International Diabetic Nephropathy Study Group: Effects of duration and age at onset of type 1 diabetes on preclinical manifestations of nephropathy. Diabetes 2003;52:1818–1824.

17 Mauer SM, Steffes MW, Ellis EN, Sutherland DE, Brown DM, Goetz FC: Structural-functional relationships in diabetic nephropathy. J Clin Invest 1984;74:1143–1155.

18 Ellis EN, Steffes MW, Goetz FC, Sutherland DE, Mauer SM: Glomerular filtration surface in type I diabetes mellitus. Kidney Int 1986;29:889–894.

19 Caramori ML, Kim Y, Huang C, Fish AJ, Rich SS, Miller ME, Russell G, Mauer M: Cellular basis of diabetic nephropathy. 1. Study design and renal structural-functional relationships in patients with long-standing type 1 diabetes. Diabetes 2002;51:506–513.

20 Steinke JM, Sinaiko AR, Kramer MS, Suissa S, Chavers BM, Mauer M, International Diabetic Nephopathy Study Group: The early natural history of nephropathy in type 1 diabetes. III. Predictors of 5-year urinary albumin excretion rate patterns in initially normoalbuminuric patients. Diabetes 2005;54:2164–2171.

21 Bilous RW, Mauer SM, Sutherland DE, Steffes MW: Mean glomerular volume and rate of development of diabetic nephropathy. Diabetes 1989;38:1142–1147.

22 Bendtsen TF, Nyengaard JR: The number of glomeruli in type 1 (insulin-dependent) and type 2 (noninsulin-dependent) diabetic patients. Diabetologia 1992;35:844–850.

23 Katz A, Caramori ML, Sisson-Ross S, Groppoli T, Basgen JM, Mauer M: An increase in the cell component of the cortical interstitium antedates interstitial fibrosis in type 1 diabetic patients. Kidney Int 2002;61:2058–2066.

24 Torbjornsdotter TB, Perrin NE, Jaremko GA, Berg UB: Widening of foot processes in normoalbuminuric adolescents with type 1 diabetes. Pediatr Nephrol 2005;20:750–758.

25 Najafian B, Crosson JT, Kim Y, Mauer M: Glomerulotubular junction abnormalities are associated with proteinuria in type 1 diabetes. J Am Soc Nephrol 2006;17(suppl 2):S53–S60.

26 Najafian B, Mauer M: Progression of diabetic nephropathy in type 1 diabetic patients. Diabetes Res Clin Pract 2009;83:1–8.

27 Najafian B, Kim Y, Crosson JT, Mauer M: Atubular glomeruli and glomerulotubular junction abnormalities in diabetic nephropathy. J Am Soc Nephrol 2003;14:908–917.

28 Fioretto P, Caramori ML, Mauer M: The kidney in diabetes: dynamic pathways of injury and repair. The Camillo Golgi Lecture 2007. Diabetologia 2008;51:1347–1355.

29 Østerby R, Gall MA, Schmitz A, Nielsen FS, Nyberg G, Parving HH: Glomerular structure and function in proteinuric type 2 (non-insulin-dependent) diabetic patients. Diabetologia 1993;36:1064–1070.

30 Brosius FC 3rd, Alpers CE, Bottinger EP, Breyer MD, Coffman TM, Gurley SB, Harris RC, Kakoki M, Kretzler M, Leiter EH, Levi M, McIndoe RA, Sharma K, Smithies O, Susztak K, Takahashi N, Takahashi T, Animal Models of Diabetic Complications Consortium: Mouse models of diabetic nephropathy. J Am Soc Nephrol 2009;20:2503–2512.

Agnes B. Fogo, MD
Department of Pathology, Vanderbilt University Medical Center
Nashville, TN 37232 (USA)
Tel. +1 615 3223114, Fax +1 615 343 7023, E-Mail agnes.fogo@vanderbilt.edu

Lai KN, Tang SCW (eds): Diabetes and the Kidney.
Contrib Nephrol. Basel, Karger, 2011, vol 170, pp 48–56

Pathogenesis and Progression of Proteinuria

Merlin C. Thomas

Baker IDI Heart and Diabetes Institute, Melbourne, Vic., Australia

Abstract

Progressive albuminuria is the sine qua non of diabetic nephropathy. It is not only a marker of renal damage but also significantly contributes to its development and progression. However, the precise mechanisms by which escalating amounts of albumin leave the bloodstream, cross the endothelial glycocalyx, the glomerular basement membrane and the slit pores between the foot processes of the podocytes, transit through Bowman's space, bypass the resorptive mechanisms of the nephron and ultimately pass into the urine remain hotly debated. Certainly, diabetes is associated with significant dysfunction at each of these levels, and will be discussed in detail in this review. Moreover, dilation of the afferent and constriction of the efferent arterioles triggered by defective autoregulation and subsequently by loss of peritubular capillaries also act to increase the glomerular transcapillary hydrostatic pressure and facilitate a far greater transit of albumin into the urine. Importantly, none of these mechanisms exists in isolation. Indeed, the most likely reason for progression of albuminuria is the fact that dysfunction initiated by compromise of one component will inevitably modify other parts, and ultimately affect the whole nephron function. From this 'holonephric' view of albuminuria, the best treatment will be a combination that can promote regression and restore the integrity of the entire pathway, as while fixing one part may slow the process, because of the integrated nature of renal function, it will not completely prevent nephropathy.

The analysis of urine has always formed an integral part of diabetes care. The production of large amounts of sweet urine associated with polydipsia provides the derivation of the term diabetes mellitus (meaning 'a honey siphon'). The importance of the urinalysis in prognostication remains for individuals with type 1 and type 2 diabetes. It was recognized nearly 30 years ago that some individuals with diabetes had increased excretion of albumin in their urine, which preceded and predicted the development of proteinuria and 'overt' kidney disease [1]. This level of albumin excretion (30–300 mg/day or 20–200 µg/min) was

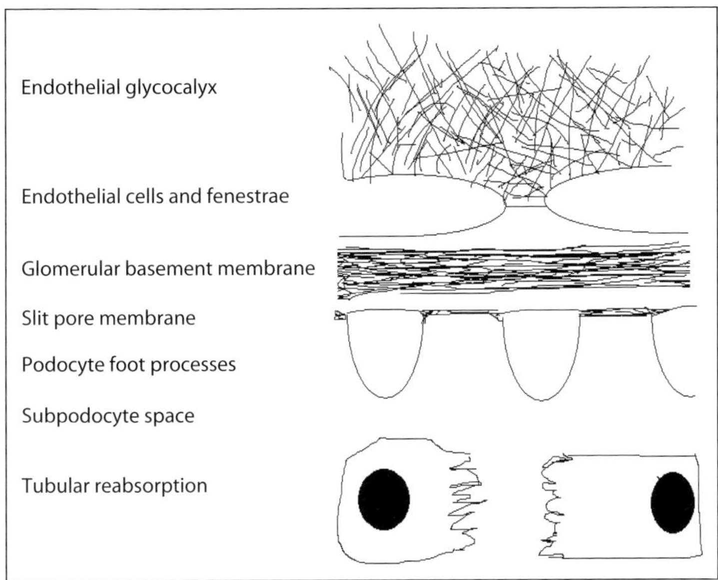

Fig. 1. Potential barriers to albuminuria.

not detectable by then standard tests for proteinuria and was coined 'microalbuminuria' or 'incipient nephropathy'. This now widely employed categorization has proved useful for identifying patients with diabetes at increased risk of a range of adverse outcomes, including renal failure and premature mortality. For example, the risk of developing end-stage kidney disease is 3- to 4-fold higher in those with microalbuminuria [2], and twice this again in those with overt nephropathy. It is also clear that the presence of protein in the primary urine has direct and cumulative effects on the nephron that lead to nephron dropout and ultimately renal impairment [3]. However, the precise mechanisms by which escalating amounts of albumin leave the bloodstream, cross the capillary endothelium, the glomerular basement membrane (GBM) and the slit pores between the foot processes of the podocytes, transit through Bowman's space, bypass the resorptive mechanisms of the nephron, and ultimately pass into the urine remain hotly debated (fig. 1) [4]. This review will examine some of the potential factors that contribute to the development of microalbuminuria and its subsequent progression to proteinuria.

Intraglomerular Hemodynamics

Although most research has focused on the role of specific structural defects for the development and progression of albuminuria, it is clear that some (modest degree of) albuminuria may be hemodynamic, and can occur without invoking any loss of integrity of the glomerular barrier. For example, a substantial proportion of healthy, fit people transiently develop microalbuminuria after a single, short bout of vigorous exercise [5]. In healthy kidneys, glomerular pressure is autoregulated by the tone of preglomerular (afferent) and postglomerular (efferent) arterioles to maintain continuous and optimal renal filtration. In response to vigorous exercise, where renal blood flow is reduced, GFR is maintained by changes in intraglomerular pressure at the expense of transient intraglomerular hypertension and albuminuria. This can be partly blocked by ACE inhibition, suggesting its hemodynamic origin [6]. For the same reasons, a reduction in nephron mass associated with kidney disease increases the glomerular transcapillary hydrostatic pressure within residual glomeruli to preserve filtration and albuminuria is also modestly elevated. Similarly in diabetes, dilation of the afferent and constriction of the efferent arterioles triggered by defective autoregulation and subsequently by loss of peritubular capillaries also act to increase the intraglomerular pressure, promoting albumin leak. To develop nephrotic range proteinuria, additional compromise of the glomerular barrier must also occur. However, in the presence of such renal injury, hemodynamic disturbances will also facilitate a far greater transit of albumin into the urine and, as a result, may be significantly more important.

Endothelial Barrier

The glomerular capillaries are lined by a unique flat endothelium, densely perforated with transcellular pores, 60–100 nm in diameter (the fenestrae), that are essential for fluid filtration [7]. Indeed, healthy fenestrae make up half of the endothelial surface area, much in the manner of a sieve (like that the glomerulus is often crudely characterized). It was originally believed that, much like a sieve, these 'wide open' fenestrae would readily allow passage of albumin (that is at most 8 nm in diameter), and as a consequence, that the glomerular endothelium did not significantly contribute to retention of albumin. However, it is now clear that a complex glycoprotein coat covers the glomerular fenestrae in a gel-like diaphragm that extends up to 500 nm from the endothelial surface to exclude macromolecules from the primary filtrate [8]. Damage to systemic endothelial glycocalyx is associated with the onset of albuminuria in patients with diabetes [9]. Similarly in mice, deterioration of glomerular endothelial surface layer induced by oxidative stress is implicated in altered permeability of macromolecules in Zucker fatty rats [10]. In vitro studies have also shown that high glucose levels modify the biosynthesis of key glycosaminoglycans within the endothelial glycocalyx to increase passage of labeled albumin across monolayers [8]. Such data are consistent with the strong clinical association between albuminuria and (the severity of) endothelial dysfunction.

Glomerular Basement Membrane

On electron microscopy, the GBM appears as the only continuous barrier between the blood and urinary space. And for a long time, it has been considered to be the principal barrier to prevent the transit of albumin into the urine. However, in diabetic glomeruli, the basement membrane is significantly thickened, both in type 1 and type 2 diabetes, with even greater thickening observed in patients with incipient nephropathy [11]. Yet the leakage of albumin through it is substantially enhanced. It has been argued that compositional changes in the GBM, including the loss of negatively charged proteoglycans, must therefore mediate proteinuria in diabetes. It has been suggested that the GBM offers little or no specific resistance to the passage of macromolecules, like albumin. Certainly, podocyte-specific knockout of proteoglycan synthesis (and as the GBM is made by podocytes, therefore also GBM-specific) does not result in proteinuria, despite significant charge defects [12]. However, patients with inherited defects in collagen or laminin fibers within the GBM do develop proteinuria. In addition, stiffening of the GBM associated with diabetes potentially contributes to the compromise of the subpodocyte space (see below) and reduced distensibility of the pericapillary wall and intraglomerular hypertension, indirectly facilitating proteinuria through hemodynamic mechanisms.

Podocytes and Their Slit Pores

Diabetic kidney disease is associated with significant podocyte injury and dysfunction, which can be observed early in the natural history of diabetic nephropathy [13]. Such changes include foot process retraction and flattening (known as effacement) that occur as mature podocytes dedifferentiate (fig. 2), losing the specialized features required for efficient glomerular function and in the process acquiring a number of profibrotic, proinflammatory and proliferative features [14]. In addition, podocyte flattening reduces the ultrafiltration coefficient, thereby exacerbating glomerular hypertension. The key role that the glomerular podocyte plays in the development and progression of albuminuria in diabetes has recently been demonstrated by studies in which mice with podocyte-specific deletion of the insulin receptor develop significant albuminuria, together with histological features that recapitulate diabetic nephropathy, but in a normoglycemic environment [14]. Such data place podocytes, and more particularly the dysregulation of their growth and differentiation, at the very heart of the pathogenesis of albuminuria. Although mature podocytes are traditionally thought of as archetypal postmitotic cells, with little or no capacity for regenerative replication, proliferating podocytes may be observed in experimental diabetes. Moreover, increased number of podocytes is detected in the urine of diabetic patients [15] long before any

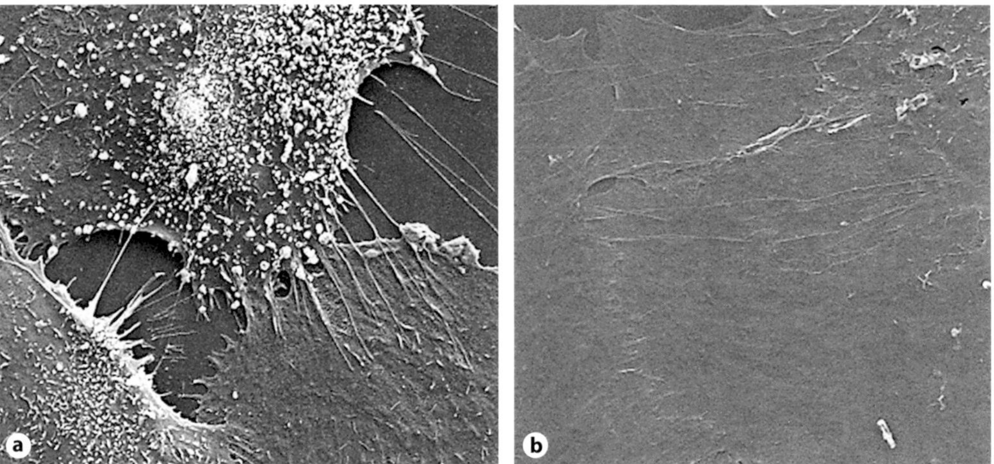

Fig. 2. Electron micrograph showing complex interdigitation of the foot process of differentiated podocytes in vitro (**a**), and their effacement following induction of dedifferentiation with TGF-β (**b**), a known pathogenic stimulus in the diabetic kidney (adapted from Herman-Edelstein et al. [25] and revised).

reduction in any glomerular podocyte numbers. This suggests that podocyte proliferation occurs in human disease alongside podocyte apoptosis. Indeed, they may be two edges of the same sword, as the reentry of dedifferentiated podocytes into the cell cycle in response to hyperglycemia allows for both outcomes, although the net effect is one of a progressive but modest podocyte loss. In the same way, re-entry into the cell cycle of terminally differentiated neuronal cells leads to apoptosis more often than proliferation, although both are always stimulated.

The slit diaphragm is a heteroporous membrane that stretches between foot processes [16]. The normal mean pore radius has been estimated to be around 12 nm, although many pores are smaller than albumin. Following the onset of diabetes, and more so with the development and progression of nephropathy, there are changes in the composition, organization, width and number of filtration slit pores. The role of slit pore dysfunction in albuminuria is well known, as specific mutations of slit pore proteins like nephrin, CD2AP and podocin are associated with nephrotic range proteinuria. In diabetes, there is a downregulation of nephrin expression [17] and other key proteins in the slit diaphragm [13]. Moreover, there is evidence that the assembly of these proteins is substantially altered by diabetes, possibly through posttranslational modification of the slit pore itself or associated cytoskeleton proteins.

Finally, recent studies have also demonstrated that podocytes are able to directly endocytose albumin [18], in keeping with the fact that podocytes have a substantial endosomal/lysosomal apparatus. The impact of diabetes on this endocytic activity is unknown, but as similarities exist with the endocytic

Thomas

functions of proximal cells, it may be anticipated that such activities may also be impaired in the diabetic kidney.

Bowman's Space

Between the podocytes and the parietal layer of Bowman's capsule exists Bowman's space or the urinary space. Rather than being a functionless void, the hydrostatic pressure in Bowman's space also appears to be important for the flow of both solute and macromolecules in much the same way that intraglomerular hypertension facilitates their flux. In diabetic kidneys, the hydrostatic pressure in Bowman's space is reduced, possibly through an increased luminal diameter, reduced tubular brush border and increased flow rate through distal nephron segments, where flow resistance is normally high. Perhaps the most important part of the urinary space with regard to albuminuria is the space between the underside of podocyte cell body/primary processes and the foot processes (known as the subpodocyte space) [4]. This labyrinthine layer covers over 60% of the entire filterable surface and confers ultrafiltration and hydraulic resistance to dynamically regulate glomerular fluid flux as well as permeability [19]. Changes in podocyte numbers and morphology associated with diabetes as well as basement membrane thickening significantly compromise the integrity of the subpodocyte space in the diabetic kidney.

Proximal Tubule

Far from being a bystander in diabetic nephropathy, changes in the proximal tubule appear important for the development of progressive diabetic kidney disease [20]. Indeed, renal function and prognosis correlate better with structural lesions in the tubuli and cortical interstitium than with classical glomerular changes in diabetic nephropathy. Maladaptive tubular hypertrophy and apoptosis, thickening of the tubular basement membrane and activation of tubuloglomerular feedback all precede proteinuria, and potentially contribute to it. For example, it has been suggested that increase in the fractional and absolute sodium reabsorption in the proximal tubule leads to activation of tubuloglomerular feedback pathways that act to increase intraglomerular pressure [21] and therefore hemodynamically facilitate albuminuria. To support this 'tubulocentric' hypothesis, selective inhibitors of tubular growth are able to attenuate albuminuria in experimental diabetes.

Tubular dysfunction may also contribute to proteinuria due to defective uptake and/or lysosomal processing of filtered protein [22]. Recent studies have demonstrated that significant quantities of albumin are filtered across the glomerular endothelium, which then undergoes complex processing in the proximal tubule involving cellular uptake, transcytosis and lysosomal processing [22]. A number of points on the endocytic pathway for albumin reabsorption may be disrupted in the diabetic kidney, contributing to decreased reabsorption of low-molecular-weight proteins (e.g. retinol-binding protein, α_2-microglobulin, N-acetyl-β-D-glucosaminidase, β_2-glycoprotein-1) in early diabetic nephropathy, and potentially albuminuria. For example, expression of the multivalent albumin transporters megalin and its cotransporter, cubilin, may be reduced in diabetes [23]. The size, flexibility and complexity of the apical brush border are reduced in diabetic kidneys, particularly in the proximal straight tubules, due to changes in the composition of the glycocalyx. Vesicular trafficking of proteins by the cytoskeleton is also modified by diabetes, possibly through the induction of protein kinase C. Finally, it has been demonstrated that protein degradation processes are also inhibited in the proximal tubule of diabetic animals, possibly via a decrease in cathepsin-mediated proteolytic activity and impaired endosomal acidification [24].

Concluding Remarks: Chicken and Egg

It will be still argued for many decades, which is the most important contributor to the development and progression of albuminuria in diabetes. Is it the podocyte, the GBM, the proximal tubule or endothelial glycocalyx that sets the ball rolling? Certainly, pathology at each of these sites may be an important determinant of progressive diabetic nephropathy in its own right. But while it may be useful to subdivide the nephron for the purposes of analysis and scientific discovery, it is likely the interactions between endothelium, glomerulus and tubule are keyed to the understanding of complex disorders like diabetic nephropathy. From this 'holonephric' point of view, the key driver of proteinuria is the progressive loss of integrity of the entire pathway that can be initiated by any or all components. For this reason different pathological appearances can result in equivalent amount of albuminuria. Equally, therapies that target hemodynamic pathways can appear as effective as those that selectively target tubular reabsorption. Ultimately, the best treatment will be a combination that can promote regression of all these lesions and restore the integrity of the entire pathway, as while fixing one part may slow the process, because of the integrated nature of renal function, it will not completely prevent nephropathy.

Acknowledgments

The author is supported by JDRF international, the Australian NHMRC, and the Bootle bequest from Kidney Health Australia.

References

1 Mogensen CE: Microalbuminuria predicts clinical proteinuria and early mortality in maturity-onset diabetes. N Engl J Med 1984; 310:356–360.
2 Newman DJ, Mattock MB, Dawnay AB, Kerry S, McGuire A, Yaqoob M, et al: Systematic review on urine albumin testing for early detection of diabetic complications. Health Technol Assess 2005;9:iii-vi, xiii-163.
3 Abbate M, Zoja C, Remuzzi G: How does proteinuria cause progressive renal damage? J Am Soc Nephrol 2006;17:2974–2984.
4 Salmon AH, Neal CR, Harper SJ: New aspects of glomerular filtration barrier structure and function: five layers (at least) not three. Curr Opin Nephrol Hypertens 2009; 18:197–205.
5 Heathcote KL, Wilson MP, Quest DW, Wilson TW: Prevalence and duration of exercise induced albuminuria in healthy people. Clin Invest Med 2009;32:E261–E265.
6 Romanelli G, Giustina A, Bossoni S, Caldonazzo A, Cimino A, Cravarezza P, et al: Short-term administration of captopril and nifedipine and exercise-induced albuminuria in normotensive diabetic patients with early-stage nephropathy. Diabetes 1990;39: 1333–1338.
7 Jarad G, Miner JH: Update on the glomerular filtration barrier. Curr Opin Nephrol Hypertens 2009;18:226–232.
8 Singh A, Satchell SC, Neal CR, McKenzie EA, Tooke JE, Mathieson PW: Glomerular endothelial glycocalyx constitutes a barrier to protein permeability. J Am Soc Nephrol 2007;18:2885–2893.
9 Nieuwdorp M, Mooij HL, Kroon J, Atasever B, Spaan JA, Ince C, et al: Endothelial glycocalyx damage coincides with microalbuminuria in type 1 diabetes. Diabetes 2006;55: 1127–1132.
10 Kuwabara A, Satoh M, Tomita N, Sasaki T, Kashihara N: Deterioration of glomerular endothelial surface layer induced by oxidative stress is implicated in altered permeability of macromolecules in Zucker fatty rats. Diabetologia 2010;53:2056–2065.
11 Dalla Vestra M, Saller A, Bortoloso E, Mauer M, Fioretto P: Structural involvement in type 1 and type 2 diabetic nephropathy. Diabetes Metab 2000;26(suppl 4):8–14.
12 Harvey SJ, Jarad G, Cunningham J, Rops AL, van der Vlag J, Berden JH, et al: Disruption of glomerular basement membrane charge through podocyte-specific mutation of agrin does not alter glomerular permselectivity. Am J Pathol 2007;171: 139–152.
13 Ziyadeh FN, Wolf G: Pathogenesis of the podocytopathy and proteinuria in diabetic glomerulopathy. Curr Diabetes Rev 2008;4:39–45.
14 Shankland SJ: The podocyte's response to injury: role in proteinuria and glomerulosclerosis. Kidney Int 2006;69:2131–247.
15 Vogelmann SU, Nelson WJ, Myers BD, Lemley KV: Urinary excretion of viable podocytes in health and renal disease. Am J Physiol Renal Physiol 2003;285: F40–F48.
16 Gagliardini E, Conti S, Benigni A, Remuzzi G, Remuzzi A: Imaging of the porous ultrastructure of the glomerular epithelial filtration slit. J Am Soc Nephrol 2010;21:2081–2089.
17 Forbes JM, Bonnet F, Russo LM, Burns WC, Cao Z, Candido R, et al: Modulation of nephrin in the diabetic kidney: association with systemic hypertension and increasing albuminuria. J Hypertens 2002;20:985–992.
18 Eyre J, Ioannou K, Grubb BD, Saleem MA, Mathieson PW, Brunskill NJ, et al: Statin-sensitive endocytosis of albumin by glomerular podocytes. Am J Physiol Renal Physiol 2007;292:F674–F681.
19 Peti-Peterdi J, Sipos A: A high-powered view of the filtration barrier. J Am Soc Nephrol 2010;21:1835–1841.
20 Thomas MC, Burns WC, Cooper ME: Tubular changes in early diabetic nephropathy. Adv Chronic Kidney Dis 2005;12: 177–186.
21 Thomson SC, Vallon V, Blantz RC : Kidney function in early diabetes: the tubular hypothesis of glomerular filtration. Am J Physiol Renal Physiol 2004;286: F8–F15.
22 Russo LM, Sandoval RM, Campos SB, Molitoris BA, Comper WD, Brown D: Impaired tubular uptake explains albuminuria in early diabetic nephropathy. J Am Soc Nephrol 2009;20:489–494.

23 Thrailkill KM, Nimmo T, Bunn RC, Cockrell GE, Moreau CS, Mackintosh S, et al: Microalbuminuria in type 1 diabetes is associated with enhanced excretion of the endocytic multiligand receptors megalin and cubilin. Diabetes Care 2009;32:1266–1268.

24 Osicka TM, Houlihan CA, Chan JG, Jerums G, Comper WD: Albuminuria in patients with type 1 diabetes is directly linked to changes in the lysosome-mediated degradation of albumin during renal passage. Diabetes 2000;49:1579–1584.

25 Herman-Edelstein M, Thomas M, Thallas-Bonke V, Saleem M, Cooper ME, Kantharidis P: De-differentiation of immortalised human podocytes in response to transforming growth factor β – a model for diabetic podocytopathy. Diabetes (in press).

Prof. Merlin C. Thomas, MBChB, PhD, FRACP
Baker IDI Heart and Diabetes Institute
St Kilda Rd Central, PO Box 6492
Melbourne, VIC 8008 (Australia)
Tel. +61 3 8532 1277, Fax +61 3 8532 1480, E-Mail mthomas@bakeridi.edu.au

Lai KN, Tang SCW (eds): Diabetes and the Kidney.
Contrib Nephrol. Basel, Karger, 2011, vol 170, pp 57–65

Epigenetic Mechanisms

Louisa M. Villeneuve · Rama Natarajan

Division of Cellular and Molecular Diabetes Research, Department of Diabetes,
Beckman Research Institute of City of Hope, Duarte, Calif., USA

Abstract
The incidence of diabetes and related complications like nephropathy is growing rapidly
and has become a major health care issue. Changes in the environment and nutritional
habits have been implicated as major players. Furthermore, it is becoming increasingly
clear that epigenetic factors may modulate the connections between genes and the
environment. While diabetes in itself is treatable to a large extent, it is still associated
with significantly increased risk for complications including chronic kidney and cardiovascular diseases. Current treatments have added preventative approaches so as to
avoid future diabetic complications. Unfortunately, diabetic patients are often plagued
with the continued development of various complications even after achieving glucose
control. This has been suggested to be attributable to a mysterious phenomenon termed
'metabolic memory' of the prior glycemic state. Recent studies have suggested that epigenetic changes to chromatin can affect gene expression in response to various stimuli,
and changes in key biochemical pathways and epigenetic histone and DNA methylation
patterns in chromatin have been observed in a diabetic milieu. These accumulating data
suggest that metabolic or hyperglycemic memory may be due to epigenetic changes in
specific target tissues altering gene expression without changing the genetic code itself.
While the genetics of diabetes has long been the focus of scientific research, much less is
known about the role of epigenetics and the related molecular pathways that might
affect the development of diabetes and the associated complications. Further studies of
epigenetic mechanisms are therefore timely and could provide valuable new insights
into the pathology of diabetic complications and also uncover much needed new therapeutic targets.   Copyright © 2011 S. Karger AG, Basel

It is well known that both type 1 and type 2 diabetes mellitus are associated
with significantly accelerated rates of macro- and microvascular diseases such
as cardiovascular complications and nephropathy. Even when blood glucose
levels are under control, diabetic patients may still develop chronic kidney

disease (CKD) leading to renal failure. Most of the treatment options for renal dysfunction rely on the use of limited available medications known to delay the progression toward renal failure with hopes of avoiding dialysis and/or kidney transplantation.

Genetic factors and key gene mutations have been implicated in the pathogenesis of diabetes. However, increasing evidence suggests that complex interactions between genes and the environment and related epigenetic factors may play major roles in many common human diseases such as diabetes and its complications. Chromatin provides a crucial interface between genetics and environment, and the epigenetic DNA methylation as well as posttranscriptional modifications (PTMs) of histone tails in chromatin can regulate gene transcription. Although numerous studies have identified key biochemical pathways triggered by hyperglycemia and diabetes in target cells related to inflammation and diabetic complications [1], the exact molecular mechanisms are still not very clear. Exciting recent research has implicated a role for epigenetic regulation in this context. Such investigations can provide new insights into the pathology of diabetes and its complications and lead to the development of sorely needed new treatment options.

Evidence for Metabolic Memory from Clinical Trials and Experimental Models

Evidence shows that some diabetic patients experience a continued development of diabetic complications even after achieving glucose control, suggesting a metabolic memory of prior glycemic exposure. Recent studies have suggested that this may be attributed to epigenetic changes in target cells without alterations in gene coding sequences. Several clinical trials have indicated that early intensive glycemic control leads to sustained beneficial effects. The Diabetes Control and Complications Trial (DCCT) demonstrated that type 1 diabetes mellitus patients under intensive insulin therapy had delayed progression of nephropathy, retinopathy, and neuropathy as compared to patients under conventional therapy [2]. The Epidemiology of Diabetic Complications and Interventions (EDIC) trial continued long-term follow-up of the DCCT and placed all participants on intensive glycemic control [3]. The EDIC trial has demonstrated that patients originally in the intensive treatment group during the DCCT and continued on intensive therapy for the EDIC trial continued to have significantly slower progression of key diabetic microvascular complications relative to patients who were in the conventional treatment group during DCCT. Better outcomes for macrovascular complications have been noted as well [4]. Clinical trials with T2D patients have also shown similar protection [5].

These major clinical trial findings demonstrate the importance of early metabolic control to reduce long-term complications and confirm that hyperglycemia can have long-lasting detrimental consequences. The mechanisms responsible for these persistent effects of the prior hyperglycemic state are still not well understood, and this phenomenon, termed 'metabolic memory' [4] has been a major challenge in the treatment of diabetic complications.

In addition, several experimental models have been used to study the molecular mechanisms of 'metabolic memory'. Early studies in dogs found that retinal complications persisted even after reversing hyperglycemia [6]. In rats, islet transplantation after 12 weeks of diabetes could not reverse the progression of retinopathy as compared to islet transplantation after 6 weeks of diabetes [7]. Studies in streptozotocin-induced diabetic rats showed that early glycemic control has protective effects in the retina, and also demonstrated a link with histone acetylation [8].

Endothelial cells cultured in high glucose (HG) showed sustained increases in key profibrotic and extracellular matrix proteins even after normalization of glucose [9]. More recently, short-term HG culture of endothelial cells displayed sustained activation of inflammatory genes and oxidant stress even after glucose normalization [10, 11]. Additional reports demonstrated the persistence of oxidant stress for up to one week after glucose normalization, which could be blocked by antioxidants or NADPH oxidase inhibitors [12]. In another model of metabolic memory, vascular smooth muscle cells (VSMC) derived from T2D obese *db/db* mice displayed sustained increases in NF-κB activation, inflammatory gene expression, migration, and oxidant stress as well as increased monocyte adhesion relative to control nondiabetic *db/+* VSMC even after in vitro culture [13].

These data suggest a metabolic memory of vascular dysfunction arising from prior hyperglycemia, and further emphasize the importance of strict glycemic control to reduce the progression of diabetic complications. They also suggest a role for oxidant stress in perpetuating this metabolic memory [12]. Hyperglycemia and oxidant stress can also increase the accumulation of advanced glycation end products (AGEs) further perpetuating local inflammation and oxidant stress to promote long-term damage to end organs, including the kidney. Thus AGEs, acting through receptors such as RAGE, could also contribute to hyperglycemic memory. However, the nuclear mechanisms responsible for the sustained 'memory' over time through multiple cell divisions at the transcriptional and epigenetic level are not fully clear and have evolved as an active area of research.

Epigenetic Regulation of Gene Expression

The basic subunit of chromatin, nucleosome, consists of DNA wrapped around a histone octamer comprised of two copies of histone H2A, H2B, H3, and H4 [14]. Transcriptional activation or repression is a dynamic process that relies on the accessibility of DNA to various transcription factors, coactivators/corepressors, and the recruitment of protein complexes that alter chromatin structure via enzymatic modifications of histone tails and nucleosome remodeling. Epigenetic changes to chromatin form an added layer of gene regulation that can be altered without altering the DNA code itself. Numerous combinations of epigenetic modifications allow for flexibility of the chromatin and transcriptional outcomes depending on the needs of the cell [15].

DNA methylation is a well-characterized epigenetic mark in the context of tumor suppressor genes and cancer [16]. However, much less is known about DNA methylation in diabetes, although a role for DNA methylation in the regulation of various genes associated with insulin expression and secretion has been implicated [17]. DNA methylation has also been shown to play a role in kidney disease. One recent study showed altered DNA methylation at key gene promoters in T1D patients with diabetic nephropathy relative to those without [18]. Changes in blood homocysteine levels affecting methyl transfer reactions by inhibiting DNA methyltransferases have also been seen in patients with CKD [19].

Histone posttranslational modifications (PTMs) such as lysine acetylation, phosphorylation, and methylation can greatly influence gene regulation. Histone acetyltransferases (HATs) and histone deacetylases (HDACs) are relatively nonselective regulators of histone lysine acetylation which can occur quite rapidly [20]. Histone methylation on the other hand is considered to be more long lasting. Histone lysine methylation is mediated by histone methyltransferases (HMTs), and is more complex since there are several HMTs and lysine residues that can be mono-, di-, or tri-methylated. HMTs are generally more specific, usually methylating only one specific lysine residue. Typically, histone H3 lysine 4 methylation (H3K4me) is associated with gene activation, while histone H3 lysine 9 methylation (H3K9me) and H3K27me3 are associated with gene repression [21]. Furthermore, histone methylation can also be dynamically regulated through the coordinated action of HMTs and lysine demethylases (KDMs). Numerous KDMs have been identified with varying specificities [22]. Overall, gene activation or repression depends on the modification, the specific residue as well as dynamic cooperation between various chromatin factors.

Although histone lysine methylation is reversible, it remains one of the more stable epigenetic modifications with some histone lysine methylation states maintaining very low turnover rates, and hence could be key factors in metabolic memory. Histone modifications are now widely accepted to play a role in epigenetics; however, there are questions as to what role they specifically play. Histone modifications are closely linked with DNA methylation, and together they can initiate or maintain transcriptional memory [23].

                                    Villeneuve · Natarajan

Epigenetic Mechanisms in Diabetic Complications and Metabolic Memory

Numerous studies have shown epigenetic changes induced by diabetic conditions. Both in vitro and in vivo studies with diabetic stimuli like HG or environmental changes have demonstrated alterations in histone lysine acetylation and methylation patterns influenced by recruitment of HATs/HDACs or HMTs [24, 25]. In addition, recent evidence shows that HG- and TGF-β_1-induced expression of fibrotic genes in renal mesangial cells was associated with changes in key histone PTMs at these gene promoters [26]. There is much interest in the biological roles of DNA methylation, histone PTMs and HMTs in metabolic memory and diabetic complication such as CKD and nephropathy that seem to persist even after treatment with drugs or glycemic control.

Emerging research has provided new insights suggesting that chromatin-based changes to histone methylation patterns may be responsible for metabolic memory. VSMC from diabetic *db/db* mice cultured in vitro for several passages continued to exhibit increased inflammatory gene expression as compared to VSMC from nondiabetic *db/+* mice [13, 27]. This corresponded to decreased H3K9me3-repressive marks at the promoters of key inflammatory genes and was associated with decreased protein levels of Suv39h1, a known HMT mediating H3K9me3. Overexpression of Suv39h1 in the diabetic *db/db* VSMC partially reversed the diabetic phenotype. *db/db* VSMC also exhibited increased TNF-α-induced inflammatory gene expression with corresponding sustained decreases in promoter H3K9me3 and Suv39h1 occupancy. Human VSMC treated with HG demonstrated similar changes in lysine methylation [27].

Interestingly, a recent study reported elevated levels of miR-125b in diabetic *db/db* VSMC. MicroRNA-125b could target Suv39h1 leading to decreased H3K9me3 and a corresponding increase in inflammatory gene expression as well as enhanced VSMC monocyte binding [28]. These results indicate a sustained loss of chromatin-repressive mechanisms in the diabetic state in part through reduction in repressive marks due to the prior hyperglycemic environment and thus may play a role in metabolic memory.

Short-term hyperglycemic conditions were also found to induce long-term changes in chromatin modifications. Endothelial cells exposed to HG for 16 h and later cultured for several days in normal glucose demonstrated a sustained increase in the expression of NF-κB p65 subunit which correlated with increased promoter H3K4me1 marks and occupancy of the HMT Set7 known to regulate this mark [10, 11]. These epigenetic changes could be prevented by blocking components of the mitochondrial electron transport chain [11]. HG could also promote epigenetic changes at the promoters of fibrotic genes in renal mesangial cells [26]. Collectively, these results indicate that prior exposure to hyperglycemia can induce epigenetic changes in target cells and alter chromatin structure resulting in long-lasting repercussions for gene expression levels associated with the pathology of diabetic micro- and macrovascular complications.

Conclusions

Epigenetic modifications have been found to play an important role in altered gene expression patterns associated with various diseases [29]. Clinical and experimental studies have clearly demonstrated the deleterious effects of hyperglycemia and the importance of maintaining good glucose control to prevent the onset or severity of diabetic complications like diabetic nephropathy. Hyperglycemia can induce epigenetic changes at genes associated with diabetic complications. DNA methylation, key histone modifications and their associated chromatin modifiers have been implicated (fig. 1), and epigenetic regulation of key inflammatory and fibrotic genes has been documented in vascular and renal cells as well as in metabolic memory. In addition to hyperglycemia, diabetes is also associated with genetic, nutritional and environmental risk factors. Each of these risk factors could in itself induce epigenetic changes to the chromatin structure ultimately altering gene expression patterns in conjunction with elevated glucose in various target tissues affected by diabetic complications. How these epigenetic changes are transmitted through multiple cell cycles is still under investigation [30].

The rates of diabetes and associated complications are increasing rapidly. Given the widespread prevalence of diabetic nephropathy, CKD and the rapid transition to ESRD, further exploration into the epigenetic causes and related treatment options is imperative. Epigenetic drugs such as inhibitors of DNA methylation, HATs and HDACs, and some KDMs are already being evaluated for cancer and other diseases [16]. In addition, recent and ongoing studies investigating epigenomic DNA and histone methylation patterns are beginning to reveal key aspects of chromatin structure, gene expression and epigenetic information under diabetic states. Such epigenomic profiling approaches will be greatly aided by the recent availability of high throughput next generation sequencing technologies. No doubt will we see a lot of research activity in this field in the upcoming years that hopefully can lead to a greater understanding of the role of epigenetics in diabetes and to the identification of sorely needed new therapeutic strategies for diabetic complications and CKD.

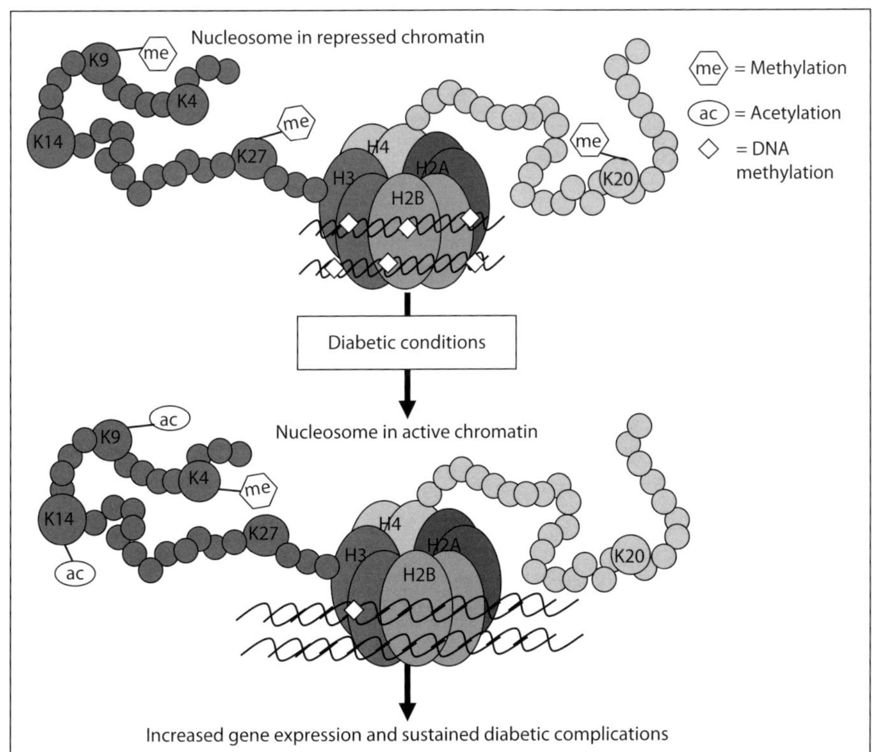

Fig. 1. Model for epigenetic regulation of pathologic gene expression associated with diabetic complications and a potential role in metabolic memory. Posttranslational modifications of N-terminal histone tails in nucleosomes of chromatin play essential roles in gene regulation and are regulated by various chromatin modifiers. As examples, promoter trimethylation of histone H3 at lysine 9 (K9) or K27 is generally associated with repression, while methylation of histone H3 at lysine 4 or acetylation at H3 lysine 9/14 are usually associated with gene activation. In addition, promoter DNA methylation may also be involved as a key repressive mark. In the proposed model, diabetic conditions such as hyperglycemia lead to a loss of repressive modifications that normally maintain strict control of gene expression, while activating histone modifications may be increased or DNA methylation may be reduced at promoters, thus leading to relaxation or opening of the chromatin structure around key pathologic genes resulting in increased transcription. Various combinations of these chromatin modifications and others not shown here are likely to be involved. These epigenetic marks can be maintained through cell division via as yet unclear mechanisms. The occurrence and persistence of these epigenetic changes might explain altered gene expression and the metabolic memory phenomenon responsible for the sustained and persistent development of diabetic complications even after glucose normalization.

Acknowledgments

The authors thank Dr. Marpadga Reddy for help with the manuscript, and also gratefully acknowledge funding from the National Institutes of Health (NIDDK and NHLBI) and the Juvenile Diabetes Research Foundation.

References

1 Brownlee M: Biochemistry and molecular cell biology of diabetic complications. Nature 2001;414:813–820.
2 Writing Team DCCT/EDIC Research Group: Effect of intensive therapy on the microvascular complications of type 1 diabetes mellitus. JAMA 2002;287:2563–2569.
3 EDIC Study: Sustained effect of intensive treatment of type 1 diabetes mellitus on development and progression of diabetic nephropathy: the Epidemiology of Diabetes Interventions and Complications (EDIC) study. JAMA 2003;290:2159–2167.
4 Nathan DM, Cleary PA, Backlund JY, et al: Intensive diabetes treatment and cardiovascular disease in patients with type 1 diabetes. N Engl J Med 2005;353:2643–2653.
5 Schernthaner G: Diabetes and cardiovascular disease: is intensive glucose control beneficial or deadly? Lessons from ACCORD, ADVANCE, VADT, UKPDS, PROactive, and NICE-SUGAR. Wien Med Wochenschr 2010;160:8–19.
6 Engerman RL, Kern TS: Progression of incipient diabetic retinopathy during good glycemic control. Diabetes 1987;36:808–812.
7 Hammes HP, Klinzing I, Wiegand S, Bretzel RG, Cohen AM, Federlin K: Islet transplantation inhibits diabetic retinopathy in the sucrose-fed diabetic Cohen rat. Invest Ophthalmol Vis Sci 1993;34:2092–2096.
8 Zhong Q, Kowluru RA: Role of histone acetylation in the development of diabetic retinopathy and the metabolic memory phenomenon. J Cell Biochem 2010;110: 1306–1313.
9 Roy S, Sala R, Cagliero E, Lorenzi M: Overexpression of fibronectin induced by diabetes or high glucose: phenomenon with a memory. Proc Natl Acad Sci U S A 1990;87: 404–408.
10 Brasacchio D, Okabe J, Tikellis C, et al: Hyperglycemia induces a dynamic cooperativity of histone methylase and demethylase enzymes associated with gene-activating epigenetic marks that coexist on the lysine tail. Diabetes 2009;58:1229–1236.
11 El-Osta A, Brasacchio D, Yao D, et al: Transient high glucose causes persistent epigenetic changes and altered gene expression during subsequent normoglycemia. J Exp Med 2008;205:2409–2417.
12 Ihnat MA, Thorpe JE, Ceriello A: Hypothesis: the 'metabolic memory', the new challenge of diabetes. Diabet Med 2007;24: 582–586.
13 Li SL, Reddy MA, Cai Q, et al: Enhanced proatherogenic responses in macrophages and vascular smooth muscle cells derived from diabetic db/db mice. Diabetes 2006;55: 2611–2619.
14 Luger K, Mader AW, Richmond RK, Sargent DF, Richmond TJ: Crystal structure of the nucleosome core particle at 2.8 A resolution. Nature 1997;389:251–260.
15 Strahl BD, Allis CD: The language of covalent histone modifications. Nature 2000;403:41–45.
16 Sharma S, Kelly TK, Jones PA: Epigenetics in cancer. Carcinogenesis 2010;31:27–36.
17 Ling C, Groop L: Epigenetics: a molecular link between environmental factors and type 2 diabetes. Diabetes 2009;58:2718–2725.
18 Bell CG, Teschendorff AE, Rakyan VK, Maxwell AP, Beck S, Savage DA: Genome-wide DNA methylation analysis for diabetic nephropathy in type 1 diabetes mellitus. BMC Med Genomics 2010;3:33.
19 Ekstrom TJ, Stenvinkel P: The epigenetic conductor: a genomic orchestrator in chronic kidney disease complications? J Nephrol 2009;22:442–449.

20 Roth SY, Denu JM, Allis CD: Histone acetyl-
    transferases. Annu Rev Biochem 2001;70:
    81–120.
21 Martin C, Zhang Y: The diverse functions of
    histone lysine methylation. Nat Rev Mol Cell
    Biol 2005;6:838–849.
22 Shi Y, Whetstine JR: Dynamic regulation of
    histone lysine methylation by demethylases.
    Mol Cell 2007;25:1–14.
23 Berger SL, Kouzarides T, Shiekhattar R,
    Shilatifard A: An operational definition of
    epigenetics. Genes Dev 2009;23:781–783.
24 Villeneuve LM, Natarajan R: The role
    of epigenetics in the pathology of
    diabetic complications. Am J Physiol
    2010;299:F14–F25.
25 Tonna S, El-Osta A, Cooper ME, Tikellis C:
    Metabolic memory and diabetic nephropa-
    thy: potential role for epigenetic mecha-
    nisms. Nature Rev Nephrol 2010;6:332–341.
26 Sun G, Reddy MA, Yuan H, Lanting L, Kato
    M, Natarajan R: Epigenetic histone methyla-
    tion modulates fibrotic gene expression.
    J Am Soc Nephrol 2010;21:2069–2080.
27 Villeneuve LM, Reddy MA, Lanting LL,
    Wang M, Meng L, Natarajan R: Epigenetic
    histone H3 lysine 9 methylation in metabolic
    memory and inflammatory phenotype of
    vascular smooth muscle cells in diabe-
    tes. Proc Natl Acad Sci U S A 2008;105:
    9047–9052.
28 Villeneuve LM, Kato M, Reddy MA, Wang
    M, Lanting L, Natarajan R: Enhanced levels
    of microRNA-125b in vascular smooth
    muscle cells of diabetic db/db mice lead to
    increased inflammatory gene expression
    by targeting the histone methyltransferase
    Suv39h1. Diabetes 2010;59:2904–2915.
29 Liu L, Li Y, Tollefsbol TO: Gene-environment
    interactions and epigenetic basis of human
    diseases. Curr Issues Mol Biol 2008;10:25–36.
30 Margueron R, Reinberg D: Chromatin
    structure and the inheritance of epigenetic
    information. Nat Rev Genet 2010;11:
    285–296.

Rama Natarajan, PhD
Department of Diabetes, Division of Cellular and Molecular Diabetes Research
Beckman Research Institute of City of Hope
Duarte, CA 91010 (USA)
Tel. +1 626 256 4673, ext. 62289, Fax +1 626 301 8136, E-Mail RNatarajan@coh.org

Lai KN, Tang SCW (eds): Diabetes and the Kidney.
Contrib Nephrol. Basel, Karger, 2011, vol 170, pp 66–74

Advanced Glycation End Products

Merlin C. Thomas

Baker IDI Heart and Diabetes Institute, Melbourne, Vic., Australia

Abstract

Prolonged hyperglycemia, dyslipidemia and oxidative stress in diabetes result in the increased production and accumulation of advanced glycation end products (AGEs) in the kidney. Covalent AGE modifications significantly influence the structure and function of key protein targets. In addition, activation of AGE receptors, alone or in combination with other ligands, is able to promote renal damage, fibrosis and inflammation associated with diabetic nephropathy. The actions of AGEs synergize and potentiate the activity of other pathogenic mediators in the diabetic kidney, including oxidative stress, protein kinase C and renin-angiotensin system activation, which subsequently promote the development and progression of kidney disease in a vicious and progressive cycle. Their importance as downstream mediators of hyperglycemia in diabetes has been amply demonstrated in studies using mechanistically different inhibitors of advanced glycation to retard the development of kidney disease without directly influencing plasma glucose levels. Furthermore, direct exposure to AGEs is able to generate lesions similar to those seen in diabetic nephropathy. The human body has a number of natural defenses against AGE accumulation, which are reduced in diabetic individuals, and in particular those with nephropathy, while the receptor for AGEs and its ligands are significantly increased. Given such data, a number of different pharmacological agents have been developed to reduce AGEs and with it prevent diabetic kidney disease. Although many have proved effective in experimental models of diabetes, their clinical utility remains unproven.

Optimal glycemic control remains the cornerstone of the prevention of the renal complications of diabetes. Indeed, diabetic nephropathy is not seen in the absence of hyperglycemia, implying that glucose is fundamental for its development and progression. Yet, a complete understanding of the mechanisms by which chronic excess of glucose causes renal dysfunction and deterioration in some (but not all) individuals remains a major challenge to biomedical research and clinical care. Although a number of hemodynamic and metabolic factors are induced by elevated glucose levels, one pivotal pathway appears to be the formation and accumulation of advanced glycation end products (AGEs), both in the kidney, and at other sites of diabetic complications [1]. AGEs have a wide range of chemical, cellular and tissue activities implicated in the pathogenesis of nephropathy. Their importance as downstream mediators of hyperglycemia in diabetes has been amply demonstrated in studies using mechanistically different inhibitors of advanced glycation to retard the development of kidney disease without directly influencing plasma glucose levels [2]. Furthermore, direct in vivo exposure to AGEs is able to generate lesions similar to those seen in diabetic nephropathy [3]. Clinical studies in patients with diabetes certainly demonstrate a strong correlation between AGE accumulation and the severity of diabetic complications. For example, in the Diabetes Control and Complications Trial, AGE levels were overall a better predictor of vascular complications than HbA1c, with over a third of the variance in complications in the study attributed to differences in AGE indices [4]. This review will examine the role of AGEs in the pathogenesis of renal complications in diabetes and explore the potential utility of interventions designed to prevent the accumulation of AGEs in diabetes.

What Are Advanced Glycation End Products?

AGEs are formed by the nonenzymatic Maillard or 'browning' reaction that begins when the aldehyde group of sugars and carbonyls reacts with amino groups of proteins, lipids or nucleic acids. The first stable adduct of this reaction is the Amadori product, like HbA1c. Its subsequent dehydration, oxidation, rearrangement and fragmentation leads to the formation of AGEs. Under physiological conditions, this reaction is slow, meaning that AGE modifications accumulate primarily in long-lived molecules such as extracellular matrix proteins. However, because glucose levels must always be present to maintain brain function, nearly all proteins carry some AGE modifications [5], in part as a way of denoting molecular age (hence the appropriate acronym and its implicated role in the ageing process). In patients with diabetes, persistent hyperglycemia, dyslipidemia and oxidative stress hasten the formation of AGEs. In addition, α-dicarbonyls (such as methylglyoxal, 3-deoxyglucosone, and glyoxal) can be formed from triose phosphate intermediates of glycolysis, lipid peroxidation, and the degradation of glycated proteins. These highly reactive compounds are able to rapidly generate AGEs, meaning that not only long-lived proteins become more heavily modified but also shorter-lived molecules become novel targets for progressive AGE modification. Moreover, this means that even transient elevations of glucose can initiate microvascular damage, while mean control (denoted by HbA1c) is not substantially abnormal.

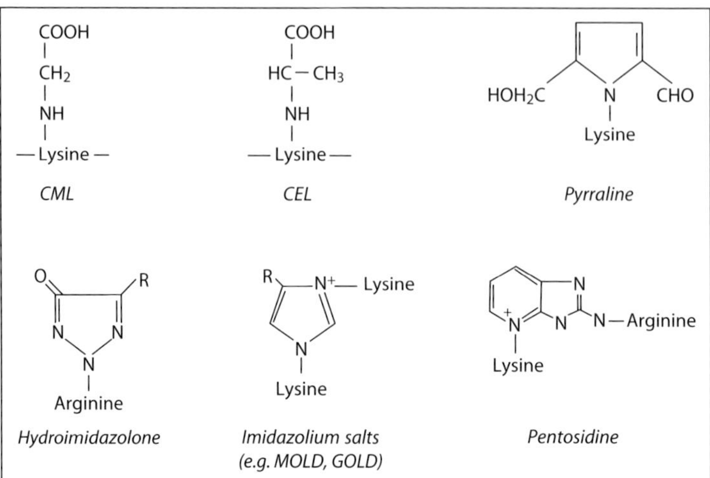

Fig. 1. Chemical structures of some AGEs that are found in increased levels within the diabetic kidney.

AGEs are a group of chemically heterogeneous compounds (fig. 1). About twenty have been chemically characterized in human tissues, the best known of which include Nε-(carboxymethyl)lysine (CML), Nε-carboxyethyllysine (CEL), the 'cross-links' pentosidine, MOLD (methylglyoxal lysine dimer) and GOLD (glyoxal lysine dimer), and the most abundant hydroimidazolone AGEs [6]. The precise identity of AGEs that contribute to diabetic complications has not been clearly determined. Although each of these specific AGEs has been shown to have toxic effects in vitro, it is likely that there is an aggregate effect in vivo. Indeed, most of the AGE-modified proteins present in the circulation of diabetic patients are large protein complexes often containing multiple ligands [7]. Moreover, given the broad chemical heterogeneity of AGEs and the fact that AGE-mediated renal injury occurs partly through the activation of AGE receptors that recognize a range of chemical structures and formations (see below), the precise chemical modification or even their abundance may be of less consequence than their ability to reach, bind and activate AGE receptors and their relevant pathogenic pathways. Nonetheless, the measurement of specific AGEs has been shown in both experimental and clinical studies to be a useful biomarker for the presence of unmeasured modifications on pathogenic sites in much the same way that HbA1c reflects glucose modification of extracellular proteins in general. For the same reasons, there appears to be a strong association between these measured AGEs and adverse outcomes in patients with diabetes, including nephropathy [4]. Indeed, because many AGE cross-link moieties have intrinsic fluorescence, even crude fluorescence can be used as a surrogate marker for the presence of AGE modifications and the severity of vascular complications in patients with diabetes [8].

How Do Advanced Glycation End Products Cause Renal Damage?

AGE modification of proteins may impact on their structural and functional integrity. For example, the AGE modification of collagen leads to structural alterations, including changes in self-assembly and packing density, manifested by increased stiffness, reduced thermal stability and resistance to digestion by metalloproteinases. Interactions between collagen and other matrix proteins and cell-matrix interactions may also be disturbed by AGE modifications, contributing to compositional changes, matrix accumulation, altered cell growth and differentiation. Cross-links also affect the permeability and charge of the extracellular matrix, which contribute to the irreversible trapping of macromolecules, manifested as amorphous hyalinization that characterizes diabetic vascular disease. In addition, AGEs have been shown to directly generate reactive oxygen intermediates [9], acting in effect as a one electron donor. Posttranslational modifications of proteins may not only lead to loss of molecular function, but also lead to gain-of-function properties and promote novel interactions [10].

Many of the renal actions of AGEs appear to be mediated by interactions with specific AGE receptors and binding proteins in the kidney, of which the receptor for AGEs (RAGE) is the best characterized [11]. RAGE is a multiligand receptor and may be activated not just by AGEs but also high mobility group box-1, S-100 calgranulins, Mac-1, amyloid-b peptide and advanced oxidation protein products. Activation of RAGE triggers the activation of key secondary messenger pathways in renal cells such as protein kinase C, NF-κB and increases production of inflammatory and fibrogenic growth factors and cytokines, including TGF-β and connective tissue growth factor. Glomerular matrix production is directly increased by exposure to AGEs [12], mediated through RAGE activation. RAGE activation also induces the expression and activity of NADPH oxidase in the kidney [13, 14], which represents an important source of oxidative stress in the diabetic kidney. In addition, RAGE activation promotes the epithelial-mesenchymal transdifferentiation of tubular cells through the induction of TGF-β and connective tissue growth factor, contributing to interstitial fibrogenesis [15].

The potential importance of RAGE in the diabetic kidney is illustrated by the prevention of indices of mesangial expansion, thickening of the glomerular basement membrane and reduced albuminuria in RAGE KO mice [14, 16] and following intervention with RAGE neutralizing antibodies in mouse models of type 1 and 2 diabetes [17, 18]. Conversely, diabetic mice which overexpress RAGE develop more advanced renal lesions following the induction of diabetes [19].

Natural Defenses against Advanced Glycation End Product Accumulation

The human body has a range of defenses that antagonize activation of the AGE/RAGE axis. Circulating soluble RAGE (sRAGE) is produced by proteolytic cleavage of the membrane-bound form via the action of metalloprotein sheddases [20], with a small additional amount derived from the alternative splicing of RAGE (so called esRAGE) [21]. sRAGE does not contain the C-terminal signaling domain, but retains the ability to bind ligands, thereby antagonizing RAGE-dependent signaling and/or facilitating ligand clearance [15]. These actions have been exploited by researchers who have used recombinant sRAGE to attenuate progressive nephropathy in experimental diabetes [22, 23]. Endogenous circulating levels of sRAGE largely reflect expression of RAGE at a tissue level, and are increased in individuals with type 1 diabetes, paralleling increased levels of circulating AGEs [24]. Kidney damage also increases sRAGE, possibly by impaired renal clearance of sRAGE. In addition, retention of AGEs and/or other RAGE ligands may lead to increased tissue RAGE expression and/or sRAGE shedding.

Lysozyme and lactoferrin are naturally occurring antibacterial defense proteins found in mucosal secretions, serum and in the lysosomes of phagocytes. Both molecules contain high-affinity binding domain for AGE-modified proteins, possibly to enhance their uptake and degradation by macrophages. In both the nonobese diabetic and db/db (+/+) mice, an intraperitoneal injection of lysozyme significantly reduces circulating AGE levels and reduced albuminuria [25]. Although the utility of this pathway in a clinical setting remains to be established, a number of common interventions are known to increase serum lysozyme activity, including caffeine intake.

Other endogenous AGE receptors may also act as competitive antagonists of RAGE, either by clearing ligands or offering alternative binding sites. In addition, some of these receptors may have direct cytoprotective effects, possibly by negatively regulating oxidant stress-dependent signaling. In diabetic individuals, and in particular those with nephropathy, the expression of these 'good receptors' is reduced, while RAGE and its ligands are increased, further shifting the balance of effects towards inflammation and injury.

Pharmacological Inhibition of Glycation as a Renoprotective Intervention

A number of agents have been specifically developed to reduce the accumulation of AGEs in diabetes (table 1). The fact that the majority of these agents have been shown to be effective in experimental models of diabetes, despite their disparate mechanisms of actions, clearly supports the pathogenic role of AGEs in diabetic kidney disease. Nonetheless, the clinical utility of AGE inhibition remains to be established. Notably, large clinical studies of pimagedine

Table 1. Some pharmacological agents specifically developed to reduce AGEs and/or RAGE activation

---

Aminoguanidine
ALT-946 [N-(2-acetamidoethyl)hydrozinecarboximidamide hydrochloride]
Pyridoxamine
Pyridoxamine-aminoguanidine conjugate
Benfotiamine
Alagebrium chloride
LR-90 [methylene bis [4,4'-(2-chlorophenylureidopheonoxyisobutyric acid)]]
PF-04494700
Kremezin
Pentoxyfylline
Aryl ureido and aryl carboxaminido phenoxy isobutyric acids

(aminoguanidine) in patients with type 1 diabetes (ACTION 1) and type 2 diabetes (ACTION 2) were terminated, due to safety concerns and apparent lack of efficacy. Nonetheless, pimagedine was associated with a reduction in urine protein excretion in a subpopulation of patients with moderate renal impairment [26]. More recent attention has been focused on the potential effects of the AGE-inhibiting vitamins, thiamine and pyridoxamine. Supplementation with benfotiamine, a lipid-soluble form of thiamine, is able to prevent renal damage associated with chronic hyperglycemia in experimental diabetes [27]. However, the only large randomized controlled clinical trials to be conducted failed to demonstrate any effect of benfotiamine on albuminuria or markers of tubular injury [28]. Pyridoxamine (a vitamin B_6 derivative) also has AGE-inhibitory activities, and is able to attenuate the development of renal disease in animal models [29]. In phase II clinical trials, pyridoxamine has proven to be safe and well tolerated. Although not designed to test this endpoint, post-hoc analyses of these studies also noted significant reductions in the rise of serum creatinine and urinary albumin excretion in the treatment group [30]. Another group of compounds effective in preventing diabetic kidney disease are the so-called 'cross-link breakers' (N-phenacylthiazolium bromide and alagebrium chloride). The designation of these compounds as 'AGE breakers' is probably a misnomer as these compounds may more likely work as dicarbonyl scavengers, antioxidants and stimulators of transketolase activity. Nonetheless, these agents are effective in the inhibition of AGE accumulation and renal damage in experimental diabetic models [2].

Other agents in clinical use may have actions to reduce AGEs, partly contributing to their efficacy. These include LMW heparin, carnosine, α-lipoic acid, some antioxidants (aspirin, benzoic acid inositol and probucol), transition metal chelators, metformin and the thioglitazone hypoglycemics and even agents that block the renin-angiotensin system (RAS). For example, blockade of the RAS using ACE inhibitors or angiotensin receptor antagonists significantly attenuates the formation and accumulation of AGEs in experimental diabetes [31], possibly suppresses the formation of reactive oxygen species, thereby reducing AGEs formed as a result of glycoxidation. These actions appear synergistic to that of inhibition of AGE accumulation with aminoguandine [31].

AGEs are also formed during fermentation, during prolonged or high temperature cooking, meaning that certain (processed) foods have high levels of AGEs. These 'browning' products contribute to textures, flavors or aromas that give many foods their appeal. However, increased dietary intake of AGEs is associated with elevated circulating AGEs (probably through the ingestion of toxic intermediates like MGO or lipid peroxides). Equally, dietary restriction of AGEs in patients with type 2 diabetes appears to both reduce AGEs and suppress circulating levels of inflammatory markers [32]. In patients with type 1 diabetes, a gluten-free diet (low in processed products and consequently AGEs) is also associated with reduced AGE levels as well as a reduced albumin excretion rate [33]. Taken together, these data support the preferred consumption of fresh foods in patients with diabetes or at the least only brief applications of heat in the presence of ample water or humidity. Such regimens have been reported to decrease the intake of AGEs by more than 50% and reduce circulating AGEs by 30% within a month [34].

Conclusions

AGEs are not the only cause of diabetic nephropathy. However, there is good evidence that they not only have direct effects but also synergize with other pathogenic pathways including oxidative stress, protein kinase C and RAS activation to promote the development and progression of kidney disease in diabetes. Moreover, renal damage contributes to the accumulation of AGEs in a vicious and progressive cycle, since the kidney is the major site of catabolism and clearance of AGEs and their precursors. To this end, optimal metabolic and blood pressure control achieved early in the disease and sustained indefinitely is currently the best recourse to prevent the accumulation of AGEs until more specific agents become a clinical reality. And even then, it is likely that these agents will need to be combined with other metabolic and hemodynamic interventions for maximal efficacy.

Acknowledgments

The author is supported by JDRF international, the Australian NHMRC, and the Bootle bequest from Kidney Health Australia.

References

1 Forbes JM, Cooper ME, Oldfield MD, Thomas MC: Role of advanced glycation end products in diabetic nephropathy. J Am Soc Nephrol 2003;14(suppl 3):S254–S258.

2 Forbes JM, Thallas V, Thomas MC, Founds HW, Burns WC, Jerums G, et al: The breakdown of preexisting advanced glycation end products is associated with reduced renal fibrosis in experimental diabetes. FASEB J 2003;17:1762–1764.

3 Vlassara H, Striker LJ, Teichberg S, Fuh H, Li YM, Steffes M: Advanced glycation end products induce glomerular sclerosis and albuminuria in normal rats. Proc Natl Acad Sci U S A 1994;91:11704–11708.

4 Monnier VM, Bautista O, Kenny D, Sell DR, Fogarty J, Dahms W, et al: Skin collagen glycation, glycoxidation, and crosslinking are lower in subjects with long-term intensive versus conventional therapy of type 1 diabetes: relevance of glycated collagen products versus HbA1c as markers of diabetic complications. DCCT Skin Collagen Ancillary Study Group. Diabetes Control and Complications Trial. Diabetes 1999;48: 870–880.

5 Monnier VM, Sell DR, Nagaraj RH, Miyata S, Grandhee S, Odetti P, et al: Maillard reaction-mediated molecular damage to extracellular matrix and other tissue proteins in diabetes, aging, and uremia. Diabetes 1992;41(suppl 2):36–41.

6 Ahmed N, Dobler D, Dean M, Thornalley PJ: Peptide mapping identifies hotspot site of modification in human serum albumin by methylglyoxal involved in ligand binding and esterase activity. J Biol Chem 2005;280: 5724–5732.

7 Penfold SA, Coughlan MT, Patel SK, Srivastava PM, Sourris KC, Steer D, et al: Circulating high-molecular-weight RAGE ligands activate pathways implicated in the development of diabetic nephropathy. Kidney Int 2010;78:287–295.

8 Thomas MC, Forbes JM, MacIsaac R, Jerums G, Cooper ME: Low-molecular weight advanced glycation end products: markers of tissue AGE accumulation and more? Ann N Y Acad Sci 2005;1043:644–654.

9 Yim MB, Yim HS, Lee C, Kang SO, Chock PB: Protein glycation: creation of catalytic sites for free radical generation. Ann N Y Acad Sci 2001;928:48–53.

10 Yao D, Taguchi T, Matsumura T, Pestell R, Edelstein D, Giardino I, et al: High glucose increases angiopoietin-2 transcription in microvascular endothelial cells through methylglyoxal modification of mSin3A. J Biol Chem 2007;282:31038–31045.

11 D'Agati V, Schmidt AM: RAGE and the pathogenesis of chronic kidney disease. Nat Rev Nephrol 2010;6:352–360.

12 Yang CW, Vlassara H, Peten EP, He CJ, Striker GE, Striker LJ: Advanced glycation end products up-regulate gene expression found in diabetic glomerular disease. Proc Natl Acad Sci U S A 1994;91:9436–9440.

13 Coughlan MT, Cooper ME, Forbes JM: Renal microvascular injury in diabetes: RAGE and redox signaling. Antioxid Redox Signal 2007; 9:331–342.

14 Coughlan MT, Thorburn DR, Penfold SA, Laskowski A, Harcourt BE, Sourris KC, et al: RAGE-induced cytosolic ROS promote mitochondrial superoxide generation in diabetes. J Am Soc Nephrol 2009;20:742–752.

15 Oldfield MD, Bach LA, Forbes JM, Nikolic-Paterson D, McRobert A, Thallas V, et al: Advanced glycation end products cause epithelial-myofibroblast transdifferentiation via the receptor for advanced glycation end products (RAGE). J Clin Invest 2001;108: 1853–1863.

16 Tan AL, Sourris KC, Harcourt BE, Thallas-Bonke V, Penfold S, Andrikopoulos S, et al: Disparate effects on renal and oxidative parameters following RAGE deletion, AGE accumulation inhibition, or dietary AGE control in experimental diabetic nephropathy. Am J Physiol Renal Physiol 2010;298: F763–F770.

17 Flyvbjerg A, Denner L, Schrijvers BF, Tilton RG, Mogensen TH, Paludan SR, et al: Long-term renal effects of a neutralizing RAGE antibody in obese type 2 diabetic mice. Diabetes 2004;53:166–172.

18 Jensen LJ, Denner L, Schrijvers BF, Tilton RG, Rasch R, Flyvbjerg A: Renal effects of a neutralising RAGE-antibody in long-term streptozotocin-diabetic mice. J Endocrinol 2006;188:493–501.

19 Inagi R, Yamamoto Y, Nangaku M, Usuda N, Okamato H, Kurokawa K, et al: A severe diabetic nephropathy model with early development of nodule-like lesions induced by megsin overexpression in RAGE/iNOS transgenic mice. Diabetes 2006;55:356–366.

20 Zhang L, Bukulin M, Kojro E, Roth A, Metz VV, Fahrenholz F, et al: Receptor for advanced glycation end products is subjected to protein ectodomain shedding by metalloproteinases. J Biol Chem 2008;283:35507–35516.

21 Yonekura H, Yamamoto Y, Sakurai S, Petrova RG, Abedin MJ, Li H, et al: Novel splice variants of the receptor for advanced glycation end-products expressed in human vascular endothelial cells and pericytes, and their putative roles in diabetes-induced vascular injury. Biochem J 2003;370:1097–1109.

22 Morcos M, Sayed AA, Bierhaus A, Yard B, Waldherr R, Merz W, et al: Activation of tubular epithelial cells in diabetic nephropathy. Diabetes 2002;51:3532–3544.

23 Wendt TM, Tanji N, Guo J, Kislinger TR, Qu W, Lu Y, et al: RAGE drives the development of glomerulosclerosis and implicates podocyte activation in the pathogenesis of diabetic nephropathy. Am J Pathol 2003;162:1123–1137.

24 Challier M, Jacqueminet S, Benabdesselam O, Grimaldi A, Beaudeux JL : Increased serum concentrations of soluble receptor for advanced glycation endproducts in patients with type 1 diabetes. Clin Chem 2005;51:1749–1750.

25 Zheng F, Cai W, Mitsuhashi T, Vlassara H: Lysozyme enhances renal excretion of advanced glycation end products in vivo and suppresses adverse age-mediated cellular effects in vitro: a potential AGE sequestration therapy for diabetic nephropathy? Mol Med 2001;7:737–747.

26 Bolton WK, Cattran DC, Williams ME, Adler SG, Appel GB, Cartwright K, et al: Randomized trial of an inhibitor of formation of advanced glycation end products in diabetic nephropathy. Am J Nephrol 2004;24:32–40.

27 Babaei-Jadidi R, Karachalias N, Ahmed N, Battah S, Thornalley PJ: Prevention of incipient diabetic nephropathy by high-dose thiamine and benfotiamine. Diabetes 2003;52:2110–2120.

28 Alkhalaf A, Klooster A, van Oeveren W, Achenbach U, Kleefstra N, Slingerland RJ, et al: A double-blind, randomized, placebo-controlled clinical trial on benfotiamine treatment in patients with diabetic nephropathy. Diabetes Care 2010;33:1598–1601.

29 Degenhardt TP, Alderson NL, Arrington DD, Beattie RJ, Basgen JM, Steffes MW, et al: Pyridoxamine inhibits early renal disease and dyslipidemia in the streptozotocin-diabetic rat. Kidney Int 2002;61:939–950.

30 Williams ME, Bolton WK, Khalifah RG, Degenhardt TP, Schotzinger RJ, McGill JB: Effects of pyridoxamine in combined phase 2 studies of patients with type 1 and type 2 diabetes and overt nephropathy. Am J Nephrol 2007;27:605–614.

31 Davis BJ, Forbes JM, Thomas MC, Jerums G, Burns WC, Kawachi H, et al: Superior renoprotective effects of combination therapy with ACE and AGE inhibition in the diabetic spontaneously hypertensive rat. Diabetologia 2004;47:89–97.

32 Vlassara H, Striker G: Glycotoxins in the diet promote diabetes and diabetic complications. Curr Diab Rep 2007;7:235–241.

33 Malalasekera V, Cameron F, Grixti E, Thomas MC: Potential reno-protective effects of a gluten-free diet in type 1 diabetes. Diabetologia 2009;52:798–800.

34 Uribarri J, Woodruff S, Goodman S, Cai W, Chen X, Pyzik R, et al: Advanced glycation end products in foods and a practical guide to their reduction in the diet. J Am Diet Assoc 2010;110:911–916.e12.

Prof. Merlin C. Thomas, MBChB, PhD, FRACP
Baker IDI Heart and Diabetes Institute, St Kilda Rd Central
PO Box 6492
Melbourne, VIC 8008 (Australia)
Tel. +61 3 8532 1277, Fax +61 3 8532 1480, E-Mail mthomas@bakeridi.edu.au

Thomas

Lai KN, Tang SCW (eds): Diabetes and the Kidney.
Contrib Nephrol. Basel, Karger, 2011, vol 170, pp 75–82

Transforming Growth Factor-β and Smads

Hui Yao Lan · Arthur C.K. Chung

Department of Medicine and Therapeutics, and Li Ka Shing Institute of Health Sciences, the Chinese University of Hong Kong, Hong Kong SAR, China

Abstract

Diabetic nephropathy (DN) is a major diabetic complication. Transforming growth factor-β (TGF-β) is a key mediator in the development of diabetic complications. It is well known that TGF-β exerts its biological effects by activating downstream mediators, called Smad2 and Smad3, which is negatively regulated by an inhibitory Smad7. Recent studies also demonstrated that under disease conditions Smads act as signal integrators and interact with other signaling pathways such as the MAPK and NF-κB pathways. In addition, Smad2 and Smad3 can reciprocally regulate target genes of TGF-β signaling. Novel research into microRNA has revealed the complexity of TGF-β signaling during DN. It has been found that TGF-β and elevated glucose concentration can positively regulate miR-192 and miR-377, but negatively regulate miR-29a in a diabetic milieu. These microRNAs are found to contribute to DN. Although targeting TGF-β may exert adverse effects on immune system, therapeutic approach against TGF-β signaling during DN still draws much attention. Blocking TGF-β signaling by neutralizing antibody, anti-sense oligonucleotides, and soluble receptors have been tested, but effects are limited. Gene transfer of Smad7 into diseased kidneys demonstrates a prominent inhibition on renal fibrosis and amelioration of renal impairment. Alteration of TGF-β-regulated microRNA expression in diseased kidneys may provide an alternative therapeutic approach against DN. In conclusion, TGF-β/Smad signaling plays a critical role in DN. A better understanding of the role of TGF-β/Smad signaling in the development of DN should provide an effective therapeutic strategy to combat DN.

As one of major diabetic complications, diabetic nephropathy (DN) has become the leading cause of end-stage renal disease in most developed countries. The pathogenesis of DN is multifactorial, and the precise mechanisms remain uncertain. Several mechanisms have been proposed which include increased production of advanced glycation end products (AGEs), enhanced polyol pathway, activation of protein kinase C, and enhanced oxidative stress [reviewed in 1, 2]. Of them, activation of transforming growth factor-β (TGF-β) signaling plays a key role in DN development [3]. Several lines of evidence support this notion. In vitro studies demonstrate that high glucose and AGEs induce the production of extracellular matrix (ECM), while reducing the degradation of ECM [3], which is TGF-β-dependent [3]. Results from animal models of type 1 or type 2 diabetes mellitus further demonstrate TGF-β as an important mediator of diabetic kidney disease [3]. Transcript and protein levels of TGF-β are upregulated in both the glomerular and tubular compartments of various murine models of experimental diabetes [3]. In addition, activation of Smad2/3 has been observed in both experimental and human diabetic kidney [3, 4], implicating that activation of the renal TGF-β system plays a critical role in the pathogenesis of DN. Taken together, results from all these studies demonstrate the importance of TGF-β signaling in the pathogenesis of DN.

TGF-β/Smad Signaling in Diabetic Nephropathy

Among its three isoforms (TGF-$β_1$, -$β_2$ and -$β_3$), TGF-$β_1$, the central player in the fibrogenic process, is a key mediator in promoting ECM production while inhibiting its degradation [5]. All these biological effects are mediated by the Smad-dependent pathway via the activation of its downstream mediators, called Smad2 and Smad3 [6]. After binding to its receptor II (TβRII), TGF-$β_1$ activates the TGF-β receptor type I (TβRI) kinase. This process results in the phosphorylation of Smad2 and Smad3. Subsequently, phosphorylated Smad2 and Smad3 form oligomeric complexes with Smad4, the common Smad. This Smad complex then translocates into the nucleus and regulates the target gene transcription [6]. TGF-β/Smad signaling is strictly regulated to maintain homeostasis within cells. This safeguard mechanism to protect cells from an unnecessary TGF-β response is negatively regulated by inhibitory Smads [7], through which the access of Smad2/3 to TβRI is blocked and the TβRI is degraded [7]. Immunohistochemical studies in patients developing DN reveal highly activated TGF-β/Smad signaling (fig. 1). Gene transfer studies from our laboratory and others demonstrate that overexpression of Smad7 can suppress the activation of Smad2/3 and inhibit renal fibrosis and inflammation [7]. This is also found in a rodent diabetic model of type 1 DN, demonstrating a protective role of Smad7 in the development of DN [8]. This is further supported by the finding that Smad7 knockout mice develop more severe diabetic kidney injury because these mice have significantly higher levels of urinary albumin excretion, renal fibrosis (collagen I, IV, and fibronectin), and renal inflammation (IL-1β, TNF-α, MCP-1, ICAM-1, and macrophages) when compared to wild-type mice [8]. Enhanced activation of both TGF-β and NF-κB signaling

Lan · Chung

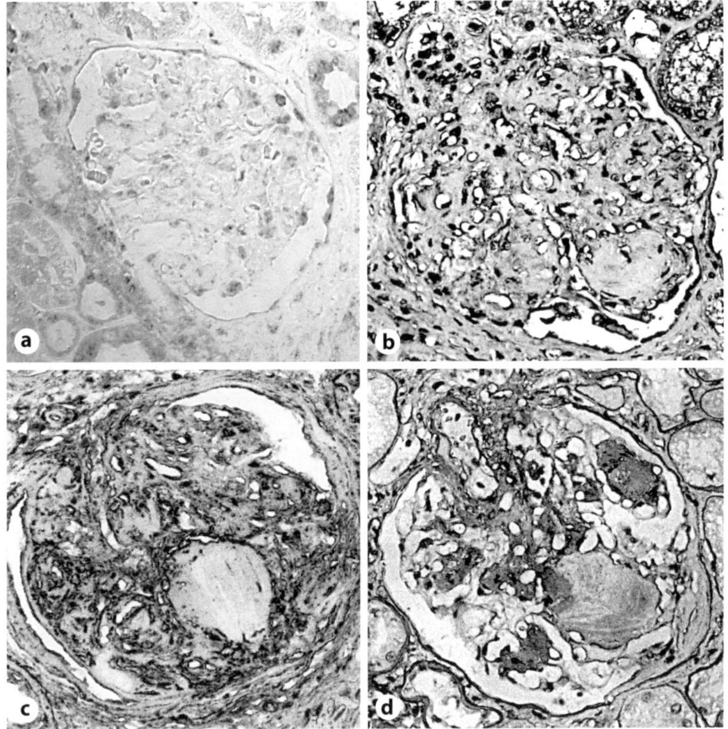

Fig. 1. Activation of the TGF-β Smad signaling pathway in DN. **a** Activation of Smad2/3 in a normal human kidney. **b** Activation of Smad2/3 in a diabetic kidney from a patient with DN. **c** Expression of TGF-β_1. **d** Collagen I in a diabetic kidney from a patient with DN. Note that Smad2/3 is activated in the DN as demonstrated by its nucleated location in the area with upregulation of TGF-β_1 and severe fibrosis. ×400.

pathways is the mechanism by which enhanced renal fibrosis and inflammation occurs in the diabetic mice deficient in Smad7. In contrast, delivery of Smad7 gene into the kidneys of diabetic rats by an ultrasound microbubble-mediated technique significantly reduces the development of microalbuminuria, TGF-β/Smad-mediated renal fibrosis, and NF-κB-driven renal inflammation [8].

Our recent findings also shed light on the precise roles of Smad2 and Smad3 in TGF-β-mediated renal fibrosis. Deletion of Smad2 from renal tubular epithelial cells (TECs) significantly promotes fibrosis in response to unilateral ureteral obstruction [9]. Knockdown of Smad2 expression in TECs also promotes the expression of fibrotic markers in response to TGF-β_1. Similarly, renal and cardiovascular fibrosis under diabetic and hypertensive conditions is mediated by Smad3, but inhibited by Smad2 [10, 11], suggesting that Smad2 and Smad3 may play reciprocal roles in regulating TGF-β target genes. Inhibition of diabetic kidney injury in a rodent model of type 1 diabetes that lacks Smad3 gene expression supports the notion that Smad3 may play a critical role in diabetic complications [12, 13].

Smads also act as signal integrators and interact with other signaling pathways, such as the MAPK, NF-κB, and Wnt signaling pathways [14, 15]. Recent bioinformatics research suggests that an interaction of β-catenin with CDH5 and TβRI through Smad molecules may contribute to endothelial dysfunction in type 2 diabetes [15]. Our earlier study in mesangial cells (MCs), TECs, and vascular smooth muscle cells shows that AGEs are able to markedly induce Smad2/3 phosphorylation in TβRI and TβRII mutant cell lines via the ERK/p38 MAP kinase-dependent mechanism [4], demonstrating a critical role for the TGF-β-independent Smad pathway in AGE-mediated fibrotic response. This may explain the finding that blockade of TGF-β_1 with specific shRNA and a neutralizing antibody is unable to significantly inhibit AGE induced connective tissue growth factor (CTGF) mRNA expression [16]. Our recent study further demonstrates that AGEs are able to induce CTGF expression via the TGF-β-independent Smad3 signaling pathway [11]. Firstly, AGEs stimulate CTGF expression in TECs lacking TGF-β_1 gene or expressing dominant TβRII. This observation is further confirmed by the finding that conditioned knockout of TβRII gene from the kidney does not prevent AGE-induced renal CTGF and collagen I expression. In addition, AGEs upregulate CTGF expression via the RAGE-ERK/p38 MAP kinase-Smad crosstalk pathway because this upregulation is suppressed by the anti-RAGE antibody, the specific inhibitors, or dominant negative adenovirus to ERK1/2 and p38. Furthermore, overexpression of Smad7 abrogates AGE-induced Smad3 phosphorylation and CTGF expression, suggesting that activation of Smad signaling is necessary for AGE-mediated CTGF expression, which is negatively regulated by Smad7. Finally, knockdown of Smad3, but not Smad2, expression attenuates AGE-induced CTGF expression. All these studies suggest that hyperglycemia may also regulate gene expression through a TGF-β-independent Smad-dependent mechanism.

Increasing evidence has shown that DN is also an inflammatory disease, and loss of Smad7 contributes to the development of diabetic inflammation as a result of activation of the NF-κB signaling pathway. Under pathological conditions, Smad7 is lost due to the activation of E3 ubiquitin ligases including Smurf1 and Smurf2 (Smad ubiquitin regulatory factors) as well as arkadia [7, 17]. Loss of renal Smad7 not only causes TGF-β/Smad3-mediated renal fibrosis, but also enhances renal inflammation by activating the NF-κB-dependent inflammatory response [8, 18].

Recent research into microRNAs has demonstrated that TGF-β may regulate several microRNAs to influence DN. Some miRNAs, such as miR-192, -194, -204, -215 and -216, are highly expressed in the kidney compared with other organs [19]. Recent reports from our laboratory and others describe upregulation of miR-192 in diabetic and obstructive nephropathy, as well as in kidney cells in response to TGF-β_1 [20, 21]. In MCs, miR-192 regulates TGF-β-induced collagen expression by downregulating SIP1 expression [20]. In addition, miR-377 is found to be upregulated in cultured MCs by high glucose concentration and TGF-β and in animal models of type 1 diabetes [22]. Overexpression of miR-377 in MCs induces fibronectin expression [22]. On the other hand, high glucose concentration and TGF-β downregulate miR-29a expression in human proximal TECs (HK-2 cells) [23]. Furthermore, miR-29a is able to suppress collagen IV expression [23]. In addition, miR-200a prevents TGF-β-dependent EMT by downregulating expression of TGF-β_2 and matrix proteins [24]. These findings suggest that miRNAs are critical mediators of TGF-β-driven renal fibrosis during DN.

Therapeutic Potential for Diabetic Nephropathy by Targeting TGF-β Signaling

As DN is the leading cause of end-stage renal disease, angiotensin-converting enzyme inhibitors, angiotensin-II AT1 receptor blockers and antioxidants have widely been used to ameliorate the dysfunction of diabetic kidney and delay the progression of renal injury [25]. However, the effectiveness of these drugs is only restricted in early stages of DN. With recent advances leading to a much better understanding of the role of TGF-β signaling in the development of DN, the search for a successful strategy that suppresses TGF-β-induced renal fibrosis in DN is ongoing. Here, we review some of the major therapeutic strategies to combat TGF-β-induced renal fibrosis in DN.

In the last decade, the use of a neutralizing antibody to suppress TGF-β signaling during DN has been proposed. The experimental use of anti-TGF-β_2 IgG4 in diabetic rats with nephropathy is associated with inhibition of renal fibrosis and reduction in albuminuria [26]. Moreover, blocking TGF-β signaling by antisense TGF-β oligodeoxynucleotides or soluble human TGF-β type II receptor produces renoprotective effects by reducing fibrosis and urinary albumin excretion through inhibiting TGF-β signaling [27]. Administration of GW788388 and IN-1130, inhibitors of TGF-β type receptor kinases, decreases renal fibrosis in rodent models [28, 29]. However, blockade of these general effects of TGF-β may also raise concern over increasing renal inflammation because TGF-β_1 is a potent anti-inflammation mediator. Thus, we proposed to target the downstream TGF-β signaling pathway as well as the downstream NF-κB pathway by exploring the therapeutic potential of Smad7 (fig. 2). It has been well documented [30] that gene transfer of Smad7 using adenovirus or an ultrasound-mediated technique is able to substantially inhibit Smad3 activation and, hence, renal fibrosis in a number of experimental models of chronic kidney diseases, including DN [8], obstructive nephropathy, remnant kidney disease, and autoimmune crescentic glomerulonephritis [reviewed in 7]. Most recently, we also proposed to specifically target the TGF-β/Smad3-dependent

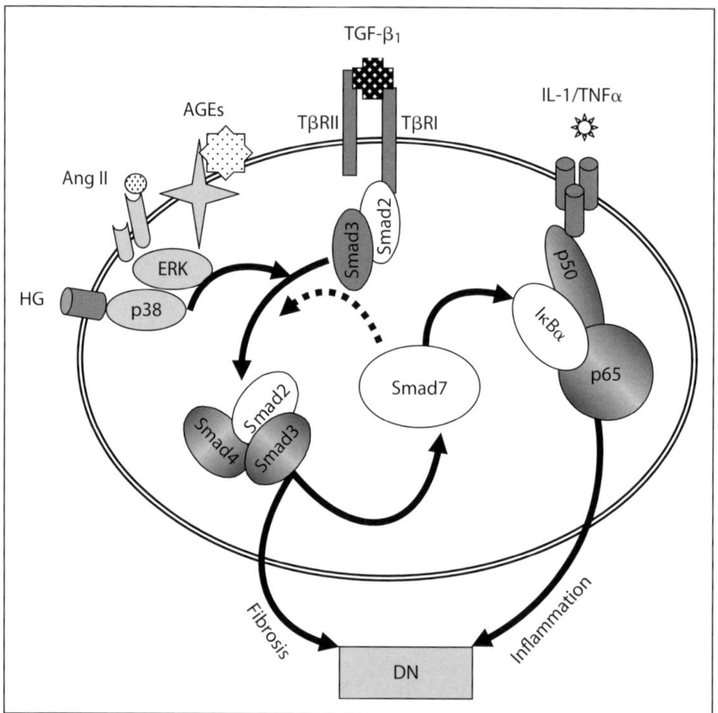

Fig. 2. Pathways of TGF-β/Smad as signal integrators in diabetic complications. Note that high glucose (HG), angiotensin II (Ang II), and AGEs can activate Smads via the ERK/p38/MAPK crosstalk pathway. Smad7 is able to block TGF-β/Smad-dependent fibrosis and also induces IκBα to inhibit NF-κB-driven inflammatory response. Empty symbols denote protective; solid lines denote positive; dotted line denotes negative.

microRNAs that directly regulate the fibrosis genes. By using an ultrasound-mediated technique, we found that overexpression of miR-29b or knockdown of miR-21 is capable of ameliorating renal fibrosis and halting the progression of renal fibrosis in an established rodent model of obstructive nephropathy [unpubl. data]. Application of microRNAs in the treatment of DN may represent a novel, specific, and effective therapeutic strategy for DN.

Conclusion

A growing body of evidence confirms that TGF-β, a fibrogenic cytokine, plays a critical key role in the development of DN. Hyperglycemia is able to induce TGF-β expression and activate Smad3 signaling via both TGF-β-dependent and TGF-β-independent pathways. Recent studies demonstrate that microRNAs regulated by TGF-β also play a critical role in renal fibrosis during DN.

Lan · Chung

Overexpression of Smad7 and alteration of microRNAs may have a therapeutic potential against DN.

Acknowledgements

This work was supported by grants from the Research Grants Council of Hong Kong (RGC GRF 768207; CUHK5/CRF09).

References

1 Kanwar YS, Wada J, Sun L, Xie P, Wallner EI, Chen S, Chugh S, Danesh FR: Diabetic nephropathy: mechanisms of renal disease progression. Exp Biol Med (Maywood) 2008;233:4–11.

2 Soldatos G, Cooper ME: Diabetic nephropathy: important pathophysiologic mechanisms. Diabetes Res Clin Pract 2008;82 (suppl 1):S75–S79.

3 Ziyadeh FN: Mediators of diabetic renal disease: the case for TGF-beta as the major mediator. J Am Soc Nephrol 2004;15(suppl 1): S55–S57.

4 Li JH, Huang XR, Zhu HJ, Oldfield M, Cooper M, Truong LD, Johnson RJ, Lan HY: Advanced glycation end products activate Smad signaling via TGF-beta-dependent and independent mechanisms: implications for diabetic renal and vascular disease. FASEB J 2004;18:176–178.

5 Schnaper HW, Hayashida T, Hubchak SC, Poncelet AC: TGF-beta signal transduction and mesangial cell fibrogenesis. Am J Physiol Renal Physiol 2003;284:F243–F252.

6 Roberts AB: Molecular and cell biology of TGF-beta. Miner Electrolyte Metab 1998;24: 111–119.

7 Lan HY: Smad7 as a therapeutic agent for chronic kidney diseases. Front Biosci 2008; 13:4984–4992.

8 Chen H, Huang XR, Wang W, Li J, Heuchel RL, Chung AC, Lan HY: The protective role of Smad7 in diabetic kidney disease: mechanism and therapeutic potential. Diabetes 2011;60:590–601.

9 Meng XM, Huang XR, Chung AC, Qin W, Shao X, Igarashi P, Ju W, Bottinger EP, Lan HY: Smad2 protects against TGF-beta/ Smad3-mediated renal fibrosis. J Am Soc Nephrol 2010;21:1477–1487.

10 Yang F, Chung AC, Huang XR, Lan HY: Angiotensin II induces connective tissue growth factor and collagen I expression via transforming growth factor-beta-dependent and -independent Smad pathways: the role of Smad3. Hypertension 2009;54:877–884.

11 Chung AC, Zhang H, Kong YZ, Tan JJ, Huang XR, Kopp JB, Lan HY: Advanced glycation end-products induce tubular CTGF via TGF-beta-independent Smad3 signaling. J Am Soc Nephrol 2010;21:249–260.

12 Fujimoto M, Maezawa Y, Yokote K, Joh K, Kobayashi K, Kawamura H, Nishimura M, Roberts AB, Saito Y, Mori S: Mice lacking Smad3 are protected against streptozotocin-induced diabetic glomerulopathy. Biochem Biophys Res Commun 2003;305:1002–1007.

13 Isono M, Chen S, Hong SW, Iglesias-de la Cruz MC, Ziyadeh FN: Smad pathway is activated in the diabetic mouse kidney and Smad3 mediates TGF-beta-induced fibronectin in mesangial cells. Biochem Biophys Res Commun 2002;296:1356–1365.

14 Derynck R, Zhang YE: Smad-dependent and Smad-independent pathways in TGF-beta family signalling. Nature 2003;425: 577–584.

15 Sengupta U, Ukil S, Dimitrova N, Agrawal S: Expression-based network biology identifies alteration in key regulatory pathways of type 2 diabetes and associated risk/complications. PLoS One 2009;4:e8100.

16 Zhou G, Li C, Cai L: Advanced glycation end-products induce connective tissue growth factor-mediated renal fibrosis predominantly through transforming growth factor beta-independent pathway. Am J Pathol 2004;165:2033–2043.

17 Inoue Y, Imamura T: Regulation of TGF-beta family signaling by E3 ubiquitin ligases. Cancer Sci 2008;99:2107–2112.

18 Wang W, Huang XR, Li AG, Liu F, Li JH, Truong LD, Wang XJ, Lan HY: Signaling mechanism of TGF-beta1 in prevention of renal inflammation: role of Smad7. J Am Soc Nephrol 2005;16:1371–1383.

19 Tian Z, Greene AS, Pietrusz JL, Matus IR, Liang M: MicroRNA-target pairs in the rat kidney identified by microRNA microarray, proteomic, and bioinformatic analysis. Genome Res 2008;18:404–411.

20 Kato M, Zhang J, Wang M, Lanting L, Yuan H, Rossi JJ, Natarajan R: MicroRNA-192 in diabetic kidney glomeruli and its function in TGF-beta-induced collagen expression via inhibition of E-box repressors. Proc Natl Acad Sci USA 2007;104:3432–3437.

21 Chung AC, Huang XR, Meng X, Lan HY: miR-192 mediates TGF-beta/Smad3-driven renal fibrosis. J Am Soc Nephrol 2010;21: 1317–1325.

22 Wang Q, Wang Y, Minto AW, Wang J, Shi Q, Li X, Quigg RJ: MicroRNA-377 is up-regulated and can lead to increased fibronectin production in diabetic nephropathy. FASEB J 2008;22:4126–4135.

23 Du B, Ma LM, Huang MB, Zhou H, Huang HL, Shao P, Chen YQ, Qu LH: High glucose down-regulates miR-29a to increase collagen IV production in HK-2 cells. FEBS Lett 2010; 584:811–816.

24 Wang B, Koh P, Winbanks C, Coughlan MT, McClelland A, Watson A, Jandeleit-Dahm K, Burns WC, Thomas MC, Cooper ME, Kantharidis P: miR-200a prevents renal fibrogenesis through repression of TGF-beta2 expression. Diabetes 2011;60:280–287.

25 Balakumar P, Arora MK, Ganti SS, Reddy J, Singh M: Recent advances in pharmacotherapy for diabetic nephropathy: current perspectives and future directions. Pharmacol Res 2009;60:24–32.

26 Hill C, Flyvbjerg A, Rasch R, Bak M, Logan A: Transforming growth factor-beta2 antibody attenuates fibrosis in the experimental diabetic rat kidney. J Endocrinol 2001;170:647–651.

27 Russo LM, del Re E, Brown D, Lin HY: Evidence for a role of transforming growth factor (TGF)-beta1 in the induction of postglomerular albuminuria in diabetic nephropathy: amelioration by soluble TGF-beta type II receptor. Diabetes 2007;56:380–388.

28 Petersen M, Thorikay M, Deckers M, van Dinther M, Grygielko ET, Gellibert F, de Gouville AC, Huet S, ten Dijke P, Laping NJ: Oral administration of GW788388, an inhibitor of TGF-beta type I and II receptor kinases, decreases renal fibrosis. Kidney Int 2008;73:705–715.

29 Moon JA, Kim HT, Cho IS, Sheen YY, Kim DK: IN-1130, a novel transforming growth factor-beta type I receptor kinase (ALK5) inhibitor, suppresses renal fibrosis in obstructive nephropathy. Kidney Int 2006;70: 1234–1243.

30 Lan HY, Mu W, Tomita N, Huang XR, Li JH, Zhu HJ, Morishita R, Johnson RJ: Inhibition of renal fibrosis by gene transfer of inducible Smad7 using ultrasound-microbubble system in rat UUO model. J Am Soc Nephrol 2003;14:1535–1548.

Prof. Hui Yao Lan, MD, PhD
Department of Medicine & Therapeutics, and Li Ka Shing Institute of Health Sciences
601 LiHS, Prince of Wales Hospital, the Chinese University of Hong Kong
Shatin, Hong Kong, SAR (China)
Tel. +852 3763 6077, Fax +852 2145 7190, E-Mail hylan@cuhk.edu.hk

Lai KN, Tang SCW (eds): Diabetes and the Kidney.
Contrib Nephrol. Basel, Karger, 2011, vol 170, pp 83–92

Angiogenic Factors

Charbel C. Khoury[a] · Fuad N. Ziyadeh[b]

[a]Feinberg School of Medicine, Northwestern University, Chicago, Ill., USA; [b]Departments of
Internal Medicine and Biochemistry, Faculty of Medicine, American University of Beirut,
Beirut, Lebanon

Abstract

Diabetic nephropathy (DN) is in essence a microvascular disease that develops as a result
of a confluence of hemodynamic and metabolic perturbations. Angiogenic factors are
prime candidates to explain the vascular and pathologic findings of DN; however, analysis
of their pathophysiology shows that they have a constellation of effects on the glomerulus
that go beyond angiogenesis. Vascular endothelial growth factor (VEGF) is an exemplary
candidate for fulfilling the criteria for Koch's postulate as an etiologic agent of the glomerulopathy in diabetes. Its expression and signaling in the kidney are amplified early on in
the diabetic state. Moreover, counteracting its effects reverses the albuminuria and other
hemodynamic and structural features of experimental DN. Finally, experimental overexpression of VEGF in adult mice replicates several aspects of diabetic kidney disease. Under
the influence of a variety of diabetic mediators, the podocyte becomes the main source of
increased expression of VEGF in the kidney. The cytokine then exerts its multitude of
effects in an autocrine fashion on the podocyte itself, on the endothelial cell in a paracrine
manner, and finally contributes to macrophage recruitment acting as a chemokine. The
angiopoietins consist primarily of two main factors acting in contrast to each other: Ang1
– an antiangiogenic ligand, and Ang2 – its competitive inhibitor. Both, however, seem to
have important roles in the maintenance of glomerular homeostasis. Diabetes disrupts
the tight balance that controls angiopoietin expression and functions and decreases the
Ang1/Ang2 ratio. The end physiologic result seems to be dependent on the concomitant
VEGF changes in the kidney. Because of the intricacy of their control, angiogenic factors
are difficult to manipulate therapeutically. However, they remain valid target points for
the treatment of DN.

Diabetes mellitus is characterized by multiorgan complications that cause the
mortality and morbidity of this major epidemic of the century. For the most
part, diabetic complications are diseases of the blood vessels: macrovascular in
coronary artery disease, peripheral arterial disease and cerebrovascular disease,

and microvascular in retinopathy, neuropathy and nephropathy. In diabetes, the kidney develops a constellation of pathologic changes: afferent and efferent artery arteriolosclerosis and hyalinosis, thickening of the glomerular basement membrane (GBM) and mesangial matrix expansion. Early on, angiogenic features may also be noted as new and abnormal capillaries develop in the vicinity of the glomerulus [1]. Perhaps the most characteristic pathology of diabetic nephropathy (DN), the 'Kimmelstiel-Wilson lesion', is a nodular sclerosis that results from microaneurysmal dilation of glomerular capillaries and mesangial lysis. In view of all these microangiopathic changes, investigators have hypothesized and examined the role of angiogenic factors in the diabetic kidney. Vascular endothelial growth factor (VEGF) and the angiopoietins are some of the main angiogenic factors that have been linked to renal disease. A growing body of work shows that their target is more than the endothelial cell, and that they are involved in intricate signaling pathways that affect the entire glomerular filtration barrier. This chapter will try to give a brief overview of the current evidence regarding the involvement of angiogenic factors in the development or progression of DN.

Vascular Endothelial Growth Factor

VEGF-A is the most widely studied isoform of VEGF and has been the target of a multitude of cancer chemotherapeutic protocols as well as treatments for diabetic retinopathy. But its original description as a vascular permeability factor was the main drive behind VEGF's candidacy for mediator of DN. VEGF widens interendothelial cell junctions and increases the cell's fenestrations. It also induces the formation of caveolae, which allow a vesicle-based transport across the cytoplasm. Moreover, it stimulates the fusion of multiple cytoplasmic vesicles to form transcytoplasmic channel known as the vesiculovacuolar organelle [2]. Theoretically then, VEGF could prime the glomerular filtration barrier for leakiness across the capillary wall.

A total of 12 isoforms of the cytokine have been described and, based on the pattern of alternative splicing, they can be classified as the angiogenic VEGF_{xxx} and the antiangiogenic $\text{VEGF}_{xxx}\text{b}$ isoforms [3]. VEGF-A can bind and signal through two main tyrosine kinase receptors: VEGF-R1 or fms-like tyrosine kinase (Flt-1), VEGF-R2 or fetal liver kinase 1 (Flk-1/KDR). In normal kidneys, VEGF-A is predominantly expressed in glomerular podocytes. It has also been detected in epithelial cells of the distal tubules, collecting ducts, and at smaller levels in some proximal tubules [4]. Both signaling receptors are expressed in the glomerular endothelial cells, podocytes, mesangial cells, proximal tubular cells, and pre- and postglomerular vessels [5]. While the focus has been on a role of VEGF-A in DN, it should be noted that other members of the VEGF family have been described in the kidney. VEGF-B and -C are both expressed in the podocyte, even though it seems that only VEGF-C is functionally important to the cell [5]. VEGF-D, on the other hand, appears to be specifically expressed by the parietal epithelial cell [6]. Finally, placental-derived growth factor (PlGF) is detected in mesangial and tubular cells grown in culture [7].

Several studies have tried to link VEGF to human DN. Whether assaying renal, serum, or urine levels of the cytokine, results have wandered between increased and decreased expression. While the animal models may not be correctly reproducing human disease, it is more likely that the inconsistency is due to the large variability in human studies. Human DN progresses at different rates in patients. VEGF may play a role in earlier stages of the disease, its levels and effect could dwindle as podocyte loss ensues. We will focus our discussion on the animal experiments; human studies are reviewed in detail elsewhere [8].

Experimental Evidence Suggests a Role for VEGF in Diabetic Nephropathy

To prove that VEGF is implicated in DN, one would have to demonstrate that: (a) its expression varies between control and diseased state, (b) reversing this abnormality can improve renal disease, and (c) elements of the renal pathology can be reproduced by replicating the VEGF changes. Indeed, different groups have shown that VEGF is generally upregulated in the kidneys of diabetic rodents. The increase in VEGF expression occurred in type 1 and type 2 diabetic models in both rats and mice. Consequently, the cytokine's activity must be amplified in the kidney, especially that there is an associated increase in the expression of VEGF-R1 and -R2 and their phosphorylation [9].

The earliest functional experiments involved neutralizing VEGF with a targeted antibody. In the streptozotocin-treated rat, a model of type 1 diabetes, anti-VEGF treatment resulted in 50% less albuminuria [10]. Similar results were obtained when the antibody was administered to mice and rats with type 2 diabetes such as the *db/db* mouse [11, 12].

Another method that researchers have used to antagonize VEGF in animal models is the soluble VEGF-R1 (s-Flt1), which acts as a natural competitive inhibitor. This version of the receptor lacks a transmembrane domain and can circulate freely. Soluble VEGF-R1 interferes with the binding of VEGF-A, -B and PlGF to functional receptors. Inducible overexpression of sFlt-1 in podocytes resulted in a decrease in albuminuria, mesangial expansion, GBM thickening and podocyte foot-process fusion [13]. Systemic overexpression of s-Flt1 using adeno-associated virus-1 also led to a reduction in albuminuria. However, treated mice were found to have a more severe tubulointerstitial disease [14]. This could indicate a differential role of the VEGF system in the glomeruli vs. the tubules, and complicates the perspective of using VEGF targeted therapies in human disease.

In order to target the downstream signaling pathways of the VEGF system, our group treated diabetic mice with SU5416, an inhibitor of both VEGF receptors. This compound resulted in significant inhibition of VEGF-R1 phosphorylation in the *db/db* type 2 diabetic mouse. The treatment did not alter hyperglycemia but had a specific renal protective effect. Albumin excretion rate was brought down to the level of nondiabetic controls. Mesangial matrix expansion was not prevented, but histologic examination by electron microscopy showed prevention of GBM thickening and improvement in the density of open slit pores between podocyte foot processes [15].

Recently, researchers from the Tufro and Quaggin groups have succeeded in showing that inducing VEGF expression in adult normal mice can reproduce features of diabetic renal disease [16, 17]. Their conditional podocyte-specific VEGF overexpressor mice were albuminuric, had thickened GBM, foot process effacement and mesangial expansion. These significant pathologic changes were reversible, indicating that VEGF is required for the initiation and maintenance of renal disease.

Molecular and Cellular Mechanisms of VEGF

When trying to elucidate the cellular and molecular mechanisms in the diabetic glomerulus, the podocyte takes center stage. A constellation of factors including hypoxia-inducible factor, transforming growth factor-β (TGF-β), angiotensin II and advanced glycation end products induce the cell to produce more VEGF [18]. But the induction of the cytokine is also mediated by hyperglycemia itself. Recent work by Long et al. [19] shows that glomeruli from diabetic mice and podocytes grown in high glucose media have decreased levels of micro-RNA 93. By using multiple in silico, in vivo, and in vitro approaches, the authors prove that this small RNA suppresses VEGF expression, and accordingly, its modulation by hyperglycemia contributes to VEGF induction in diabetes. Another important pathway is via Notch-1 signaling, that is known to mediate podocyte disease in DN. Inhibition of Notch-1, using DAPT or shRNA, decreases VEGF levels in diabetic kidneys and podocytes, indicating a possible interaction between the two pathways [20].

The modus operandi of podocyte-derived VEGF is tripartite: autocrine, paracrine, and chemokine. The autocrine effects are shown by our finding that treating cultured podocytes with TGF-β and angiotensin II results in upregulation of α_3-collagen type IV in a dose-dependent manner [21, 22]. The deposition of α_3-collagen type IV is projected to alter the composition of the GBM and contribute to its thickening. This effect was inhibited by the VEGF receptor inhibitor SU5416, indicating the presence of a VEGF autocrine loop in the podocyte. In their transgenic mouse, Veron et al. [16] confirmed that GBM thickening was dependent on podocyte-derived VEGF. However, while their real-time qPCR did not show any significant changes in collagen genes expression, it

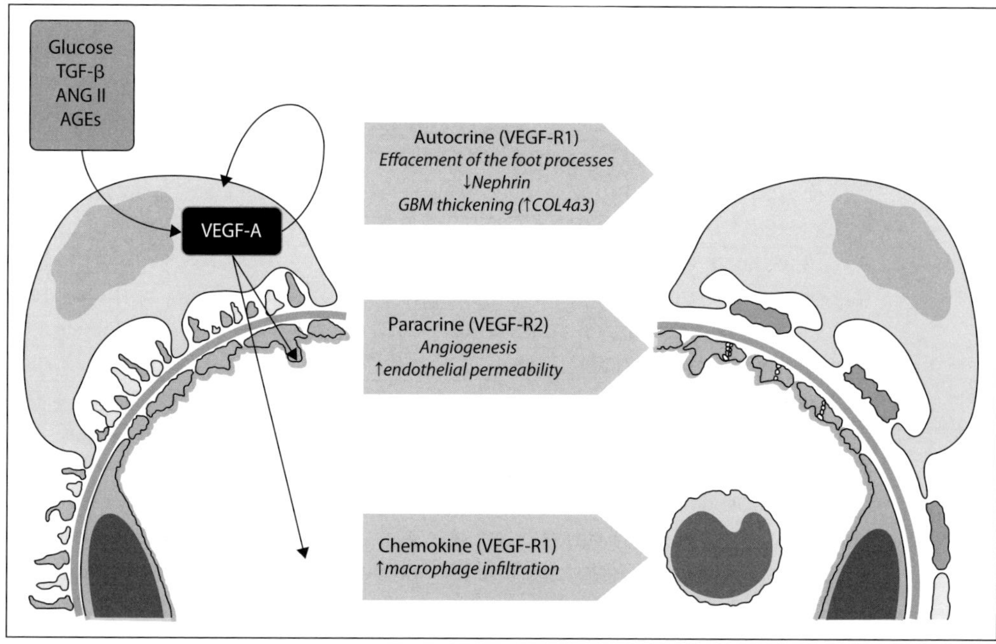

Fig. 1. Molecular and cellular mechanisms of VEGF in DN.

suggested that the VEGF autocrine loop may be responsible for nephrin and MMP-9 downregulation. Which of the VEGF-receptor types completes this autocrine loop? Veron et al. [16] noted the expression and phosphorylation of VEGF-R2 in the podocytes of their transgenic mouse. In addition, they showed that the receptor colocalizes and coimmunoprecipitates with nephrin, making it the most plausible candidate for them. On the other hand, Sison et al. [17] failed to detect VEGF-R2 in renal podocytes despite using multiple methodologies, and identified VEGF-R1 as the dominant receptor in the cell. Moreover, the latter group showed that the knockout of VEGF-R2 in adult mice is asymptomatic, and cannot rescue the glomerulopathy resulting from VEGF overexpression [17]. Like Sison et al. [17], our group could not detect a significant level of VEGF-R2 expressed in podocytes in vivo or in vitro. We have however shown that VEGF-R1 is expressed and phosphorylated in cultured podocytes and in diabetic kidneys [15]. Treating podocytes and diabetic mice with SU5416 inhibited the activation of VEGF-R1 while reversing VEGF's autocrine effects as well as albuminuria, GBM thickening, slit pore density, and nephrin quantity. These findings suggest that a VEGF/VEGF-R1 autocrine loop is activated in the diabetic podocyte, leading to disruption of the slit diaphragm, disorganization of the GBM and subsequently albuminuria (fig. 1). Further experiments using VEGF-R1 transgenic mice are needed to certify these findings.

Podocyte-derived VEGF also exerts a paracrine effect on glomerular endothelial cells. This is essential for endothelial cell homeostasis and is normally under tight regulation [23]. All VEGF isoforms, except for $VEGF_{121}$, are capable of binding to a variety of matrix proteins. This could allow the cytokine to travel across the GBM to the endothelial cell. Interestingly matrix-anchored VEGF may actually outperform its soluble counterpart. It has been shown to induce a prolonged activation of VEGF-R2 with differential phosphorylation of Y1214 and extended activation of p38 [24]. In DN, endothelial dysfunction and angiogenesis result from the amplification of this paracrine effect in the context of its 'uncoupling' from the nitric oxide (NO) axis [1]. Increased VEGF signaling should normally result in upregulation of NO; however, a multitude of diabetic factors keep renal NO levels low. The eNOS knockout mouse in the study by Nakagawa et al. [1] shows that the uncoupling of the VEGF-endothelial NO axis is essential for the development of a full-blown endothelial phenotype in the diabetic kidney. The recent description of the podocyte-specific $VEGF_{165}b$ overexpressor mouse suggests that there may be alternative hypotheses for the regulation of the podocyte-endothelial cell crosstalk [25]. $VEGF_{165}b$, generally referred to as the antiangiogenic isoform, did not affect the mouse's viability. There was only reduced normalized glomerular ultrafiltration fraction associated with less numerous and smaller endothelial fenestrations, as well as moderate thinning of the GBM. It would be interesting to examine whether diabetes results in an imbalance between $VEGF_{xxx}$ and $VEGF_{xxx}b$ in the glomerulus. A disease favoring $VEGF_{xxx}$ would be expected to result in more severe endothelial damage.

DN is induced, at least in part, by inflammatory mechanisms. In fact, diseased kidneys are infiltrated by macrophages. These cells most probably add to the local slew of cytokines, further exacerbating the sclerosis. Podocyte-derived VEGF seems to contribute to the chemotaxis of immune cells into the glomerulus. Sison's transgenic mouse had evidence of periglomerular macrophage infiltration within 4 weeks of VEGF overexpression in the podocyte [17]. In diabetes, this inflammatory response could be further promoted with 'VEGF-NO uncoupling'. Local decrease in NO seems to lead to increased endothelial expression of intercellular adhesion molecule 1, and permits the increased expression of VEGF-R1 in macrophages, thus turning VEGF into a potent chemokine [26].

Angiopoietins

The angiopoietins are a family of vascular growth factors that have recently been examined in the context of DN (fig. 2). The two main angiopoietins, Ang1 and Ang2, bind and signal through a tyrosine kinase receptor referred to as Tie2 (Tyrosine kinase with Ig and EGF homology domain 2). The other receptor, Tie1, remains to be fully characterized, but appears to function in fine-tuning

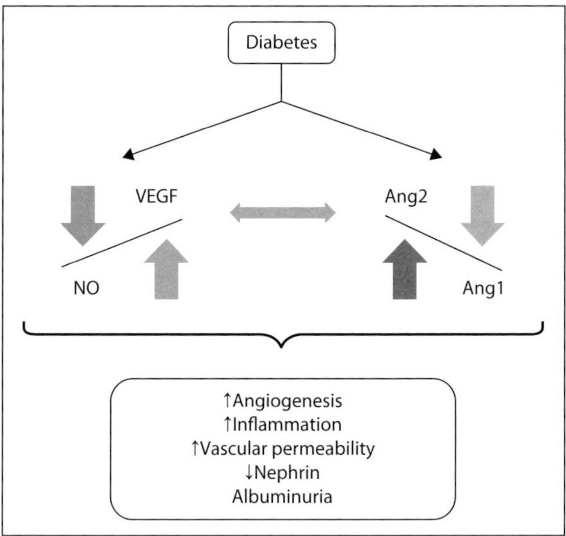

Fig. 2. Angiogenic factors in DN: a 'balance act'.

Tie-2 signaling [27]. In a healthy adult kidney, Ang1 is generally expressed in the podocyte. Ang2 seems to be generally localized to glomerular endothelial cells in mice, but has been detected in rat podocytes. Both receptors are expressed in endothelial cells, glomerular and extraglomerular; however, only Tie2 can be detected in the rat podocyte [28].

Ang1 is generally known to have antiangiogenic properties; it promotes endothelial cell survival, decreases vascular permeability, and counters VEGF-induced effects. Also, Ang1 has anti-inflammatory functions, and inhibits leukocyte capillary transmigration [27]. Yet, in vitro experiments in microvessels and glomeruli indicate that this growth factor has very pertinent glomerular homeostatic functions. Ang1 can decrease water permeability and limit albumin seepage. Moreover, it has the potential to increase the thickness of the glycolayx layer covering microvessels [28].

On the other hand, Ang2 acquires most of its proangiogenic and proinflammatory effects by competitive inhibition of Ang1. It has partial agonist properties only in the absence of Ang1 [29]. Podocyte-specific conditional overexpression of Ang2 in mice resulted in albuminuria and glomerular endothelial apoptosis. While podocytes were structurally unaffected, both VEGF-A and nephrin expression were downregulated. Ang1 levels were unaffected [30].

Multiple murine models of type 1 and type 2 diabetes show that renal Ang1/Ang2 ratio is decreased in diseased kidneys [31–34]. Hypoxia, hyperglycemia, and TNF-α all have been known to upregulate Ang2, and may be involved in tilting the ratio in diabetes [27]. Because of the interdependence of these factors in their signaling, it is very difficult to decipher which of the two could be a better therapeutic target. The current evidence suggests that it is the imbalance between the two angiopoietins that leads to glomerular disease. Yet, since the transgenic Ang2 mouse does not fully replicate DN, the angiopoietin system most likely contributes to other signaling pathways in diabetes, and is in turn modulated by them. In fact, it is well known that the angiopoietins induce much of their functions in the context of VEGF signaling. Ang2 activates apoptotic signaling in vascular structures in the absence of VEGF, while it promotes angiogenesis in a VEGF-rich milieu [28]. From the current experimental evidence, it appears that in diabetes Ang1 expression is prominent in kidney tubules, whereas Ang2 is mostly expressed in glomerular endothelial cells and podocytes. Further experiments are needed to elucidate whether, just like with VEGF, autocrine and paracrine pathways are in place in the glomerulus.

Conclusion

The VEGF and angiopoietin systems are the two main systems of angiogenic factors that have been studied in DN. A growing body of evidence is starting to implicate other receptors and factors in this microvascular disease. The neuropilins could be involved secondary to their role as coreceptors for VEGF, or as a result of pathways involving their ligands the semaphorins. Members of the ephrin family have been localized to glomerular slit diaphragm and may as such have angiogenic and local effects in diabetes. A number of inhibitors of angiogenesis have been experimentally tested in DN, showing promising results [31, 32, 34]. Whether or not these compounds or other more targeted therapies will be used to treat or prevent human DN should be carefully considered. Because of the tight control under which VEGF and the angiopoietins are kept, manipulating them therapeutically could lead to serious side effects. Further understanding of their mechanism of action may open windows of opportunity by elucidating a safer approach. Could it be the antiangiogenic $VEGF_{xxx}b$? or Ang1? It remains to be seen.

References

1 Nakagawa T, Kosugi T, Haneda M, Rivard CJ, Long DA: Abnormal angiogenesis in diabetic nephropathy. Diabetes 2009;58:1471–1478.

2 Weis SM, Cheresh DA: Pathophysiological consequences of VEGF-induced vascular permeability. Nature 2005;437:497–504.

3 Harper SJ, Bates DO: VEGF-A splicing: the key to anti-angiogenic therapeutics? Nat Rev Cancer 2008;8:880–887.

4 Schrijvers BF, Flyvbjerg A, De Vriese AS: The role of vascular endothelial growth factor (VEGF) in renal pathophysiology. Kidney Int 2004;65:2003–2017.

5 Muller-Deile J, Worthmann K, Saleem M, Tossidou I, Haller H, Schiffer M: The balance of autocrine VEGF-A and VEGF-C determines podocyte survival. Am J Physiol Renal Physiol 2009;297:F1656–F1667.

6   Foster RR, Satchell SC, Seckley J, Emmett MS, Joory K, Xing CY, Saleem MA, Mathieson PW, Bates DO, Harper SJ: VEGF-C promotes survival in podocytes. Am J Physiol Renal Physiol 2006;291: 196–207.

7   Wu Q, Du Y, Yang N, Liang Y, Li Y: Microvasculature change and placenta growth factor expression in the early stage of a rat remnant kidney model. Am J Nephrol 2006;26:97–104.

8   Chen S, Ziyadeh FN: Vascular endothelial growth factor and diabetic nephropathy. Curr Diab Rep 2008;8:470–476.

9   Chou E, Suzuma I, Way KJ, Opland D, Clermont AC, Naruse K, Suzuma K, Bowling NL, Vlahos CJ, Aiello LP, King GL: Decreased cardiac expression of vascular endothelial growth factor and its receptors in insulin-resistant and diabetic states: a possible explanation for impaired collateral formation in cardiac tissue. Circulation 2002; 105:373–379.

10 de Vriese AS, Tilton RG, Elger M, Stephan CC, Kriz W, Lameire NH: Antibodies against vascular endothelial growth factor improve early renal dysfunction in experimental diabetes. J Am Soc Nephrol 2001;12:993–1000.

11 Flyvbjerg A, Dagnaes-Hansen F, De Vriese AS, Schrijvers BF, Tilton RG, Rasch R: Amelioration of long-term renal changes in obese type 2 diabetic mice by a neutralizing vascular endothelial growth factor antibody. Diabetes 2002;51:3090–3094.

12 Schrijvers BF, Flyvbjerg A, Tilton RG, Lameire NH, De Vriese AS: A neutralizing VEGF antibody prevents glomerular hypertrophy in a model of obese type 2 diabetes, the Zucker diabetic fatty rat. Nephrol Dial Transplant 2006;21:324–329.

13 Ku CH, White KE, Dei Cas A, Hayward A, Webster Z, Bilous R, Marshall S, Viberti G, Gnudi L: Inducible overexpression of sFlt-1 in podocytes ameliorates glomerulopathy in diabetic mice. Diabetes 2008;57:2824–2833.

14 Kosugi T, Nakayama T, Li Q, Chiodo VA, Zhang L, Campbell-Thompson M, Grant M, Croker BP, Nakagawa T: Soluble Flt-1 gene therapy ameliorates albuminuria but accelerates tubulointerstitial injury in diabetic mice. Am J Physiol Renal Physiol 2010;298:F609–F616.

15 Sung SH Ziyadeh FN, Wang A, Pyagay PE, Kanwar YS, Chen S: Blockade of vascular endothelial growth factor signaling ameliorates diabetic albuminuria in mice. J Am Soc Nephrol 2006;17:3093–3104.

16 Veron D, Reidy KJ, Bertuccio C, Teichman J, Villegas G, Jimenez J, Shen W, Kopp JB, Thomas DB, Tufro A: Overexpression of VEGF-A in podocytes of adult mice causes glomerular disease. Kidney Int 2010;77: 989–999.

17 Sison K, Eremina V, Baelde H, Min W, Hirashima M, Fantus IG, Quaggin SE: Glomerular structure and function require paracrine, not autocrine, VEGF-VEGFR-2 signaling. J Am Soc Nephrol 2010;21: 1691–1701.

18 Wolf G, Chen S, Ziyadeh FN: From the periphery of the glomerular capillary wall toward the center of disease: podocyte injury comes of age in diabetic nephropathy. Diabetes 2005;54:1626–1634.

19 Long J, Wang Y, Wang W, Chang BH, Danesh FR: Identification of microRNA-93 as a novel regulator of vascular endothelial growth factor in hyperglycemic conditions. J Biol Chem 2010;285:23457–23465.

20 Lin CL, Wang FS, Hsu YC, Chen CN, Tseng MJ, Saleem MA, Chang PJ, Wang JY: Modulation of notch-1 signaling alleviates vascular endothelial growth factor-mediated diabetic nephropathy. Diabetes 2010;59: 1915–1925.

21 Chen S. Lee JS, Iglesias-de la Cruz MC, Wang A, Izquierdo-Lahuerta A, Gandhi NK, Danesh FR, Wolf G, Ziyadeh FN: Angiotensin II stimulates alpha3(IV) collagen production in mouse podocytes via TGF-beta and VEGF signaling: implications for diabetic glomerulopathy. Nephrol Dial Transplant 2005;20:1320–1328.

22 Chen S, Kasama Y, Lee JS, Jim B, Marin M, Ziyadeh FN: Podocyte-derived vascular endothelial growth factor mediates the stimulation of alpha3(IV) collagen production by transforming growth factor-beta1 in mouse podocytes. Diabetes 2004;53:2939–2949.

23 EreminaV, Baelde HJ, Quaggin SE: Role of the VEGF-A signaling pathway in the glomerulus: evidence for crosstalk between components of the glomerular filtration barrier. Nephron Physiol 2007;106:32–37.

24 Chen TT, Luque A, Lee S, Anderson SM, Segura T, Iruela-Arispe ML: Anchorage of VEGF to the extracellular matrix conveys differential signaling responses to endothelial cells. J Cell Biol 2010;188:595–609.

25 Qiu Y, Ferguson J, Oltean S, Neal CR, Kaura A, Bevan H, Wood E, Sage LM, Lanati S, Nowak DG, Salmon AH, Bates D, Harper SJ: Overexpression of VEGF165b in podocytes reduces glomerular permeability. J Am Soc Nephrol 2010;21:1498–509.

26 Nakagawa T: Uncoupling of the VEGF-endothelial nitric oxide axis in diabetic nephropathy: an explanation for the paradoxical effects of VEGF in renal disease. Am J Physiol Renal Physiol 2007;292:F1665–F1672.

27 Woolf AS, Gnudi L, Long DA: Roles of angiopoietins in kidney development and disease. J Am Soc Nephrol 2009;20:239–244.

28 Woolf AS: Angiopoietins: vascular growth factors looking for roles in glomeruli. Curr Opin Nephrol Hypertens 2010;19:20–25.

29 Yuan HT, Khankin EV, Karumanchi SA, Parikh SM: Angiopoietin 2 is a partial agonist/antagonist of Tie2 signaling in the endothelium. Mol Cell Biol 2009;29: 2011–2022.

30 Davis B, Dei Cas A, Long DA, White KE, Hayward A, Ku CH, Woolf AS, Bilous R, Viberti G, Gnudi L: Podocyte-specific expression of angiopoietin-2 causes proteinuria and apoptosis of glomerular endothelia. J Am Soc Nephrol 2007;18:2320–2329.

31 Ichinose K, Maeshima Y, Yamamoto Y, Kinomura M, Hirokoshi K, Kitayama H, Takazawa Y, Sugiyama H, Yamasaki Y, Agata N, Makino H: 2-(8-hydroxy-6-methoxy-1-oxo-1h-2-benzopyran-3-yl) propionic acid, an inhibitor of angiogenesis, ameliorates renal alterations in obese type 2 diabetic mice. Diabetes 2006;55:1232–1242.

32 Ichinose K, Maeshima Y, Yamamoto Y, Kitayama H, Takazawa Y, Hirokoshi K, Sugiyama H, Yamasaki Y, Eguchi K, Makino H: Antiangiogenic endostatin peptide ameliorates renal alterations in the early stage of a type 1 diabetic nephropathy model. Diabetes 2005;54:2891–2903.

33 Rizkalla B, Forbes JM, Cao Z, Boner G, Cooper ME: Temporal renal expression of angiogenic growth factors and their receptors in experimental diabetes: role of the renin-angiotensin system. J Hypertens 2005;23:153–164.

34 Yamamoto Y, Maeshima Y, Kitayama H, Kitamura S, Takazawa Y, Sugiyama H, Yamasaki Y, Makino H: Tumstatin peptide, an inhibitor of angiogenesis, prevents glomerular hypertrophy in the early stage of diabetic nephropathy. Diabetes 2004;53:1831–1840.

Prof. Fuad N. Ziyadeh, MD
Departments of Internal Medicine and Biochemistry, Faculty of Medicine, American University of Beirut
Bliss Street
Beirut (Lebanon)
Tel. +961 1 350000, ext. 5353, Fax +961 1 370814, E-Mail fz05@aub.edu.lb

                                                   Khoury · Ziyadeh

Lai KN, Tang SCW (eds): Diabetes and the Kidney.
Contrib Nephrol. Basel, Karger, 2011, vol 170, pp 93–101

Endothelial Nitric Oxide Synthase

Takahiko Nakagawa · Richard J. Johnson

Division of Renal Diseases and Hypertension, University of Colorado Denver,
Aurora, Colo., USA

Abstract

Diabetic nephropathy remains a leading cause for end-stage renal disease indicating a failure of current therapeutic strategies. One factor that impairs our ability to make advances has been the inadequacy of most animal models in manifesting advanced diabetic renal disease. Since these animal models develop marked hyperglycemia, one could assume that hyperglycemia is not enough for the development of advanced nephropathy and thereby additional factors are likely involved. Recently, our research group and others have discovered a new mouse model in which diabetes is induced in mice lacking endothelial nitric oxide synthase (eNOS). Diabetic eNOS knockout mice develop severe renal injuries resembling advanced human diabetic nephropathy. This model thereby suggests a key role for reduced nitric oxide levels in the pathogenesis of diabetic nephropathy. In this article, we summarize recent clinical and experimental evidence for the role of eNOS in diabetic nephropathy.

Nitric oxide (NO), a free radical in the form of a highly diffusible gas, regulates renal hemodynamics. The kidney expresses three different forms of NO synthase (NOS), including neuronal NOS (nNOS), inducible NOS and endothelial NOS (eNOS). Expressions of these three isoforms are cell specific and exhibit distinct mechanisms. In general, nNOS and eNOS are constitutively expressed in the normal kidney. In particular, nNOS is predominantly expressed in tubular epithelial cells of the macula densa and inner medullary collecting duct [1] whereas eNOS is restricted to endothelial cells in the kidney [2]. On the other hand inducible NOS is known to be an inducible type. While NO derived from each enzyme plays a key role in the kidney, endothelial NO has been shown to be protective in the vascular system due to its antithrombogenic effects, by blocking endothelial cell activation/injury induced by cytokines, and by promoting vasodilatation.

Role of eNOS in Diabetic Nephropathy

Komers and Anderson [3] recently reviewed a large number of reports regarding NO levels in experimental diabetic nephropathy and concluded that there was a controversy over this issue. They assumed that these contradictory findings might be accounted for by the different experimental and clinical settings. Certainly, there are diverse mechanisms by which NO could be low in diabetic nephropathy [4, 5] and one such mechanism could relate to eNOS gene polymorphisms. However, some [6, 7], but not all studies [8, 9], identified a positive association of specific eNOS polymorphisms with low NO levels in diabetic nephropathy.

Despite the fact that NO bioavailability is reduced in diabetes, several studies have paradoxically documented that eNOS expression is increased in the diabetic kidney. For example, Hohenstein et al. [10] investigated eNOS expression in type 2 diabetic patients with immunohistochemistry and found that eNOS protein levels were increased in glomeruli. Similarly, streptozotocin (STZ)-induced diabetic rats have been reported to exhibit an increase in eNOS expression in endothelial cells in the afferent arterioles and in the glomerulus [11]. However, it is important to note that eNOS protein levels may not directly correspond to its activity. Basically, eNOS needs to be 'coupled' to form a dimer to produce NO, whereas the 'uncoupled' eNOS generates superoxide as opposed to NO. Komers et al. [12] demonstrated that eNOS in the diabetic kidney is present primarily in the uncoupled form and localizes to the cytosolic fraction. Since eNOS activation also requires translocation to the plasma membrane in the coupled form [12], it is likely that the upregulated eNOS in diabetes might be an inactivated form. This finding is also consistent with a previous report showing that high glucose can 'uncouple' eNOS [13].

Diabetic Mouse Models with eNOS Deficiency

In 1995, Huang et al. [14] first described a mouse model lacking a functional eNOS gene. This mouse model was found to be unable to produce endothelial NO, leading to endothelial dysfunction and hypertension. Using this model, the role of eNOS in *db/db* type 2 diabetic mice and STZ-induced type 1 diabetic mice have been recently studied (table 1).

db/db-eNOS–/– Mice
The *db/db* mouse is a model for type 2 diabetes. In 2006, Zhao et al. [15] crossed this *db/db* mouse (C57BSK/J background) with eNOS–/– mice and examined renal histology and function in the *db/db*-eNOS–/– mice at 26 weeks. Although the severity of hyperglycemia in the *db/db*-eNOS–/– mice was almost identical to that in the parent *db/db* mice, loss of eNOS expression caused more severe

Table 1. Histological features in diabetic eNOS–/– mice

	Kidney	Other organs
STZ-diabetic eNOS–/– mice [18, 19]	Glomerular hypertrophy Mesangial expansion Glomerular BM thickening Mesangiolysis Glomerular capillary microaneurysm Kimmelstiel-Wilson-like nodule Arteriolar hyalinosis Tubulointerstitial injury Abnormal angiogenesis	Retina: • Capillary BM thickness • Vessel leakage • Gliosis • Acellular retinal capillaries
db/db-eNOS–/– mice [15,16]	Glomerular hypertrophy Mesangial expansion Glomerular BM thickening Mesangiolysis Arteriolar hyalinosis Tubulointerstitial injury	Pancreas • Hypertrophic islets Liver • Lipid droplets in hepatocyte Aorta • Impairment of remodeling after arterial injury

BM = Basement membrane.

renal injury including dramatic albuminuria, arteriolar hyalinosis, increased glomerular basement membrane thickness, mesangial expansion, mesangiolysis, and focal segmental and early nodular glomerulosclerosis. In addition, eNOS deficiency resulted in a 50% decline in glomerular filtration rate (GFR) in this mouse model as compared to normal *db/db* mice.

In 2008, Mohan et al. [16] investigated more precisely this model regarding general characteristics and other organs in addition to the renal disease. They found that *db/db*-eNOS–/– mice exhibit nonrenal abnormalities. For instance, body weight, plasma insulin, and serum cholesterol are significantly increased compared to C57BKS/J, normal *db/db*, or non-diabetic eNOS–/– mice. In contrast, hyperglycemia was significantly milder in *db/db*-eNOS–/– compared to that in normal *db/db* mice. Interestingly, eNOS deficiency likely contributes to the development of hypertrophic islets in the pancreas as well as fat droplets in hepatocytes in this mouse model. It is likely that these systemic abnormalities could modulate the renal injury observed in this model.

STZ-Induced Diabetic eNOS–/– Mice
An NIH group (The Animal Models of Diabetic Complications Consortium, AMDCC) has proposed guidelines for an ideal diabetic mouse model which includes the following criteria: (1) progressive renal insufficiency in the setting

of hyperglycemia, more specifically characterized as 50% decline in GFR during the lifetime of the animal; (2) albuminuria (10-fold increase compared with age-, gender-, and strain-matched controls), and (3) characteristic pathologic changes including basement membrane thickening by electron microscopy, advanced mesangial matrix expansion with or without mesangiolysis and nodular mesangial sclerosis, interstitial fibrosis, and any degree of arteriolar hyalinosis [17].

STZ has been widely used to induce type 1 diabetes. In 2007, our group and others first examined renal injury in diabetic eNOS–/– mice, in which diabetes was induced by STZ [18, 19]. Although the STZ regimen was slightly different between the two research groups, diabetic eNOS–/– mice developed a similar phenotype [18, 19]. According to the AMDCC guidelines, the STZ-diabetic eNOS–/– mouse meets two criteria including albuminuria and histological findings. However, Kanetsuna et al. [18] found that GFR was not reduced in diabetic eNOS–/– mice until 20 weeks after diabetic onset, whereas both GFR decline and blood nitrogen urea elevation was found in our study [19, 20]. This discrepancy might be attributed to the different protocol of STZ injection. Interestingly, mortality is also somewhat higher in our model [19]. The precise mechanism for such higher mortality remains unknown, although such finding corresponds to the clinical evidence that life expectancy is low in type 1 diabetes when patients developed advanced nephropathy [21].

The most impressive feature of this model is the histological manifestation. Most other diabetic mouse models predominantly exhibit only early features, including mesangial expansion and glomerular basement membrane thickening, yet fail to develop advanced lesions of diabetic nephropathy. However, diabetic eNOS–/– mice exhibit not only early lesions but also advanced lesions similar to human diabetic nephropathy including mesangiolysis, Kimmelstiel-Wilson-like nodules, and arteriolar hyalinosis [19]. In addition, this model also develops abnormal angiogenesis, which is considered as another important histological manifestation of human diabetic nephropathy [22]. Finally, diabetic eNOS–/– mice also develop tubulointerstitial injury which is, interestingly, more severe than that observed in diabetic wild-type mice [23].

Retinopathy in Streptozotocin-Diabetic eNOS–/– Mice

In contrast to human subjects with diabetic nephropathy, most diabetic animal models do not develop features of severe retinal injury observed in human diabetic retinopathy. For example, STZ-induced diabetic C57BL6 or Ins2^{Akita} mice exhibit lesions of early stage of diabetic retinopathy, including an increase in vascular permeability, a modest increase in acellular capillaries, leukostasis, and thinner layers of the inner plexiform [24, 25]. However, these models do not have advanced lesions, such as capillary obliteration, capillary dropout, retinal edema and preretinal neovascularization. Lately, Li et al. [26] examined retinopathy in STZ-induced diabetic eNOS–/– mice and found that this animal model displays increased retinal capillary basement membrane thickness, increased retinal vessel leakage, gliosis, and increased number of acellular retinal capillaries. Although some features are also observed in other types of diabetic mice, these features are more pronounced in the setting of eNOS deficiency. Therefore, it is likely that a lack of endothelial NO could be also a key factor in the pathogenesis of diabetic retinopathy.

Mechanisms by Which eNOS Deficiency Accelerates Diabetic Nephropathy

Endothelial NO in the kidney helps to maintain physiological function in several ways. An important function is likely the control of glomerular pressure through adjusting vascular resistance of afferent and efferent arterioles [27, 28]. Endothelial NO is also known to negatively regulate the release of von Willebrand factor by inhibiting its exocytosis from the Weibel-Palade bodies in endothelial cells, which presumably blocks the formation of platelet thrombi [29]. In addition, NO inhibition, even low grade, may cause endothelial cell detachment from basement membrane [30]. Given these facts, eNOS deficiency might functionally and structurally potentiate renal injury in the diabetic condition.

Vascular endothelial growth factor (VEGF) is also a key factor in the kidney. While VEGF is protective in the diseased kidney, it is likely due in part to its ability to stimulate NO production in endothelial cells. However, it has been demonstrated that VEGF is not protective, but rather deleterious in diabetic nephropathy [22, 31, 32]. While the precise mechanisms for such diversity of VEGF effect remain an area of study, perhaps it can be accounted for by the diverse levels of NO in distinct renal diseases. Since NO production/bioavailability is low in diabetes, we have hypothesized that the deleterious effect of VEGF may be augmented in the setting of NO deficiency. In particular, we assume that this condition might contribute to the development of abnormal angiogenesis observed in diabetic nephropathy. Our research group has recently documented that (a) proliferation of cultured endothelial cells in response to VEGF is enhanced in the setting of endothelial NO deficiency [33], (b) VEGF expression is associated with glomerular disease and tubulointerstitial injury in diabetic eNOS–/– mice [19], (c) VEGF-induced macrophage migration is prevented by NO [34], and (d) renal VEGFR2 stimulation induces abnormal angiogenesis and glomerular and tubulointerstitial injury, all of which are augmented in uninephrectomized eNOS–/– mice [35]. Given these facts, NO deficiency could enhance the deleterious effect of VEGF, which might be a mechanism for severe renal injury in the mouse diabetic nephropathy.

Therapeutic Approaches in Streptozotocin-Diabetic eNOS–/– Mice

Insulin
Controlling blood glucose by insulin was able to reduce glomerular and tubulo-interstitial injury in the diabetic eNOS–/– mice [19, 23]. The protective effect was accompanied by the lowering of blood pressure, a reduction in albuminuria and by inhibiting VEGF upregulation in the kidney. One might argue that tubulointerstitial injury could stem from STZ toxicity. However, the protective effect of insulin indicates that the diabetic milieu, and not STZ toxicity, is a key factor for the tubulointerstitial injury in diabetic eNOS–/– mice.

Hydralazine
Hydralazine is an antihypertensive agent and its action is likely independent of the renin-angiotensin system (RAS) inhibition. Hydralazine administered at 6 weeks in diabetic eNOS–/– mice significantly reduced blood pressure in this model [23]. Glomerular injuries, including both early and advanced lesions, were significantly ameliorated despite the presence of eNOS deficiency, although no favorable effects were observed on albuminuria. Tubulointerstitial injury was not prevented by lowering blood pressure, and was likely due to the effects of hyperglycemia.

Inhibition of Renin-Angiotensin System
Enalapril (angiotensin-converting enzyme inhibitor) and telmisartan (angiotensin receptor blockade) treatment 6 weeks after STZ administration in the diabetic eNOS–/– mice provided therapeutic benefit in diabetic wild-type mice [20]. Surprisingly, such treatments gave less benefit in diabetic eNOS–/– mice. The blood pressure was only transiently reduced by these treatments in diabetic eNOS–/– mice and later returned to levels similar to that of untreated diabetic eNOS–/– mice. In this study, neither glomerular damage nor tubulointerstitial injury was prevented, while albuminuria was slightly reduced by these treatments. In addition, serum aldosterone concentration was reduced only in diabetic wild-type mice, but not in diabetic eNOS–/– mice, suggesting that sustained high levels of aldosterone may account for the RAS blockade refractory in this model [20]. Future studies on the role of eNOS dysfunction in aldosterone production are warranted.

Aldosterone Antagonist
Spironolactone was found to be the most protective treatment in diabetic eNOS–/– mice. In fact, spironolactone reduced blood pressure, albuminuria and renal injury in this model [20].

Antihypertensive Therapy in Retinopathy of STZ-Diabetic eNOS–/– Mice
The effect of hydralazine, enalapril and telmisartan on retinopathy were examined in the mouse model in which hydralazine and RAS inhibitor was

Table 2. Therapeutic effects on kidney and retina of STZ-diabetic eNOS–/– mice

	Target	Insulin	Hydralazine	RAS inhibitors
Kidney	Glomerular injury	o	o	×
	Tubulointerstitial injury	o	×	×
	Albuminuria	o	×	Δ
Retina	Acellular capillary	NE	o	o

o = Protective; × = not protective; Δ = mildly protective; NE = not examined.

administered [26]. As shown in table 2, these treatments could prevent acellular capillary development in this model. The distinct effects of RAS inhibitors on the kidney and retina suggest different pathogenic mechanisms underlying retinopathy and nephropathy.

Conclusion

A new animal model has been employed to demonstrate that eNOS deficiency accelerates renal and retinal injury resembling advanced lesions of human diabetic patients. These studies indicate a key role for eNOS in diabetic nephropathy and retinopathy. Since the effects of RAS inhibitors seem to be weaker in this model, better identification of mechanisms and therapeutic strategy is required in the presence of eNOS deficiency-associated diabetic nephropathy.

Acknowledgment

The authors acknowledge the enormous contribution of Dr. Katsuyuki Tanabe, Dr. Wataru Kitagawa and Dr. Christopher J. Rivard of the Division of Renal Diseases and Hypertension, University of Colorado, Dr. Tomoki Kosugi and Dr. Waichi Sato of the Division of Nephrology, University of Florida, Dr. Quihong Li of the Department of Ophthalmology, University of Florida, and Dr. Byron P. Croker of the Department of Pathology, Immunology and Laboratory Medicine, University of Florida for the research work.

References

1 Bachmann S, Bosse HM, Mundel P: Topography of nitric oxide synthesis by localizing constitutive no synthases in mammalian kidney. Am J Physiol 1995;268:F885–F898.

2 Han KH, Lim JM, Kim WY, Kim H, Madsen KM, Kim J: Expression of endothelial nitric oxide synthase in developing rat kidney. Am J Physiol Renal Physiol 2005;288:F694–F702.

3 Komers R, Anderson S: Paradoxes of nitric oxide in the diabetic kidney. Am J Physiol Renal Physiol 2003;284:F1121–F1137.

4 Brodsky SV, Morrishow AM, Dharia N, Gross SS, Goligorsky MS: Glucose scavenging of nitric oxide. Am J Physiol Renal Physiol 2001;280:F480–F486.

5 Ceriello A, Mercuri F, Quagliaro L, Assaloni R, Motz E, Tonutti L, Taboga C: Detection of nitrotyrosine in the diabetic plasma: evidence of oxidative stress. Diabetologia 2001; 44:834–838.

6 Liu Y, Burdon KP, Langefeld CD, Beck SR, Wagenknecht LE, Rich SS, Bowden DW, Freedman BI: T-786c polymorphism of the endothelial nitric oxide synthase gene is associated with albuminuria in the diabetes heart study. J Am Soc Nephrol 2005;16:1085–1090.

7 Shin Shin Y, Baek SH, Chang KY, Park CW, Yang CW, Jin DC, Kim YS, Chang YS, Bang BK: Relations between eNOS Glu298Asp polymorphism and progression of diabetic nephropathy. Diabetes Res Clin Pract 2004;65:257–265.

8 Rippin JD, Patel A, Belyaev ND, Gill GV, Barnett AH, Bain SC: Nitric oxide synthase gene polymorphisms and diabetic nephropathy. Diabetologia 2003;46:426–428.

9 Shimizu T, Onuma T, Kawamori R, Makita Y, Tomino Y: Endothelial nitric oxide synthase gene and the development of diabetic nephropathy. Diabetes Res Clin Pract 2002;58:179–185.

10 Hohenstein B, Hugo CP, Hausknecht B, Boehmer KP, Riess RH, Schmieder RE: Analysis of no-synthase expression and clinical risk factors in human diabetic nephropathy. Nephrol Dial Transplant 2008;23:1346–1354.

11 Sugimoto H, Shikata K, Matsuda M, Kushiro M, Hayashi Y, Hiragushi K, Wada J, Makino H: Increased expression of endothelial cell nitric oxide synthase (ecNOS) in afferent and glomerular endothelial cells is involved in glomerular hyperfiltration of diabetic nephropathy. Diabetologia 1998;41:1426–1434.

12 Komers R, Schutzer WE, Reed JF, Lindsley JN, Oyama TT, Buck DC, Mader SL, Anderson S: Altered endothelial nitric oxide synthase targeting and conformation and caveolin-1 expression in the diabetic kidney. Diabetes 2006;55:1651–1659.

13 Brodsky SV, Gao S, Li H, Goligorsky MS: Hyperglycemic switch from mitochondrial nitric oxide to superoxide production in endothelial cells. Am J Physiol Heart Circ Physiol 2002;283:H2130–H2139.

14 Huang PL, Huang Z, Mashimo H, Bloch KD, Moskowitz MA, Bevan JA, Fishman MC: Hypertension in mice lacking the gene for endothelial nitric oxide synthase. Nature 1995;377:239–242.

15 Zhao HJ, Wang S, Cheng H, Zhang MZ, Takahashi T, Fogo AB, Breyer MD, Harris RC: Endothelial nitric oxide synthase deficiency produces accelerated nephropathy in diabetic mice. J Am Soc Nephrol 2006;17:2664–2669.

16 Mohan S, Reddick RL, Musi N, Horn DA, Yan B, Prihoda TJ, Natarajan M, Abboud-Werner SL: Diabetic eNOS knockout mice develop distinct macro- and microvascular complications. Lab Invest 2008;88:515–528.

17 Brosius FC, 3rd, Alpers CE, Bottinger EP, Breyer MD, Coffman TM, Gurley SB, Harris RC, Kakoki M, Kretzler M, Leiter EH, Levi M, McIndoe RA, Sharma K, Smithies O, Susztak K, Takahashi N, Takahashi T: Mouse models of diabetic nephropathy. J Am Soc Nephrol 2009;20:2503–2512.

18 Kanetsuna Y, Takahashi K, Nagata M, Gannon MA, Breyer MD, Harris RC, Takahashi T: Deficiency of endothelial nitric-oxide synthase confers susceptibility to diabetic nephropathy in nephropathy-resistant inbred mice. Am J Pathol 2007;170:1473–1484.

19 Nakagawa T, Sato W, Glushakova O, Heinig M, Clarke T, Campbell-Thompson M, Yuzawa Y, Atkinson M, Johnson RJ, Croker B: Diabetic eNOS knockout mice develop advanced diabetic nephropathy. J Am Soc Nephrol 2007;18:539–550.

20 Kosugi T, Heinig M, Nakayama T, Matsuo S, Nakagawa T: eNOS knockout mice with advanced diabetic nephropathy have less benefit from renin-angiotensin blockade than from aldosterone receptor antagonists. Am J Pathol 2010;49:51–54.

21 Borch-Johnsen K, Kreiner S, Deckert T: Mortality of type 1 (insulin-dependent) diabetes mellitus in Denmark: a study of relative mortality in 2930 Danish type 1 diabetic patients diagnosed from 1933 to 1972. Diabetologia 1986;29:767–772.

22 Nakagawa T, Kosugi T, Haneda M, Rivard CJ, Long DA: Abnormal angiogenesis in diabetic nephropathy. Diabetes 2009;58:1471–1478.

23 Kosugi T, Heinig M, Nakayama T, Connor T, Yuzawa Y, Li Q, Hauswirth WW, Grant MB, Croker BP, Campbell-Thompson M, Zhang L, Atkinson MA, Segal MS, Nakagawa T: Lowering blood pressure blocks mesangiolysis and mesangial nodules, but not tubulointerstitial injury, in diabetic eNOS knockout mice. Am J Pathol 2009;174:1221–1229.

24 Feit-Leichman RA, Kinouchi R, Takeda M, Fan Z, Mohr S, Kern TS, Chen DF: Vascular damage in a mouse model of diabetic retinopathy: relation to neuronal and glial changes. Invest Ophthalmol Vis Sci 2005;46:4281–4287.

25 Gastinger MJ, Kunselman AR, Conboy EE, Bronson SK, Barber AJ: Dendrite remodeling and other abnormalities in the retinal ganglion cells of Ins2 Akita diabetic mice. Invest Ophthalmol Vis Sci 2008;49:2635–2642.

26 Li Q, Verma A, Han PY, Nakagawa T, Johnson RJ, Grant MB, Campbell-Thompson M, Jarajapu YP, Lei B, Hauswirth WW: Diabetic eNOS knockout mice develop accelerated retinopathy. Invest Ophthalmol Vis Sci 2010;51:5240–5246.

27 Edwards RM, Trizna W: Modulation of glomerular arteriolar tone by nitric oxide synthase inhibitors. J Am Soc Nephrol 1993;4:1127–1132.

28 Patzak A, Kleinmann F, Lai EY, Kupsch E, Skelweit A, Mrowka R: Nitric oxide counteracts angiotensin II induced contraction in efferent arterioles in mice. Acta Physiol Scand 2004;181:439–444.

29 Matsushita K, Morrell CN, Cambien B, Yang SX, Yamakuchi M, Bao C, Hara MR, Quick RA, Cao W, O'Rourke B, Lowenstein JM, Pevsner J, Wagner DD, Lowenstein CJ: Nitric oxide regulates exocytosis by S-nitrosylation of N-ethylmaleimide-sensitive factor. Cell 2003;115:139–150.

30 Stoessel A, Paliege A, Theilig F, Addabbo F, Ratliff B, Waschke J, Patschan D, Goligorsky MS, Bachmann S: Indolent course of tubulointerstitial disease in a mouse model of subpressor, low-dose nitric oxide synthase inhibition. Am J Physiol Renal Physiol 2008;295:F717–F725.

31 Flyvbjerg A, Schrijvers BF, De Vriese AS, Tilton RG, Rasch R: Compensatory glomerular growth after unilateral nephrectomy is VEGF dependent. Am J Physiol Endocrinol Metab 2002;283:E362–E366.

32 de Vriese AS, Tilton RG, Elger M, Stephan CC, Kriz W, Lameire NH: Antibodies against vascular endothelial growth factor improve early renal dysfunction in experimental diabetes. J Am Soc Nephrol 2001;12:993–1000.

33 Nakagawa T, Sato W, Sautin YY, Glushakova O, Croker B, Atkinson MA, Tisher CC, Johnson RJ: Uncoupling of vascular endothelial growth factor with nitric oxide as a mechanism for diabetic vasculopathy. J Am Soc Nephrol 2006;17:736–745.

34 Sato W, Kosugi T, Zhang L, Roncal CA, Heinig M, Campbell-Thompson M, Yuzawa Y, Atkinson MA, Grant MB, Croker BP, Nakagawa T: The pivotal role of VEGF on glomerular macrophage infiltration in advanced diabetic nephropathy. Lab Invest 2008;88:949–961.

35 Sato W, Nakagawa T: Selective stimulation of VEGFR2 accelerates progressive renal disease, in revision.

Takahiko Nakagawa, MD, PhD
Division of Renal Diseases and Hypertension
University of Colorado Denver, C281
Aurora, CO 80045 (USA)
Tel. +1 303 724 3346, Fax +1 303 724 4868, E-Mail Takahiko.Nakagawa@ucdenver.edu

Lai KN, Tang SCW (eds): Diabetes and the Kidney.
Contrib Nephrol. Basel, Karger, 2011, vol 170, pp 102–112

Reactive Oxygen Species and Oxidative Stress

Hyunjin Noh[a] · Hunjoo Ha[b]

[a]Hyonam Kidney Laboratory, Division of Nephrology, Department of Internal Medicine, Soon Chun Hyang University, and [b]Division of Life and Pharmaceutical Sciences, Department of Bioinspired Science, College of Pharmacy, Ewha Womans University, Seoul, South Korea

Abstract

Oxidative stress defined as an excessive production of reactive oxygen species (ROS) surpassing existing antioxidative defense mechanisms plays a critical role in the development and progression of diabetic vascular complications including nephropathy. Overproduction of ROS in diabetic milieu is both a direct consequence of hyperglycemia and an indirect consequence through advanced glycation end products (AGEs) or mediators of glucotoxicity such as cytokines and growth factors. Among many pathways, nicotinamide adenosine dinucleotide phosphate (NADPH) oxidase and mitochondrial dysfunction have been recognized as two major sources of ROS generation in diabetic kidneys, and NADPH oxidase-derived ROS has been shown to facilitate renal mitochondrial superoxide production in hyperglycemia. Low antioxidant bioavailability promotes cellular oxidative stress leading to additional cellular damage. Although large-scale clinical trials using classical antioxidants have failed to show a significant effect on the development of vascular complications in diabetes, new strategies targeting NF-E2-related factor 2, the primary transcription factor that controls the antioxidant response, mitochondrial dysfunction, or NADPH oxidase might provide a potential approach for the prevention and treatment of diabetic nephropathy.

Diabetic nephropathy (DN) is the leading cause of end-stage renal disease worldwide and an independent risk factor for all-cause and cardiovascular mortalities in diabetic patients [1]. A causal relationship between chronic sustained hyperglycemia and diabetic microvascular complications including nephropathy has been demonstrated conclusively by large randomized clinical studies. The Diabetes Control and Complications Trial in type 1 diabetes [2] and the United Kingdom Prospective Diabetes Study in type 2 diabetes [3] demonstrated that intensive blood glucose control successfully delayed the onset and retarded the progression of diabetic microvascular complications, suggesting that hyperglycemia-induced factors are responsible for diabetic vascular complications.

Oxidative stress defined as a tissue injury induced by excessive amounts of reactive oxygen species (ROS) surpassing various endogenous antioxidative defense mechanisms is one of the major factors in the development of diabetic nephropathy [4]. Several pathways including auto-oxidation of glucose, transition metal-catalyzed Fenton reactions, advanced glycation, polyol pathway, mitochondrial dysfunction, xanthine oxidase, nitric oxide synthase, and nicotinamide adenosine dinucleotide phosphate (NADPH) oxidase all contribute to oxidative stress in diabetic nephropathy. Among these, it appears that mitochondrial dysfunction [5] and NADPH oxidase [6] play a key role in the pathogenesis of diabetic nephropathy. This chapter reviews the current knowledge base regarding the pathogenic role for the oxidative stress in DN focusing on mitochondrial dysfunction and NADPH oxidase and potential new therapeutic strategies involving more targeted antioxidant approaches.

Oxidative Stress and Tissue Injury

In a steady state, formation of ROS is finely balanced by a similar rate of their consumption by antioxidants [7]. Oxidative stress occurs when the generation of ROS exceeds the capacity of antioxidant defense mechanism in terms of extent, timing, and location. Within certain boundaries, the generation of ROS is essential to maintain homeostasis. However, a rise in intracellular oxidant levels has deleterious effects by both oxidation of various cell components including proteins, lipids, carbohydrates, and DNA and triggering of the activation of specific signaling pathways [7]. Both of these effects can influence a number of cellular stress-sensitive pathways linked to the development of diseases such as diabetes and ageing [4, 7].

The ROS encompass a family of molecules including superoxide anion (O_2^-), hydroxyl radical, hydrogen peroxide, peroxynitrite, hypochlorous acid, nitric oxide (NO), and lipid radicals (fig. 1). Some of these, such as superoxide anion and hydroxyl radicals, are extremely unstable, whereas others such as hydrogen peroxide are freely diffusible and relatively long lived [8]. ROS are generated as a result of normal intracellular metabolism in mitochondria and peroxisomes, as well as other enzyme systems including NADPH oxidase. In addition, a variety of exogenous sources such as ultraviolet light, ionizing radiation, environmental toxins, inflammatory cytokine, and chemotherapeutics can trigger ROS production [7]. Evidence indicates that the majority of intracellular ROS generation is derived from the mitochondria, where in normal conditions approximately 4% of oxygen is converted to radicals [9]. The burden of ROS is largely counterbalanced by antioxidant defense systems, mainly enzymatic scavengers such

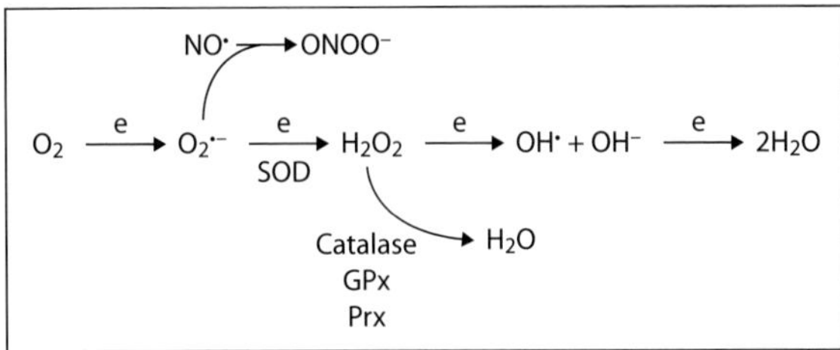

Fig. 1. ROS and antioxidant enzymes. Sequential reduction in molecular oxygen (partial reduction in oxygen) in our body produces superoxide anion, hydrogen peroxide, and hydroxyl radical before converting to water. Overproduction of superoxide anion along with increased NO favors the formation of peroxynitrite, a strong oxidant. GPx = Glutathione peroxidase; Prx = peroxiredoxin.

as superoxide dismutase (SOD), catalase, glutathione peroxidase, and peroxiredoxin (fig. 1). SOD converts superoxide to hydrogen peroxide, whereas catalase, glutathione peroxidase, and peroxiredoxin convert hydrogen peroxide to water. Other important antioxidant enzymes include superoxide reductase, glutathione reductase and thioredoxin/thioredoxin reductase. Vitamin C, vitamin E, glutathione, carotenoids, and flavonoids form other nonenzymatic antioxidants [7].

In addition to the modification of macromolecules, ROS can function as signaling molecules to activate a number of signal transduction cascades and transcription factors including mitogen-activated protein kinases, tyrosine kinases, nuclear factor-κB, activator protein-1, and specificity protein-1 leading to the abnormal regulation of inflammation, cell proliferation, fibrosis, endothelial dysfunction, and angiogenesis. Furthermore, ROS appear to act as a signal amplifier in the development of diabetic nephropathy. Protein kinase C (PKC), transforming growth factor (TGF)-β_1, and angiotensin II (Ang II) stimulated by hyperglycemia-induced ROS, in turn, generate and signal through ROS [10].

Measurement of ROS

The problem faced in the study of oxidative stress arises from the lack of reliable and reproducible methods to measure ROS because of their short lifetime, very low steady state concentrations, and their highly reactive nature. In addition, characterization of the cellular functions of each ROS molecule requires measurement of its concentration selectively in the presence of other oxygen metabolites. Although a number of methods using surrogate markers of ROS-mediated damage such as oxidized low-density lipoprotein, F_2-isoprostane, and 8-hydroxydeoxyguanosine (8-OHdG) and determination of ROS in biological fluids based on horseradish peroxidase and artificial substrates have been available, none of these is applicable for real-time measurement in different sub-cellular compartments. Recently, genetically encoded probes have been developed as a tool for detection of hydrogen peroxide inside and outside of cells [11] yet to be validated in various experimental and clinical settings.

Increased Oxidative Stress in Diabetic Nephropathy

Multiple studies have indicated that oxidative stress contributes to the pathogenesis of diabetic nephropathy. High glucose rapidly increasing dichlorofluorescein (DCF)-sensitive cytosolic ROS has been demonstrated in rat and mouse mesangial cells [12]. Notably, 8-OHdG is increased in streptozotocin (STZ)-induced diabetic kidney [13]. Koya et al. [14] also reported increased DCF-sensitive ROS in STZ-induced diabetic rat glomeruli which were effectively inhibited by antioxidants vitamin E and probucol. Further, plasma level and urinary excretion of thiobarbituric acid-reactive substances, a marker of lipid peroxidation, have been shown to be increased in STZ-induced diabetic rat [15]. Similarly, the increased oxidative stress associated with nephropathy has been shown in type 2 diabetic experimental models. An increase in ROS generation has been observed in proximal tubular cells from *db/db* mice as compared to *db/m* control [16] and increased urinary 8-OHdG and decreased NO have been shown in obese ZSF1 rats, a new experimental model for type 2 diabetes [17]. Several clinical studies also confirmed an increased oxidative stress in diabetic nephropathy. Fluorescent products of lipid peroxidation and malondialdehyde have been found to be enhanced in type 1 diabetic patients, especially in those with microalbuminuria when compared with control subjects [18]. Protein carbonyl and 4-hydroxy-2-nonenal have been shown to be significantly increased in plasma and urine from type 2 diabetic patients with nephropathy with respect to control subjects [19].

Mitochondrial ROS Overproduction in Diabetic Milieu

Mitochondria-derived ROS are byproducts of oxidative phosphorylation across the electron transport chain. Normally, electron transfer through complex I (NADH dehydrogenase), III (ubiquinone-cytochrome c reductase), and IV (cytochrome c oxidase) extrudes protons across the mitochondrial inner membrane, generating a proton gradient that derives ATP synthase (complex V) as protons pass back into the matrix. However, in diabetic milieu with high intracellular glucose, there is an overproduction of electron donors (NADH and $FADH_2$) from the tricarboxylic acid cycle. As a result, the voltage gradient across the mitochondrial inner membrane increases until a critical threshold is reached. At this point, electron transport at complex III is inhibited, causing the electrons to back up to coenzyme Q, which reduces oxygen to superoxide by donation of the electron [5].

Mitochondrial ROS in hyperglycemia have been suggested as the key molecules linking four major pathways causing hyperglycemic tissue damage: polyol pathway, hexosamine pathway, PKC pathway, and advanced glycation end products (AGEs) pathway [5]. Although there is no direct evidence showing that this unifying hypothesis is true in individual renal cell types, a role for mitochondrial ROS production in diabetic nephropathy is further supported by recent observations that almost half of children with mitochondrial diseases have renal dysfunction [20] and that a newly described inherited mitochondriopathy involving a deficiency in coenzyme Q10 also has primary renal pathology [21]. It is interesting to note that interaction of AGEs and receptor for AGE-induced cytosolic ROS promote mitochondrial superoxide generation in diabetes, suggesting that mitochondria is not only a source of ROS generation but also a target of cytosolic ROS [22].

ROS Production by NADPH Oxidase in a Diabetic Milieu

NADPH oxidase is an enzyme complex that catalyzes generation of superoxide by reduction of oxygen using NADPH. Initially, it was found in neutrophils and phagocytic cells, where it plays a pivotal role in host defense and innate immunity by production of large quantities of superoxide. It is now well recognized that NADPH oxidase is also present in nonphagocytic cells including endothelial cells, vascular smooth muscle cells, fibroblasts, mesangial cells, and proximal tubular epithelial cells [23]. In these cells, the quantity of superoxide production by NADPH oxidase is much smaller than phagocytic cells. However, they play an important role by serving as second messengers and regulate vascular tone, oxygen sensing, endothelial dysfunction, inflammation, hypertrophy, apoptosis, migration, and tissue remodeling. Various growth factors such as TGF-β_1 and platelet-derived growth factor, cytokines such as tumor necrosis factor-α, or G-protein-coupled receptor agonists like Ang II can activate NADPH oxidase upon binding their cognate receptors. In addition, metabolic factors including hyperglycemia, free fatty acid, and AGEs as well as mechanical stretch also increase the activity of NADPH oxidase followed by overproduction of ROS. Since various ligands and metabolic factors alluded above are known to be elevated under diabetic condition, it appears that ROS derived from NADPH oxidase act as a potent causal factor that initiates and accelerates DN.

NADPH oxidase consists of multiple subunits: membrane-associated p22^{phox} and Nox isoforms (based on gp91^{phox}), cytosolic subunit p47^{phox}, p67^{phox}, p40^{phox}, and GTPase Rac1 or Rac2. The Nox isoforms are the catalytic subunits for ROS production that are differently expressed and regulated in various tissues. Nox1, 2, 3, 4, and 5 have been found to be expressed in the kidney [24, 25].

The expression of various subunits of NADPH oxidase has been reported to be increased in experimental diabetic nephropathy. Furthermore, we and others have shown that pharmacological inhibition of NADPH oxidase prevents high glucose or TGF-β_1-induced ROS generation [26], p47 and gp91 overexpression, and mesangial matrix expansion [27]. More specific inhibition using antisense oligonucleotide to Nox4 inhibited NADPH oxidase-dependent ROS production in renal cortex and glomeruli [28]. These observations support the pathogenic role of NADPH oxidase-dependent ROS generation in DN.

Finally, it is worth mentioning that ROS such as hydrogen peroxide can activate NADPH oxidase, thereby amplifying the vascular injury [29]. Observations from our laboratory also suggested that hydrogen peroxide increases mRNA expression of Nox4 and translocation of cytosolic subunits in mouse mesangial cells [unpubl. data].

Antioxidant Defense Mechanism

Although it is generally accepted that excessive generation of ROS induce antioxidant enzyme expression and activity as a compensatory mechanism to keep cellular redox homeostasis, decreases in expression or activity of antioxidant enzymes have been shown in type 1 diabetic patients with nephropathy [30]. Antioxidant enzymes can be glycated, oxidized, or nitrated under hyperglycemia-induced oxidative stress, which result in loss of their activity. Diabetic renal injury is accelerated in the SOD knockout mice, and Tempol, a SOD mimetic, has shown to suppress albuminuria, TGF-β_1, collagen synthesis, and oxidative stress [31]. In addition, a functional polymorphism of MnSOD is associated with the development of DN [32]. Chronic therapy with ebselen, a glutathione mimic, has demonstrated a scavenging effect of peroxynitrite associated with an improvement in tubulointerstitial pathology and inflammation in early diabetic nephropathy [33]. Furthermore, overexpression of catalase in *db/db* mice has been demonstrated to attenuate interstitial fibrosis and tubular apoptosis [34]. This effect was attributed to decreased oxidative stress resulting in a reduced activation of TGF-β_1 induced by Ang II. In contrast, glutathione peroxidase 1-deficient mice have not been shown to have a protective effect against STZ-induced diabetic nephropathy [35], most likely because of redundancy with respect to other isoforms.

The heme oxygenase (HO) system is another crucial antioxidant mechanism in the kidney [36]. HO converts heme, a pro-oxidant, into iron, carbon monoxide, and biliverdin. Biliverdin is then transformed into a potent antioxidant, bilirubin, by biliverdin reductase. The inducible HO-1 has been shown to protect against oxidative stress in the kidney, and also the constitutively expressed HO-2 is getting recognized for its antioxidative role. Given the protective role of HO-1 in the kidney, the studies showing that hyperglycemia decreases HO activity in vascular cells suggest a pathogenic role of the HO system in the development of DN [37].

NF-E2-related factor 2 (Nrf2) is the primary transcription factor that regulates intracellular antioxidants, phase II detoxifying enzymes [38]. It has been demonstrated that the glomeruli of diabetic patients with nephropathy had elevated Nrf2 levels as a protective mechanism [39]. Further strengthening a potential role for Nrf2, Nrf2 knockout mice have higher production of ROS and suffer from greater oxidative stress and renal damage in STZ-induced diabetic nephropathy, which is inhibited by overexpression of Nrf2 [39].

Together, these data strongly suggest that low antioxidant bioavailability along with an increased ROS generation promotes oxidative damage associated with DN.

Therapies Targeting Oxidative Stress

Since hyperglycemia is the main driver of oxidative stress, strict glucose control is the basis of antioxidant therapy. In addition, many of the current standard therapies such as angiotensin-converting enzyme inhibitors, angiotensin receptor blockers, and 3-hydroxy-3-methylglutaryl-coenzyme A reductase inhibitors may also help to attenuate oxidative stress. In animal models, administration of antioxidant and overexpression of CuZnSOD [40], catalase [16], or thioredoxin 1 [41] clearly show remarkable effects in preventing DN. On the other hand, large-scale clinical trials with classic antioxidants such as vitamin C and vitamin E have failed to demonstrate beneficial effects. The lack of efficiency might be attributed to the low doses used or to the failure of antioxidant efficacy in human. The ideal antioxidant therapy would be more targeted with minimal adverse effects. Several potentially beneficial interventions targeting mitochondrial ROS generation such as poly(ADP-ribose) polymerase inhibitors, SOD/catalase mimetics, and mitoQ, NADPH oxidase, or Nrf-2 are currently under investigation in the experimental phase and the effect in humans remains to be further examined. Although experimental studies have shown some beneficial effects [42], a recent clinical study using high-dose benfotiamine, a thiamine derivative, for 12 weeks in addition to Ang II blockade has failed to demonstrate any significant positive effect on the development of albuminuria and renal injury [43].

Conclusion

Experimental and clinical evidence suggests that oxidative stress play a major role in the pathogenesis of diabetic nephropathy. Several mitochondrial and cellular pathways including NADPH oxidase are implicated in the increased

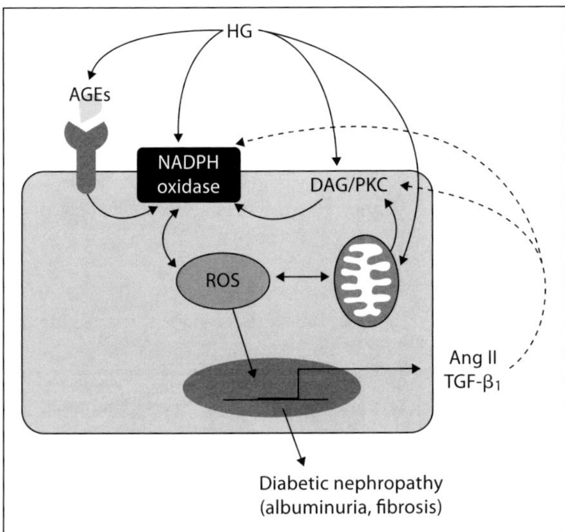

Fig. 2. ROS-mediated glucotoxicity in diabetic kidney. High glucose promotes ROS production either through increased mitochondrial activity or through increased formation of DAG, an endogenous activator of PKC, leading to the activation of NADPH oxidase. AGEs, by-products of high glucose, interact with its receptor (the receptor for AGEs, RAGE) and produce ROS through NADPH oxidase. In addition, Ang II and TGF-β_1 stimulated by hyperglycemia-induced ROS, in turn, generate and signal through ROS after binding to each of their receptors. ROS, thus, appear to act as a signal amplifier in the development of diabetic nephropathy. While mitochondrial ROS in hyperglycemia has been suggested as the key molecule activating all other pathways involved in diabetic vascular complications, NADPH oxidase-derived ROS has been shown to facilitate renal mitochondrial superoxide production in hyperglycemia and ROS such as hydrogen peroxide activates NADPH oxidase in mesangial cells [Ha, unpubl. data]. DAG = Diacylglycerol; HG = high glucose.

generation of ROS (fig. 2), which is associated with reduced antioxidant defense systems. Combination of targeted strategies to prevent overproduction of ROS and to enhance antioxidant mechanisms may prove to be effective in preventing the development and progression of diabetic nephropathy.

Acknowledgements

This work was supported, in part, by grants R01-2006-000-10829, R15-2006-020, and R31-2008-000-10010-0 (H.H.) and 2009-0064424 (H.N.) from the National Research Foundation of Korea.

References

1 Sarnak MJ, Levey AS, Schoolwerth AC, Coresh J, Culleton B, Hamm LL, McCullough PA, Kasiske BL, Kelepouris E, Klag MJ, Parfrey P, Pfeffer M, Raij L, Spinosa DJ, Wilson PW: Kidney disease as a risk factor for development of cardiovascular disease: a statement from the American Heart Association councils on kidney in cardiovascular disease, high blood pressure research, clinical cardiology, and epidemiology and prevention. Circulation 2003;108: 2154–2169.

2 The Diabetes Control and Complications Trial Research Group: The effect of intensive treatment of diabetes on the development and progression of long-term complications in insulin-dependent diabetes mellitus. N Engl J Med 1993;329:977–986.

3 UK Prospective Diabetes Study Group: Intensive blood-glucose control with sulphonylureas or insulin compared with conventional treatment and risk of complications in patients with type 2 diabetes (UKPDS 33). Lancet 1998;352:837–853.

4 Forbes JM, Coughlan MT, Cooper ME: Oxidative stress as a major culprit in kidney disease in diabetes. Diabetes 2008;57:1446–1454.

5 Brownlee M: Biochemistry and molecular cell biology of diabetic complications. Nature 2001;414:813–820.

6 Kakehi T, Yabe-Nishimura C: NOX enzymes and diabetic complications. Semin Immunopathol 2008;30:301–314.

7 Finkel T, Holbrook NJ: Oxidants, oxidative stress and the biology of ageing. Nature 2000; 408:239–247.

8 Sies H: Oxidative stress: from basic research to clinical application. Am J Med 1991;91:31S–38S.

9 Lenaz G, Bovina C, D'Aurelio M, Fato R, Formiggini G, Genova ML, Giuliano G, Pich MM, Paolucci UGO, Castelli GP, Ventura B: Role of mitochondria in oxidative stress and aging. Ann N Y Acad Sci 2002;959:199–213.

10 Ha H, Hwang IA, Park JH, Lee HB: Role of reactive oxygen species in the pathogenesis of diabetic nephropathy. Diabetes Res Clin Pract 2008;82:S42–S45.

11 Rhee S, Chang T-S, Jeong W, Kang D: Methods for detection and measurement of hydrogen peroxide inside and outside of cells. Mol Cells 2010;29:539–549.

12 Ha H, Lee HB: Reactive oxygen species as glucose signaling molecules in mesangial cells cultured under high glucose. Kidney Int Suppl 2000;77:S19–S25.

13 Ha H, Kim C, Son Y, Chung MH, Kim KH: DNA damage in the kidneys of diabetic rats exhibiting microalbuminuria. Free Radic Biol Med 1994;16:271–274.

14 Koya D, Hayashi K, Kitada M, Kashiwagi A, Kikkawa R, Haneda M: Effects of antioxidants in diabetes-induced oxidative stress in the glomeruli of diabetic rats. J Am Soc Nephrol 2003;14:S250–S253.

15 Lee EA, Seo JY, Jiang Z, Yu MR, Kwon MK, Ha H, Lee HB: Reactive oxygen species mediate high glucose-induced plasminogen activator inhibitor-1 up-regulation in mesangial cells and in diabetic kidney. Kidney Int 2005;67:1762–1771.

16 Brezniceanu ML, Liu F, Wei CC, Tran S, Sachetelli S, Zhang SL, Guo DF, Filep JG, Ingelfinger JR, Chan JSD: Catalase overexpression attenuates angiotensinogen expression and apoptosis in diabetic mice. Kidney Int 2007;71:912–923.

17 Prabhakar S, Starnes J, Shi S, Lonis B, Tran R: Diabetic nephropathy is associated with oxidative stress and decreased renal nitric oxide production. J Am Soc Nephrol 2007; 18:2945–2952.

18 Chiarelli F, Cipollone F, Mohn A, Marini M, Iezzi A, Fazia M, Tumini S, De Cesare D, Pomilio M, Pierdomenico SD, Di Gioacchino M, Cuccurullo F, Mezzetti A: Circulating monocyte chemoattractant protein-1 and early development of nephropathy in type 1 diabetes. Diabetes Care 2002;25:1829–1834.

19 Calabrese V, Mancuso C, Sapienza M, Puleo E, Calafato S, Cornelius C, Finocchiaro M, Mangiameli A, Di Mauro M, Stella AMG, Castellino P: Oxidative stress and cellular stress response in diabetic nephropathy. Cell Stress Chaperones 2007;12:299–306.

20 Martín-Hernández E, García-Silva MT, Vara J, Campos Y, Cabello A, Muley R, del Hoyo P, Martín MA, Arenas J: Renal pathology in children with mitochondrial diseases. Pediatr Nephrol 2005;20:1299–1305.

21 Diomedi-Camassei F, Di Giandomenico S, Santorelli FM, Caridi G, Piemonte F, Montini G, Ghiggeri GM, Murer L, Barisoni L, Pastore A, Muda AO, Valente ML, Bertini E, Emma F: COQ2 nephropathy: a newly described inherited mitochondriopathy with primary renal involvement. J Am Soc Nephrol 2007;18:2773–2780.

22 Coughlan MT, Thorburn DR, Penfold SA, Laskowski A, Harcourt BE, Sourris KC, Tan ALY, Fukami K, Thallas-Bonke V, Nawroth PP, Brownlee M, Bierhaus A, Cooper ME, Forbes JM: RAGE-Induced cytosolic ROS promote mitochondrial superoxide generation in diabetes. J Am Soc Nephrol 2009;20: 742–752.

23 Lassegue B, Griendling KK: NADPH oxidases: functions and pathologies in the vasculature. Arterioscler Thromb Vasc Biol 2010;30:653–661.

24 Kashihara N, Haruna Y, Kondeti VK, Kanwar YS: Oxidative stress in diabetic nephropathy. Curr Med Chem 2010;17:4256–4269.

25 Gao L, Mann GE: Vascular NAD(P)H oxidase activation in diabetes: a double-edged sword in redox signalling. Cardiovasc Res 2009;82:9–20.

26 Rhyu DY, Yang Y, Ha H, Lee GT, Song JS, Uh ST, Lee HB: Role of reactive oxygen species in TGF-β1-induced mitogen-activated protein kinase activation and epithelial-mesenchymal transition in renal tubular epithelial cells. J Am Soc Nephrol 2005;16: 667–675.

27 Asaba K, Tojo A, Onozato ML, Goto A, Quinn MT, Fujita T, Wilcox CS: Effects of NADPH oxidase inhibitor in diabetic nephropathy. Kidney Int 2005;67:1890–1898.

28 Gorin Y, Block K, Hernandez J, Bhandari B, Wagner B, Barnes JL, Abboud HE: Nox4 NAD(P)H oxidase mediates hypertrophy and fibronectin expression in the diabetic kidney. J Biol Chem 2005;280:39616–39626.

29 Li W-G, Miller FJ, Zhang HJ, Spitz DR, Oberley LW, Weintraub NL: H_2O_2-induced O_2^- production by a non-phagocytic NAD(P) H oxidase causes oxidant injury. J Biol Chem 2001;276:29251–29256.

30 Ceriello A, Morocutti A, Mercuri F, Quagliaro L, Moro M, Damante G, Viberti GC: Defective intracellular antioxidant enzyme production in type 1 diabetic patients with nephropathy. Diabetes 2000;49: 2170–2177.

31 Frederick RD, Patricia AC, Mona FM: Acceleration of diabetic renal injury in the superoxide dismutase knockout mouse: effects of tempol. Metabolism 2007;56: 1256–1264.

32 Möllsten A, Marklund SL, Wessman M, Svensson M, Forsblom C, Parkkonen M, Brismar K, Groop P-H, Dahlquist G: A functional polymorphism in the manganese superoxide dismutase gene and diabetic nephropathy. Diabetes 2007;56:265–269.

33 Chander PN, Gealekman O, Brodsky SV, Elitok S, Tojo A, Crabtree M, Gross SS, Goligorsky MS: Nephropathy in Zucker diabetic fat rat is associated with oxidative and nitrosative stress: prevention by chronic therapy with a peroxynitrite scavenger ebselen. J Am Soc Nephrol 2004;15:2391–2403.

34 Brezniceanu M-L, Liu F, Wei C-C, Chénier I, Godin N, Zhang S-L, Filep JG, Ingelfinger JR, Chan JSD: Attenuation of interstitial fibrosis and tubular apoptosis in db/db transgenic mice overexpressing catalase in renal proximal tubular cells. Diabetes 2008;57:451–459.

35 de Haan JB, Stefanovic N, Nikolic-Paterson D, Scurr LL, Croft KD, Mori TA, Hertzog P, Kola I, Atkins RC, Tesch GH: Kidney expression of glutathione peroxidase-1 is not protective against streptozotocin-induced diabetic nephropathy. Am J Physiol Renal Physiol 2005;289:F544–F551.

36 Abraham NG, Kappas A: Heme oxygenase and the cardiovascular-renal system. Free Radic Biol Med 2005;39:1–25.

37 Di Noia MA, Van Driesche S, Palmieri F, Yang L-M, Quan S, Goodman AI, Abraham NG: Heme oxygenase-1 enhances renal mitochondrial transport carriers and cytochrome c oxidase activity in experimental diabetes. J Biol Chem 2006;281:15687–15693.

38 Zhang DD: Mechanistic studies of the Nrf2-Keap1 signaling pathway. Drug Metab Rev 2006;38:769–789.

39 Jiang T, Huang Z, Lin Y, Zhang Z, Fang D, Zhang DD: The protective role of Nrf2 in streptozotocin-induced diabetic nephropathy. Diabetes 2010;59:850–860.

40   Craven PA, Melhem MF, Phillips SL, DeRubertis FR: Overexpression of Cu^{2+}/Zn^{2+} superoxide dismutase protects against early diabetic glomerular injury in transgenic mice. Diabetes 2001;50:2114–2125.

41   Hamada Y, Miyata S, Nii-Kono T, Kitazawa R, Kitazawa S, Higo S, Fukunaga M, Ueyama S, Nakamura H, Yodoi J, Fukagawa M, Kasuga M: Overexpression of thioredoxin1 in transgenic mice suppresses development of diabetic nephropathy. Nephrol Dial Transplant 2007;22:1547–1557.

42   Karachalias N, Babaei-Jadidi R, Rabbani N, Thornalley P: Increased protein damage in renal glomeruli, retina, nerve, plasma and urine and its prevention by thiamine and benfotiamine therapy in a rat model of diabetes. Diabetologia 2010;53:1506–1516.

43   Alkhalaf A, Klooster A, van Oeveren W, Achenbach U, Kleefstra N, Slingerland RJ, Mijnhout GS, Bilo HJG, Gans ROB, Navis GJ, Bakker SJL: A double-blind, randomized, placebo-controlled clinical trial on benfotiamine treatment in patients with diabetic nephropathy. Diabetes Care 2010;33:1598–1601.

Hunjoo Ha, PhD
Department of Bioinspired Science, Division of Life and Pharmaceutical Sciences,
College of Pharmacy, Ewha Womans University
11-1 Daehyun Dong
Seodaemun Ku, Seoul 120-750 (South Korea)
Tel. +82 2 3277 4075, Fax +82 2 3277 2851, E-Mail hha@ewha.ac.kr

Lai KN, Tang SCW (eds): Diabetes and the Kidney.
Contrib Nephrol. Basel, Karger, 2011, vol 170, pp 113–123

Inflammatory Pathways

Juan F. Navarro-González[a,b] · Carmen Mora-Fernández[b]

[a]Nephrology Service and [b]Research Unit, University Hospital Nuestra Señora de Candelaria,
Santa Cruz de Tenerife, Spain

Abstract
Diabetes mellitus and its complications have become one of the most important health problems in the world. Nowadays, diabetic nephropathy is the main cause of end-stage renal failure and need for renal substitutive therapy. The exact mechanisms leading to the development and progression of renal damage in diabetes are not yet completely known. Growing evidence indicates that activation of innate immunity with the development of a chronic low-grade inflammatory response is a recognized factor in the pathogenesis of this disease. Inflammatory molecules and pathways, including metabolic routes, oxidative stress, growth factors, chemokines, adhesion molecules and inflammatory cytokines, interact in manifold ways leading to renal injury responsible for the development and progression of this complication. The increasing knowledge and understanding of the role of these inflammatory mechanisms, with an integrative comprehension of this network, will facilitate the identification of new therapeutic targets and the development of new strategies that can be translated successfully into clinical applications.

Diabetes mellitus, especially type 2 diabetes, and its complications are major health problems worldwide. According to data from the International Diabetes Federation, the number of diabetics in adult population older than 20 years will be higher than 400 million in 2030. Therefore, target organ complications secondary to diabetes will become a critical medical concern in the near future. Diabetic nephropathy (DN), one of the most relevant diabetic complications, has become the main cause of end-stage renal disease in the western world in the last decade [1].

Inflammation and Diabetes Mellitus

Nowadays, growing evidence indicates that activation of the innate immune system and the development of a systemic low-grade chronic inflammation are closely involved in the pathogenesis of type 2 diabetes mellitus. The innate immune system modulates the effects of many factors including genes, ethnicity, fetal programming, nutrition and age on the later development of metabolic sequelae associated with insulin resistance [2].

Several cross-sectional studies in the general population, as well as in non-diabetic subjects, in individuals with impaired glucose tolerance/impaired fasting glucose, and in newly diagnosed or established type 2 diabetic patients have shown that diverse inflammatory molecules, such as C-reactive protein, sialic acid, tumor necrosis factor-α (TNF-α), and interleukin-6 (IL-6), are positively correlated with measures of insulin resistance [3], with diabetic patients exhibiting elevated levels of these parameters compared with nondiabetic subjects [4]. In addition, different prospective epidemiological studies have confirmed that circulating inflammatory markers, acute-phase reactants, and proinflammatory cytokines are strongly associated with the risk of developing type 2 diabetes [5, 6].

Inflammation and Diabetic Nephropathy

From a pathophysiologic point of view, the intimate mechanisms leading from chronic hyperglycemia to the development of renal injury in diabetes mellitus are complex and in many aspects are not well known. Both experimental and clinical studies have demonstrated the significant role of diverse inflammatory molecules and pathways in the setting on DN (table 1). The interplay between inflammatory and metabolic abnormalities leads to tissue damage in diabetes, and therefore the identification of the molecules and pathways implicated in these harmful processes will offer new opportunities for prevention, early diagnosis and targeted therapy.

The Polyol-Aldose Reductase Pathway

Hyperglycemia is a key point in the pathogenesis of diabetic microvascular complications, with poor glycemic control being an independent predictor of development and progression of DN. The polyol pathway of glucose metabolism becomes active when intracellular glucose levels are elevated. Aldose reductase (AR) is the first and rate-limiting enzyme in the pathway, which reduces glucose to sorbitol using NADPH as a cofactor; sorbitol is then metabolized to fructose by sorbitol dehydrogenase that uses NAD+ as a cofactor.

Table 1. Inflammatory molecules and mediators involved in the development and progression of DN

Chemokines
MCP-1
Fractalkine
Regulated upon activation, normal T cell expressed and secreted (RANTES)
Monokine induced by interferon-γ
Interferon-γ-inducible protein (IP-10)

Adhesion molecules
ICAM-1
VCAM-1
Endothelial cell-selective adhesion molecule
E-selectin (CD62E) and P-selectin (CD62P)
α-Actinin-4

Transcription factors
Nuclear factor-κB
Upstream stimulatory factors 1 and 2
Activator protein 1
cAMP response element binding protein
Nuclear factor of activated T cells
Stimulating protein 1

Inflammatory cytokines
IL-1, -6 and -18
TNF-α
Suppressors of cytokine signalling

Growth factors
TGF-β
CTGF
Hepatocyte growth factor
VEGF
Platelet-derived growth factor
Pigment epithelium-derived factor
Midkine

In the hyperglycemic state, the affinity of AR for glucose rises, causing much sorbitol to accumulate, and using much more NADPH, leaving less NADPH for other processes of cellular metabolism. Activation of AR enzyme itself causes damage, as well as through other mechanisms such as activation of protein kinase C (PKC) and protein glycosylation. Moreover, excessive activation of the polyol pathway increases intracellular and extracellular sorbitol levels, concentration of reactive oxygen species (ROS), and decreases concentrations of nitric oxide (NO) and glutathione [7]. AR, in addition to a glucose-reducing enzyme via polyol pathway, is believed to be an important component of the antioxidant defense system, as well as a key mediator of oxidative stress-induced inflammation, with the participation of downstream inflammatory molecular signaling resulting in final cytotoxic events [8].

Protein Kinase C

The PKCs are a family of serine/threonine kinases that act in the regulation of signal transduction in a wide variety of cell types, regulating vascular function and contractility, flow, cell proliferation and vascular permeability. Among the different isoforms, PKCβ is important in endothelial cells and is activated by diacylglycerol under conditions of increased glucose and fatty acid concentrations. Activation of this isoform has been suggested as the link between inflammation, endothelial dysfunction, and insulin resistance in diabetes mellitus [9].

Effects of PKCβ activation are diverse: it contributes to endothelial dysfunction by reducing NO synthase phosphorylation, activates nuclear factor-κB, one of the most relevant transcription factors linked to inflammation, and stimulates the production of prostaglandin E_2 and thromboxane A_2, which alter the permeability and the response to angiotensin II in vascular smooth muscle, leading to kidney damage [9, 10].

Advanced Glycosylation End Products

Advanced glycation end products (AGEs) is a class of complex products that result from a reaction between carbohydrates and free amino group of proteins. The AGEs are the result of glycoxidation, and represent very unstable and reactive compounds. The formation of AGEs, also called the Maillard reaction, is a succession of complex chemical reactions with dehydration and molecular rearrangement which are linked in a complicated network, finally resulting in cross-linked proteins.

Formation of AGEs alters the structure and function of proteins in plasma, arterial wall, mesangium and glomerular basement membrane (GBM). Their binding to specific receptors (RAGEs) on the podocytes, endothelial and smooth muscle cells, mesangial and tubular epithelial cells determines the activation of intracellular signaling pathways that leads to the generation of ROS, activation of transcription factors, release of inflammatory cytokines, and the expression of adhesion molecules and growth factors [11]. Detailed discussion of AGEs is found in chapter 8 [see chapter by Thomas, this vol., pp. 66–74].

Oxidative Stress

Oxidative stress, a situation that occurs when production of oxidants or ROS exceeds local antioxidant capacity, is strongly implicated as a mediator of multiple diabetes-induced microvascular complications, including the development and progression of DN [12].

ROS seem to be closely related to various pathways, including inflammatory, which are central to the pathogenesis of hyperglycemia-induced microvascular and renal injury. Thus, oxidative stress results in oxidation of important macromolecules including proteins, lipids, carbohydrates, and DNA oxidation ensues. In the diabetic kidney, ROS derive from both enzymatic and nonenzymatic sources, including auto-oxidation of glucose, advanced glycation, polyol pathway flux, mitochondrial respiratory chain deficiencies, xanthine oxidase activity, peroxidases, NO synthase and NADPH oxidase [13]. In this setting, a direct relationship between the severity of renal injury and the degree of oxidative stress has been demonstrated histologically with the presence of glyco- and lipo-oxidation products in the mesangial matrix and nodular lesions of DN [14].

Growth Factors

Progression of DN is characterized by excessive amassing of extracellular matrix with thickening of glomerular and tubular basement membranes and increased amount of mesangial matrix, which ultimately progress to glomerulosclerosis and tubulointerstitial fibrosis with gradual scarring of both the renal glomerulus and tubulointerstitial region. Diabetic milieu results in the activation of growth factor pathways, which are relevant mediators of renal injury progression. These growth factors, which are generally expressed in the normal kidney and are implicated in the control of renal matrix composition, cell hypertrophy, proliferation and survival, modulation of cells of the immune system, and enzymes involved in glucose metabolism, and whose levels increase in relation to diabetes, have been implicated in the pathogenesis of DN through complex intrarenal systems [15].

Transforming Growth Factor-β
Transforming growth factor-β (TGF-β) remains the most important cytokine for renal fibrogenesis, and it has been identified as a major stimulus in the process of epithelial-mesenchymal transition of tubular epithelial cells. Hyperglycemia induces sustained overexpression of TGF-β, whose action following inflammatory responses is characterized by increased production of extracellular matrix components, as well as mesenchymal cell proliferation, migration, and accumulation. Thus, TGF-β is a key factor in the induction of fibrosis often associated with chronic phases of inflammatory diseases [16].

The role of TGF-β in DN is based on the induction and maintenance of interstitial fibrosis due to its regulatory effect on cell proliferation and the synthesis and degradation of the extracellular matrix. The effect of TGF-β is mediated by the connective tissue growth factor (CTGF), a downstream mediator of TGF-β which promotes glomerular damage through increased production of extracellular matrix proteins and induction of changes in cytoskeletal structure [17].

In addition to these effects in fibrosis, current evidence shows that TGF-β also plays a main role in signaling abnormalities in diabetic podocytes. Reduced attachment of the podocytes to the GBM and activation of apoptotic pathways have been shown as two mechanisms related to the central effect of TGF-β in the development of podocytopenia in DN [18]. Detailed discussion of TGF-β is found in the chapter by Lan et al. [this vol., pp. 75–82].

Connective Tissue Growth Factor
CTGF is overexpressed in mesangial, epithelial, tubular and podocyte cells in response to diverse factors, such as hyperglycemia, AGEs, mechanical strain, inflammatory cytokines, TGF-β, as well as CTGF itself. Glomerular CTGF mRNA levels have been reported to be upregulated in both diabetic subjects with microalbuminuria or overt proteinuria, with mRNA expression levels significantly correlated with the degree of urinary albumin excretion.

After CTGF pathway activation, it is possible to detect changes in extracellular matrix composition and dynamics, including enhanced expression of fibronectin, increased fibronectic assembly into an insoluble matrix, increased collagen production (types I, III and IV), and facilitation of the deposition and assembly of extracellular matrix proteins. Other significant effects regarding kidney damage in DN are rearrangement of the actin cytoskeleton, induction of plasminogen activator inhibitor 1, and chemoattraction of peripheral blood mononuclear cells, which in turn may induce inflammation and fibrosis [19].

Vascular Endothelial Growth Factor
Vascular endothelial growth factor (VEGF) is an endothelium-specific growth factor whose primary function is to maintain the integrity and viability of the endothelium, and critically contributes to the physiological maintenance of the glomerular filtration barrier. VEGF has been related to endothelial cell proliferation, differentiation and survival, endothelium-dependent vasodilatation, and interstitial matrix remodeling. VEGF acts through specific membrane receptors (VEGR1 and VEGR2), which are found on periglomerular, glomerular, and peritubular endothelial cells, as well as on mesangial cells and podocytes, whereas VEGF expression in the kidney is most prominent in podocytes, tubular epithelial cells, and mesangial cells. Podocytes are the major renal source of VEGF. In addition to its central effect on endothelial cell physiology, VEGF also exerts an effect on podocytes. Secreted VEGF activates podocytes in an autocrine manner and, possibly through VEGFR1, induces collagen synthesis [20].

                                                  Navarro-González · Mora-Fernández

Inflammatory cytokines, elevated glucose, AGEs, and TGF-β represent a stimulus for VEGF production. In human DN, VEGF levels and downstream signaling are increased in the early stages of this complication, but then appear to be reduced below normal with progressive DN. The increase in the expression and activity of VEGF has been suggested to be detrimental because NO bioavailability is reduced in diabetes, leading to the uncoupling of the otherwise protective VEGF-NO signaling pathways [21]. It has recently been proved that this factor participates in processes of neovascularization in diabetic vasculopathy and glomerulosclerosis.

Chemokines and Adhesion Molecules

Growing evidence indicates that immune and inflammatory cells, including monocytes, macrophages, lymphocytes and neutrophils, are involved in the development and progression of DN. A critical factor in the participation of these cells in the pathogenesis of DN is the role chemoattractant cytokines (chemokines) and adhesion molecules. These elements mediate the attraction, recruitment, adhesion and migration of the immune and inflammatory cells from the blood into the renal tissue [22].

Intrarenal accumulation of macrophages is associated with renal injury in DN. Monocyte-specific chemokines, notably the monocyte chemoattractant protein-1 (MCP-1), are the most potent factor affecting recruitment and renal infiltration by these cells [23]. Clinical studies have shown that upregulation of MCP-1 in the kidney is associated with macrophage recruitment, urinary albumin excretion, tubulointerstitial damage and disease progression. In addition to monocytes and macrophages, activated T-lymphocytes and neutrophils have also been suggested to play a role in development and progression of DN.

Regarding cell adhesion molecules, the main elements linked to DN are intercellular adhesion molecule-1 (ICAM-1) and vascular cell adhesion molecule-1 (VCAM-1). ICAM-1 is involved in the activation and transmigration of leukocytes and macrophages from vessels to sites of inflammation. Early studies showed that expression of ICAM-1 increases in parallel with the progression of DN, whereas more recent works with gene-deficient mice have demonstrated this molecule is involved in the pathogenesis of DN [24]. Concerning VCAM-1, cross-sectional clinical studies have reported an elevation in circulating levels of this molecule in patients with DN, which may result from underlying systemic endothelial dysfunction, increased VCAM-1 production in damaged renal tubular or glomerular epithelial cells and/or decreased renal clearance of this molecule, depending on the stage of nephropathy. In addition, clinical prospective studies in type 2 diabetic patients have shown that markers of endothelial dysfunction and inflammatory activity, including VCAM-1, were strongly associated with increases in urinary albumin excretion during the 10-year follow-up [25].

Proinflammatory Cytokines

Cytokines are low-molecular-weight polypeptides that possess autocrine, paracrine and juxtacrine effects, which are relevant humoral mediators in a highly complex coordinate network regulating inflammatory and immune responses. In addition, they exert important pleiotropic actions and are cardinal effectors of the inflammatory and immune systems. These molecules play important roles in many physiological responses, and they are also involved in the pathophysiology of a range of diseases. The mainly reported cytokines involved in the pathogenesis of DN are IL-1, -6, -18, and TNF-α, which can be synthesized by infiltrating as well as by intrinsic renal cells, including endothelial, mesangial, epithelial and tubular cells [26].

IL-1 is upregulated in the diabetic kidney, and it has been involved in increased vascular endothelial permeability, proliferation of mesangial cells, enhanced matrix synthesis and prostaglandin production. IL-6 is elevated in patients with DN, being related to increased GBM width, an early lesion of diabetic glomerulopathy. This cytokine stimulates mesangial cell proliferation, enhances fibronectin expression, affects extracellular matrix dynamics, increases endothelial permeability, and it has been related to the progression of renal disease. Levels of IL-18 have also been reported to be increased in patients with type 2 diabetes, and they were related to the development of increased albuminuria, which was independently associated with both serum and urinary IL-18 levels. Finally, IL-18 has been suggested as a predictor of early renal dysfunction in type 2 diabetes mellitus.

In addition to the implications of several ILs, a great deal of attention has been paid to the role of TNF-α in the setting of DN. TNF-α, which is significantly overexpressed in the diabetic kidney, has been implicated in the development of diverse abnormalities relevant for the pathogenesis of DN [27, 28], including direct cytotoxicity to renal cells, activation of apoptosis and cell necrosis pathways, dysregulation of intraglomerular hemodynamics and glomerular filtration, alteration in the distribution of adhesion receptors participating in intercellular adhesion and intercellular junction, with the subsequent loss of vascular endothelial permeability. Furthermore, TNF-α participates in the hemodynamic misbalance between vasodilatory and vasoconstrictive mediators, resulting in alterations of intraglomerular blood flow and glomerular filtration rate, and is able to promote oxidative stress through activation of NADPH oxidase, leading to local production of ROS, which favors the disruption of the glomerular capillary wall leading to injury of the barrier function and the subsequent increase in permeability to albumin.

Experimental researches have demonstrated that urinary albumin excretion significantly correlates with renal expression levels and urinary TNF-α excretion. Moreover, the increase in TNF-α concentration in urine and renal interstitial fluid preceded the increase in albuminuria. Clinical studies have found

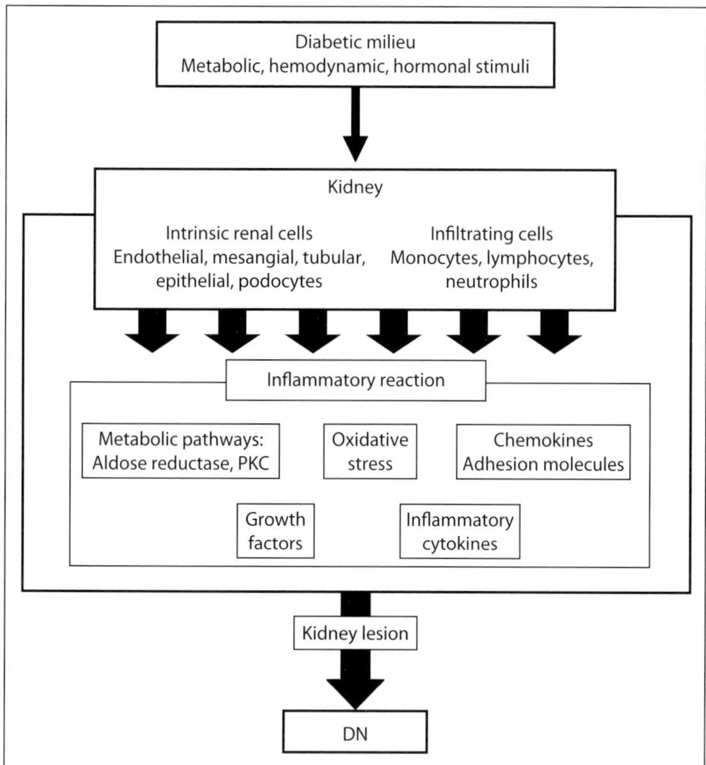

Fig. 1. Diverse factors from the diabetic milieu act on the kidney and trigger an inflammatory reaction with the participation and interplay of diverse molecules and pathways causing renal injury and leading to the development and progression of DN.

a direct and significant association between urinary TNF-α concentration and urinary albumin excretion in diabetic patients with normal renal function and microalbuminuria, as well as in subjects with overt nephropathy and renal insufficiency [29, 30].

Conclusions

Growing evidence indicates that inflammation is a key process in the development of diabetes mellitus, and that the proinflammatory milieu in diabetes contributes significantly to the development of DN (fig. 1). Different inflammatory molecules and pathways interact in different cell types finally leading to renal injury, responsible for the development and progression of DN. Better knowledge of the role of inflammatory molecules and processes in the pathogenesis of DN will facilitate the development of new and improved therapeutic targets and strategies that can be translated successfully into clinical applications for the treatment of this complication.

References

1 Atkins RC: The epidemiology of chronic kidney disease. Kidney Int 2005;67(suppl 94):S14–S18.

2 Pickup J, Crook M: Is type II diabetes mellitus a disease of the innate immune system? Diabetologia 1998;41:1241–1248.

3 Temelkova-Kurktschiev T, Henkel E, Koelher C, Karrei K, Hanefield M: Subclinical inflammation in newly detected type II diabetes and impaired glucose tolerance. Diabetologia 2002;45:151.

4 Pickup JC, Chusney GC, Thomas SM, Burt D: Plasma interleukin-6, tumor necrosis factor alpha and blood cytokine production in type 2 diabetes. Life Sci 2000;67:291–300.

5 Pradhan AD, Manson JE, Rifai N, Buring JE, Ridker PM: C-reactive protein, interleukin 6, and risk of developing type 2 diabetes mellitus. JAMA 2001;286:327–334.

6 Spranger J, Kroke A, Möhlig M, Hoffman K, Bergman MM, Ristow M, Boeing H, Pfeiffer AFH: Inflammatory cytokines and the risk to develop type 2 diabetes: results of the prospective population-based European Prospective Investigation into Cancer and Nutrition (EPIC) Postsdam study. Diabetes 2003;52:812–817.

7 Ramana KV, Friedrich B, Tammali R, West MB, Bhatnagar A, Srivastava SK: Requirement of aldose reductase for the hyperglycemic activation of protein kinase C and formation of diacylglycerol in vascular smooth muscle cells. Diabetes 2005;54: 818–829.

8 Ramana KV, Srivastrava SK: Aldose reductase: a novel therapeutic target for inflammatory pathologies. Int J Biochem Cell Biol 2010;42:17–20.

9 Das EN, King GL: The role of protein kinase C activation and the vascular complications of diabetes. Pharmacol Res 2007;55:498–510.

10 Way KJ, Katai N, King GL: Protein kinase C and the development of diabetic vascular complications. Diabetes Med 2001;18: 945–959.

11 Bierhaus A, Nawroth PP: Multiple levels of regulation determine the role of the receptor for AGE (RAGE) as common soil in inflammation, immune responses and diabetes mellitus and its complications. Diabetologia 2009;52:2251–2263.

12 Forbes JM, Coughlan MT, Cooper ME: Oxidative stress as a major culprit in kidney disease in diabetes. Diabetes 2008;57: 1446–1454.

13 Giacco F, Brownlee M: Oxidative stress and diabetic complications. Circ Res 2010;107:1058–1070.

14 Vasavada N, Agarwal R: Role of oxidative stress in diabetic nephropathy. Adv Chronic Kidney Dis 2005;12:146–154.

15 Chiarelli F, Gaspari S, Marcovecchio ML: Role of growth factors in diabetic kidney disease. Horm Metab Res 2009;41:585–593.

16 Pohlers D, Brenmoehl J, Löffler I, Müller CK, Leipner C, Schultze-Mosgau S, Stallmach A, Kinne RW, Wolf G: TGF-beta and fibrosis in different organs – molecular pathway imprints. Biochim Biophys Acta 2009; 1792:746–756.

17 Qi W, Chen X, Poronink P, Pollock CA: Transforming growth factor-beta/connective tissue growth factor axis in the kidney. Int J Biochem Cell Biol 2008;40:9–13.

18 Ziyadeh FN, Wolf G: Pathogenesis of the podocytopatia and proteinuria in diabetic glomerulopathy. Curr Diab Rev 2008;4: 39–45.

19 Schmidt-Ott KM: Unraveling the role of connective tissue growth factor in diabetic nephropathy. Kidney Int 2008;73:375–376.

20 Schrijvers BF, Flyvbjerg A, De Vriese AS: The role of vascular endothelial growth factor (VEGF) in renal pathophysiology. Kidney Int 2004;65:2003–2007.

21 Nakagawa T: Uncoupling of VEGF with NO as a mechanism for diabetic nephropathy. Diabetes Res Clin Pract 2008;13(suppl 1): S67–S69.

Navarro-González · Mora-Fernández

22 Ruster C, Wolf G: The role of chemokines and chemokine receptors in diabetic nephropathy. Front Biosci 2008;13:944–955.

23 Chow F, Nikolic-Paterson DJ, Ozols E, Atkins RC, Rollin BJ, Tesch GH: Monocyte chemoattractant protein-1 promotes the development of diabetic renal injury in streptozotocin-treated mice. Kidney Int 2006;69:73–80.

24 Chow FY, Nikolic-Paterson DJ, Ozols E, Atkins RC, Tesch GH: Intercellular adhesion molecule-1 deficiency is protective against nephropathy in type 2 diabetic db/db mice. J Am Soc Nephrol 2005;16:1711–1722.

25 Stehouwer CDA, Gall MA, Twisk JWR, Knudsen E, Emeis JJ, Parving HH : Increased urinary albumin excretion, endothelial dysfunction, and chronic low-grade inflammation in type 2 diabetes: progressive, interrelated, and independently associated with risk of death. Diabetes 2002;51:1157–1165.

26 Navarro-González JF, Mora-Fernández C: The role of inflammatory cytokines in diabetic nephropathy. J Am Soc Nephrol 2008;19:433–442.

27 Navarro JF, Mora-Fernández C: The role of TNF-α in diabetic nephropathy: pathogenic and therapeutic implications. Cytokine Growth Factor Rev 2006;17:441–450.

28 Navarro-González JF, Jarque A, Muros M, Mora C, García J: Tumor necrosis factor-α as a therapeutic target for diabetic nephropathy. Cytokine Growth Factor Rev 2009;20: 165–173.

29 Kalantarinia K, Awas AS, Siragy HM: Urinary and renal interstitial concentrations of TNF-alpha increase prior to the rise in albuminuria in diabetic rats. Kidney Int 2003;64:1208–1213.

30 Navarro JF, Mora C, Muros M, García J: Urinary tumor necrosis factor-α excretion independently correlates with clinical markers of glomerular and tubulointerstitial injury in type 2 diabetic patients. Nephrol Dial Transplant 2006;21:3428–3434.

Juan F. Navarro-González, MD, PhD, FASN
Nephrology Service and Research Unit, University Hospital Nuestra Señora de Candelaria
ES–38010 Santa Cruz de Tenerife (Spain)
Tel. +34 922602061, E-Mail jnavgon@gobiernodecanarias.org

Lai KN, Tang SCW (eds): Diabetes and the Kidney.
Contrib Nephrol. Basel, Karger, 2011, vol 170, pp 124–134

Diabetic Tubulopathy: An Emerging Entity

Sydney C.W. Tang · Joseph C.K. Leung · Kar Neng Lai

Department of Medicine, University of Hong Kong, Queen Mary Hospital,
Hong Kong SAR, China

Abstract

In chronic glomerulopathic disease, renal function correlates more with the degree of tubulointerstitial injury than that of the glomerular lesions. Proteinuria may be one of the pathologic links between these two intrarenal compartments. It is apparent that the proximal tubular epithelial cell (PTEC) assumes a proinflammatory and profibrotic role during proteinuria in which the PTEC expresses a variety of chemokines and injury signals that culminate in progressive interstitial inflammation and fibrosis. During diabetes, other substrates including advanced glycation end products (AGEs), AGE intermediates, and high glucose (HG) may provoke the PTEC even further. Glycated albumin, but not the equivalent dose of bovine serum albumin (BSA), stimulates tubular IL-8 and ICAM-1 expression via NF-κB-, MAPK- and STAT-1-dependent pathways. Human biopsies of diabetic nephropathy (DN) reveal colocalization of AGE and ICAM-1 in proximal tubules. The biologically active carbonyl intermediates methylglyoxal-BSA-AGE and AGE-BSA upregulate tubular expression of CTGF, TGF-β, and VEGF, whereas carboxymethyllysine-BSA stimulates tubular expression of IL-6, CCL-2, CTGF, TGF-β, and VEGF via RAGE activation and NF-κB signal transduction. Hyperglycemia (30 mM), but not the equivalent dose of mannitol, promotes proinflammatory (IL-6 and CCL-2), profibrotic (TGF-β) and angiogenic (VEGF) responses in tubular cells via MAPK and PKC signaling and induces epithelial mesenchymal transition, which is TGF-β_1 mediated. It has recently been shown that toll-like receptor (TLR) is implicated in the diabetic kidney. In human DN biopsies and PTEC, TLR4 is upregulated and plays a permissive role in HG-induced IL-6 and CCL-2 overexpression and monocyte transmigration. In streptozotocin-induced rat DN and PTEC, TLR2 appears to be upregulated. Other novel mediators that become activated in PTEC exposed to HG include macrophage inflammatory protein-3-α, Krüppel-like factor 6 and thioredoxin-interacting protein, which may be attenuated by peroxisome proliferator-activated receptor-γ activation. Collectively, these phenomena suggest that the renal tubules are heavily involved in the pathogenesis of DN. These pathophysiologic responses may be collectively described as diabetic tubulopathy.

Importance of the Proximal Tubule in Progressive Renal Disease

In patients with chronic kidney disease, the extent of renal dysfunction is poorly associated with changes in glomerular morphology, whereas it correlates well with chronic tubulointerstitial injury [1]. This has shifted the interest in its pathogenesis to numerous other possible causes. For example, the pathogenic link between glomerular damage and tubulointerstitial lesions may be mediated via proteinuria. This is because the severity of tubulointerstitial injury correlates with the amount of proteinuria [2], and this is supported by observations in animal models of protein overload and experimental nephrotic syndrome [3] in which tubulointerstitial lesions are a hallmark. Clinically, the risk of progression to kidney failure associated with a given level of GFR is independently increased in patients with higher levels of proteinuria [4].

The role of the proximal tubular epithelial cell (PTEC) as a proinflammatory and profibrotic cell first came into light when in vitro studies by Zoja et al. [5] showed that serum proteins activated endothelin-1 secretion in PTEC. This was later confirmed by Burton et al. [6] who showed fibronectin oversecretion by PTEC when they were exposed to serum proteins. These observations fueled further research into the putative proinflammatory role of the PTEC in orchestrating tubulointerstitial inflammation. At least three components of serum proteins have been found to stimulate PTEC expression of a variety of chemokines. These include albumin, transferrin and immunoglobulins. Albumin and/or transferrin induce overexpression of complement C3 [7, 8], IL-8 [9], MCP-1 [10, 11], IL-6 [12], ICAM-1 [12], MIF [10], and RANTES [12, 13] in PTEC. These interactions are pivotal in the development of tubulointerstitial injury and progressive renal failure, as many glomerular disorders including diabetic nephropathy (DN) are characterized by different degrees of proteinuria.

Role of the Proximal Tubular Epithelial Cell in Diabetic Nephropathy

Although DN is not generally regarded as an inflammatory disorder, this view has changed because recent studies of human biopsy specimens and animal models have established glomerular macrophage accumulation [14] and juxtaglomerular T cell infiltration as some of the cardinal features of DN. Although glomerulosclerosis is a cardinal feature of DN, tubulointerstitial fibrosis also appears early in DN, and closely correlates with renal function decline [15]. Tubulointerstitial injury contributed significantly to renal failure through secretion of inflammatory cytokines and changes in the expression of cytoskeletal proteins [16]. Infiltrating monocytes, macrophages, mast cells and T cells have all been demonstrated predominantly in the interstitium of diabetic renal disease. These cells are densely packed with vasoactive peptides and growth factors and are thought to locally contribute to tissue injury. The authors coined the term diabetic tubulopathy to describe this phenomenon. At least two mediators have been considered as important in the pathogenesis of diabetic tubulopathy, namely advanced glycation end products (AGEs) and high glucose (HG). Their independent roles in stimulating an inflammatory and profibrotic phenotypic switch in PTEC will be discussed below.

Role of Advanced Glycation End Products in Diabetic Tubulopathy

Reducing sugars may react nonenzymatically with amino groups in proteins or lipids, resulting in oxidative and non-oxidative molecular rearrangements termed the Maillard reaction. Of importance in this reaction is the formation of reactive intermediate products known as α-dicarbonyls such as methylglyoxal and 3-deoxyglucosone that may ultimately lead to stable covalent adducts known as AGEs. In human, irreversible advanced glycation is a part of the aging process, which is markedly accelerated in diabetes due to hyperglycemia.

AGEs have been implicated in the pathogenesis of diabetic tubulopathy. Gugliucci and Bendayan [17] demonstrated that AGE-containing proteins injected into rats are reabsorbed by proximal tubules, suggesting that renal handling of AGEs is effected through uptake, either specifically or nonspecifically, by PTEC. Recent animal studies have localized specific binding of AGEs in the diabetic kidney primarily to the proximal tubule of the renal cortex [18].

Glycated albumin, but not the equivalent dose of bovine serum albumin (BSA), stimulates tubular IL-8 and ICAM-1 expression via NF-κB-, MAPK- and STAT-1-dependent pathways [19]. In human kidney biopsy studies, sections from subjects with a histologic diagnosis of DN show strong IL-8 and ICAM-1 signals in the proximal tubules, particularly toward the apical membrane, and peritubular capillaries [9, 19]. Furthermore, tubular ICAM-1 signals are associated with deposition of AGE and infiltration of CD45-positive leukocytes. On the other hand, sections from subjects with nondiabetic kidney diseases such as nonproliferative glomerulopathies (minimal change nephrotic syndrome and hypertensive nephrosclerosis) display weak ICAM-1 staining, while subjects with kidney histology of no or minor abnormality have minimal ICAM-1 signals in the tubules (fig. 1). Urinary IL-8 levels in type 2 diabetic subjects correlate with the severity of diabetes [20]. Although there has not been a model of IL-8-deficient diabetic animal for analysis, the strategic role of IL-8 in vivo may be inferred indirectly from animal studies where the administration of a neutralizing anti-IL-8 antibody prevented albuminuria and glomerular infiltration of neutrophils [21]. In ICAM-1-deficient diabetic mice, there is markedly reduced interstitial leukocyte infiltration, tubular damage, interstitial fibrosis,

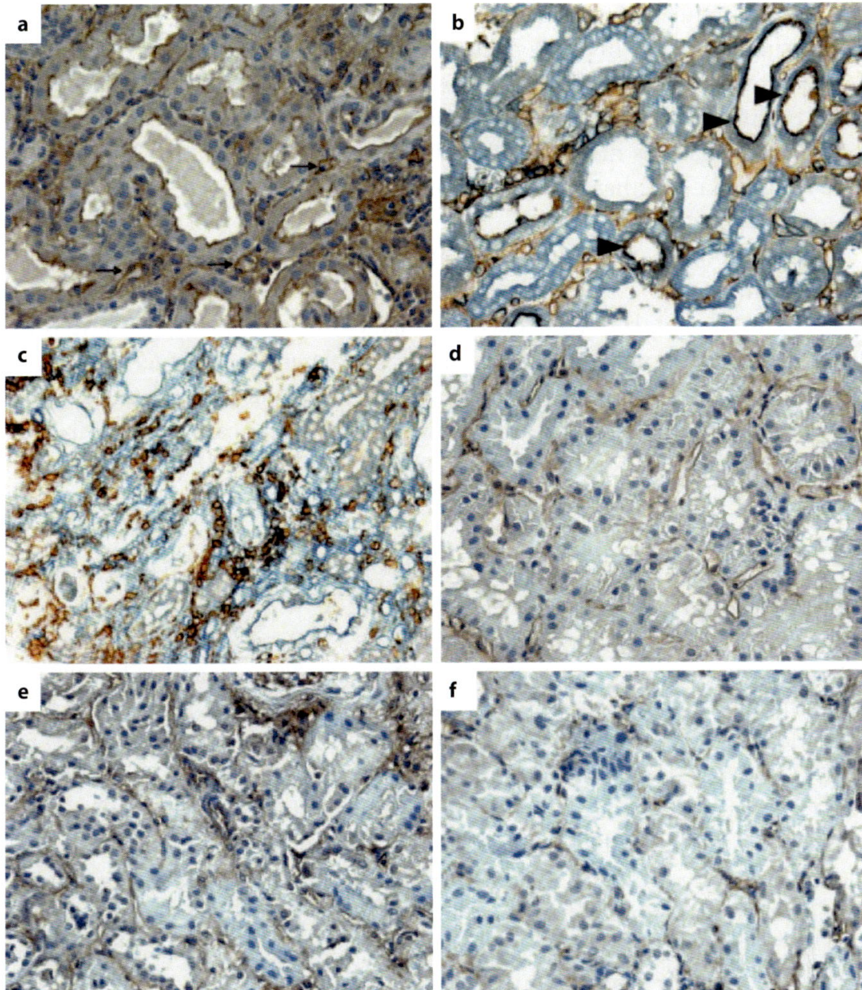

Fig. 1. Immunohistochemical staining for ICAM-1 in human renal tissue [19]. **a** Sections from subjects with DN showed intense signals in proximal tubules, particularly toward the apical membrane, and in peritubular capillaries (arrows). **b** Double immunohistochemical staining for AGE and ICAM-1 showed colocalization of AGE (in blue) and ICAM-1 (in brown) signals (arrowheads) in the tubules. **c** Double immunohistochemical staining for CD45-positive leukocytes and ICAM-1 demonstrated leukocyte infiltration (in brown) of the interstitial space in close association with ICAM-1-expressing tubules (in blue). Sections from subjects with hypertensive nephrosclerosis (**d**) and minimal change nephrotic syndrome (**e**) showed weak signals, while only minimal background staining was present in renal tissue with minor glomerular abnormality (**f**). ×100, all counterstained with hematoxylin.

albuminuria, and better renal function compared with their ICAM-1-intact counterparts [22, 23]. In *db/db* mice, there is upregulated tubular expression of ICAM-1 and MIP-2 (the murine equivalent of human IL-8) [24]. Collectively, these findings suggest that the infiltration of leukocytes into the renal interstitial space in diabetes may be mediated by sequential binding to ICAM-1 and IL-8 release that together promote rolling, arrest, firm adhesion, and ultimately, transmigration.

Apart from these molecules, IL-6 and CCL-2 are also known to play pivotal roles in DN [20, 25–27], but they were not induced by glycated albumin in PTEC. This implies that mediators other than GA may be involved. These as yet undefined mediators may include some of the biologically active carbonyl intermediates of AGEs. Indeed, methylglyoxal-BSA-AGE and AGE-BSA upregulate tubular expression of CTGF, TGF-β, and VEGF, whereas carboxymethyllysine-BSA stimulates tubular expression of IL-6, CCL-2, CTGF, TGF-β, and VEGF via RAGE activation and NF-κB signal transduction [28].

Role of High Glucose in Diabetic Tubulopathy

Hyperglycemia is the main metabolic stressor in diabetes and is believed to be an important causative factor in the pathogenesis of DN. HG exerts both proinflammatory and profibrotic effects on PTEC. In vitro, exposure of PTEC to HG (30 mM) promotes proinflammatory (IL-6 and CCL-2), profibrotic (TGF-β) and angiogenic (VEGF) responses in tubular cells (fig. 2). These effects are partly mediated via MAPK and PKC signaling [29]. In vivo, IL-6, TGF-β, and VEGF are heavily expressed in the tubules of diabetic kidney but not in kidney specimens from nondiabetic control subjects without kidney disease.

A novel receptor through which HG induces tubular IL-6 and CCL-2 expression is the toll-like receptor (TLR), a conserved family of pattern recognition receptors that play a fundamental role in the innate immune system by triggering proinflammatory signaling pathways in response to microbial pathogens. In support of the proinflammatory role of TLRs in diabetes, Dasu et al. [30] observed an upregulation and activation of TLR2 and TLR4 and their ligands in circulating monocytes of recently diagnosed type 2 diabetic subjects. Differential expression of TLR2 and TLR4 has been observed in human DN and rat models of DN. In humans, TLR4 but not TLR2 is highly expressed in renal tubules of DN kidneys compared with normal controls [31]. Tubular TLR4 expression correlated positively with infiltration of CD68+ cells in the interstitial space and HbA1c, and negatively with eGFR at the time of kidney biopsy. In primary cultured human PTEC, HG but not mannitol induces TLR4 overexpression (fig. 3), resulting in upregulation of IL-6 and CCL-2 expression via IκB/NF-κB activation. Molecular silencing of TLR4 in PTEC with siRNA attenuates HG-induced

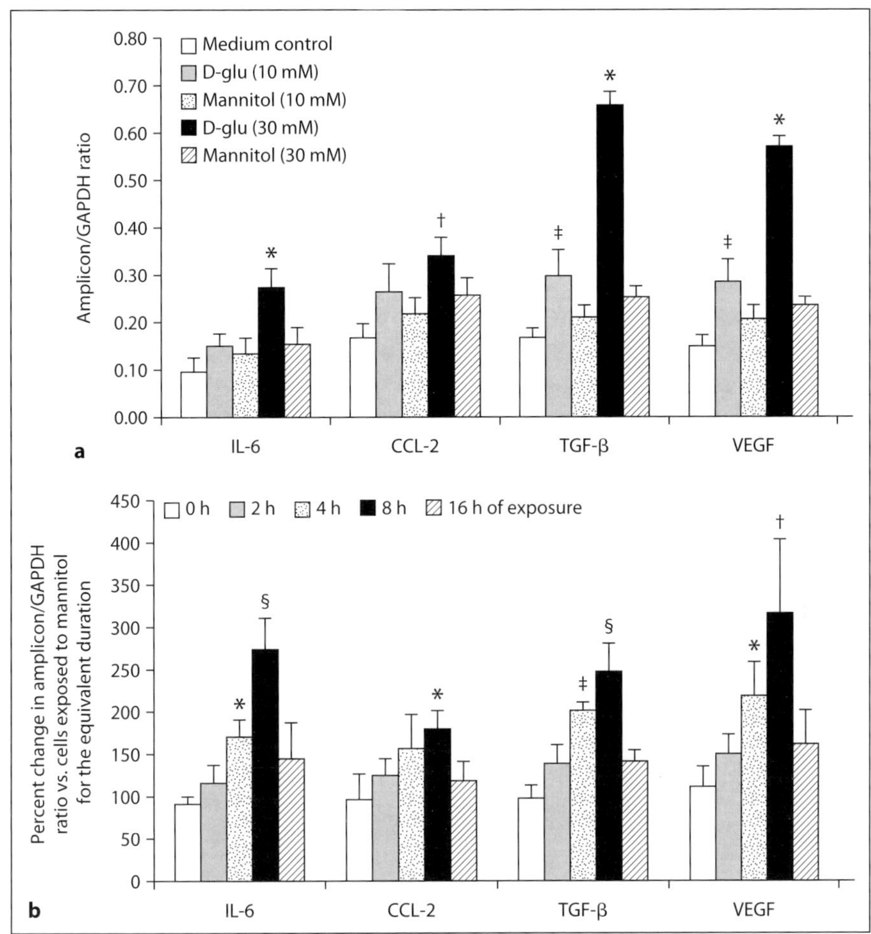

Fig. 2. Tubular expression of IL-6, CCL-2, TGF-β, and VEGF mRNA after acute exposure to D-glucose or mannitol [29]. **a** Dose response at 6 h of incubation. * p < 0.001, † p = 0.008, ‡ p = 0.003, § p = 0.004 compared with growth-arrested cells incubated in serum-free medium alone. **b** Time response after exposure to 30 mM D-glucose. * p < 0.05, † p < 0.005, ‡ p = 0.001, § p < 0.001 compared with baseline control. Results are means ± SD obtained from triplicate experiments.

IκB/NF-κB activation, the associated downstream IL-6 and CCL-2 synthesis, and impairs the ability of PBMC/U937 mononuclear cell transmigration induced by HG-treated PTEC-conditioned media. On the other hand, there is upregulation of TLR2 in the kidneys of streptozotocin (STZ)-treated rats. In rat NRK-52E tubular cell line, HG induces the expression of TLR2 mRNA [32]. Several factors may contribute to such divergent findings, including a generic difference between human and rodent DN pathologies, disease duration in human versus

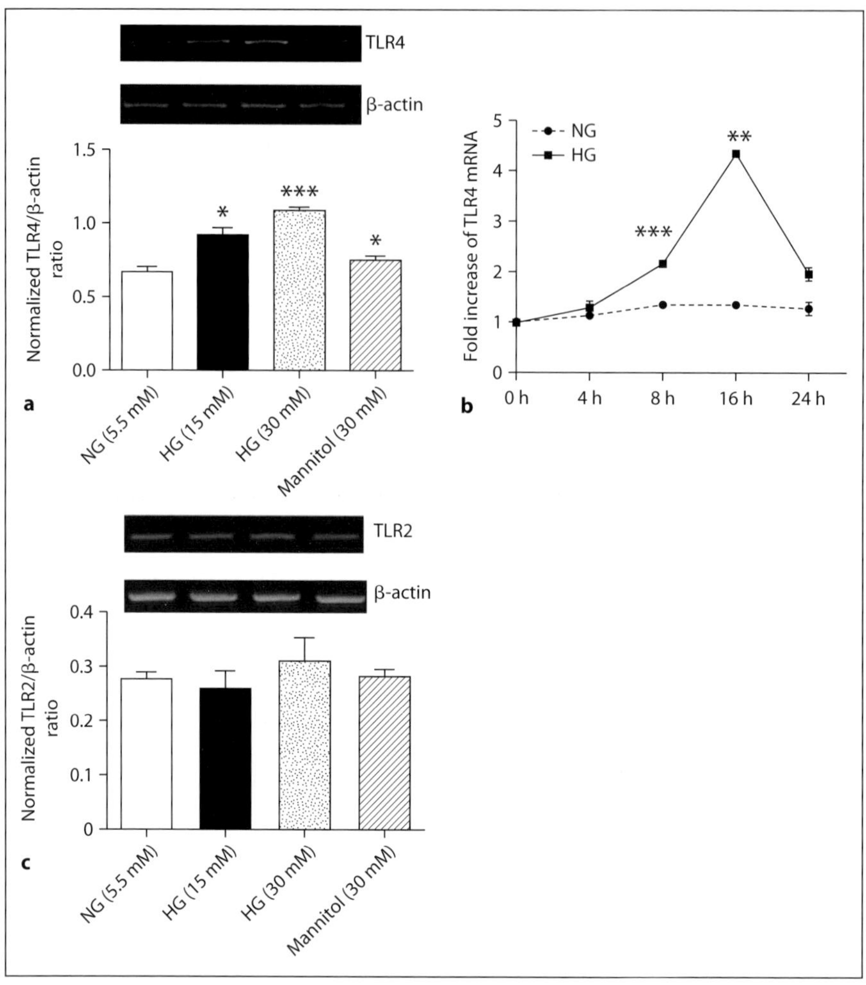

Fig. 3. Effect of high ambient glucose on TLR4 and TLR2 mRNA expression in PTEC [31]. Dose effect of glucose on mRNA expression of TLR4 (**a**) and TLR2 (**c**). PTEC were incubated with increasing doses of ambient glucose (from 5.5 to 30 mM) for 8 h. Expression of TLR4 and TLR2 was determined by semiquantitative PCR, and their values were normalized to β-actin as loading control. * $p < 0.05$, *** $p < 0.001$ versus PTEC cultured with normal glucose (NG) media. Representative images of the corresponding PCR product are shown at the top. **b** Time effect of HG on mRNA expression of TLR4. PTEC were incubated with high ambient glucose (30 mM) for 0–24 h. Expression of TLR4 was determined by real-time PCR and the values were normalized to cyclophilin A. ** $p < 0.01$, *** $p < 0.001$ versus time zero control. All results represent means ± SD obtained from five independent experiments.

animal models, type 2 DN in human versus type 1 DN in STZ-treated rats, use of primary culture versus cell lines in in vitro studies, and the semiquantitative nature of immunohistochemical staining.

Macrophage inflammatory protein-3α (MIP-3α), first identified and cloned in 1997 and also known as chemokine (C-C motif) ligand 20, attracts memory T lymphocytes. Chemokine receptor CCR6 is an exclusive receptor for MIP-3α, and the ligand-receptor (MIP-3α-CCR6) has been reported to chemoattract immature dendritic cells. Recent microarray analysis [33] suggests that HG upregulates MIP-3α expression when HK-2 cells (immortalized human PTEC cell line) are exposed to HG, and that this process is dependent on TGF-β_1. On the other hand, HG induces thioredoxin-interacting protein (Txnip), also known as vitamin D_3 and promotes intracellular oxidative stress, through a TGF-β_1-independent pathway in cultured PTEC.

Krüppel-like factor 6 (KLF6) is a DNA-binding protein containing a triple zinc-fingered motif, and plays a key role in the regulation of cell proliferation, differentiation, and development. More recently, it has been implicated in hepatic fibrosis via its binding to the TGF-β control element. HG induces KLF6 and epithelial mesenchymal transition (EMT) in PTEC, which is again TGF-β_1-mediated [34]. KLF6 overexpression in PTEC significantly promotes a phenotype consistent with EMT. In an in vivo model of DN (STZ-treated Ren-2 rats), KLF6 is also increased, suggesting a permissive role of KLF6 in promoting EMT in the diabetic kidney [34].

In terms of strategies to combat diabetic tubulopathy, peroxisome proliferator-activated receptor-γ agonists have been shown to be renoprotective. Rosiglitazone dose dependently attenuates GA-induced IL-8 and ICAM-1 signals in PTEC and completely abolishes GA-induced STAT-1 signals but has no effect on NF-κB and MAPK activation [19]. In uninephrectomized *db/db* mice, rosiglitazone but not metformin attenuates renal cortical expression of CCL2, MIP-2 and ICAM-1, inhibits p-STAT1 signal activation, and restores glomerular nephrin expression [24]. Both rosiglitazone and pioglitazone attenuate HG-induced Txnip [35] and MIP-3α [36] expression in HK-2 cells, and inhibit renal KLF6 and Txnip expression in rats with diabetes induced by STZ. In addition, both agents suppress renal MIP-3α and macrophage infiltration in these animals [36].

Conclusion

Emerging in vitro and in vivo evidence suggests that the proximal tubule plays a determining role in the pathogenesis and progression of DN. This mainly stems from the ability of the PTEC to secrete proinflammatory and profibrotic molecules and to undergo EMT upon stimulation by the various metabolic substrates of diabetes to effect a coherent succession of intracellular signaling events. Novel mediators including TLR, KLF6 and MIP-3α in PTEC have been identified. Therapeutic strategies targeting diabetic tubulopathy hold promise for the treatment of DN.

Acknowledgments

This study was supported by a General Research Fund of the Research Grants Council (grant number HKU 7764/07M) of Hong Kong, and JCKL was supported by the L&T Charitable Foundation and the House of INDOCAFE.

References

1   Nath KA: Tubulointerstitial changes as a major determinant in the progression of renal damage. Am J Kidney Dis 1992;20: 1–17.

2   Benigni A, Zoja C, Remuzzi G: The renal toxicity of sustained glomerular protein traffic. Lab Invest 1995;73:461–468.

3   Eddy AA, McCulloch L, Liu E, Adams J: A relationship between proteinuria and acute tubulointerstitial disease in rats with experimental nephrotic syndrome. Am J Pathol 1991;138:1111–1123.

4   Hemmelgarn BR, Manns BJ, Lloyd A et al: Relation between kidney function, proteinuria, and adverse outcomes. JAMA 2010;303: 423–429.

5   Zoja C, Morigi M, Figliuzzi M et al: Proximal tubular cell synthesis and secretion of endothelin-1 on challenge with albumin and other proteins. Am J Kidney Dis 1995;26: 934–941.

6   Burton CJ, Combe C, Walls J, Harris KP: Fibronectin production by human tubular cells: the effect of apical protein. Kidney Int 1996;50:760–767.

7   Tang S, Sheerin NS, Zhou W, Brown Z, Sacks SH: Apical proteins stimulate complement synthesis by cultured human proximal tubular epithelial cells. J Am Soc Nephrol 1999;10:69–76.

8   Tang S, Lai KN, Chan TM, Lan HY, Ho SK, Sacks SH: Transferrin but not albumin mediates stimulation of complement C3 biosynthesis in human proximal tubular epithelial cells. Am J Kidney Dis 2001;37:94–103.

9   Tang S, Leung JC, Abe K, et al: Albumin stimulates interleukin-8 expression in proximal tubular epithelial cells in vitro and in vivo. J Clin Invest 2003;111:515–527.

10 Tang S, Leung JC, Tsang AW, Lan HY, Chan TM, Lai KN: Transferrin up-regulates chemokine synthesis by human proximal tubular epithelial cells: implication on mechanism of tubuloglomerular communication in glomerulopathic proteinuria. Kidney Int 2002;61:1655–1665.

11 Wang Y, Chen J, Chen L, Tay YC, Rangan GK, Harris DC: Induction of monocyte chemoattractant protein-1 in proximal tubule cells by urinary protein. J Am Soc Nephrol 1997;8:1537–1545.

12 Lai KN, Leung JC, Chan LY, Guo H, Tang SC: Interaction between proximal tubular epithelial cells and infiltrating monocytes/T cells in the proteinuric state. Kidney Int 2007;71:526–538.

13 Zoja C, Donadelli R, Colleoni S, et al: Protein overload stimulates RANTES production by proximal tubular cells depending on NF-kappa B activation. Kidney Int 1998;53: 1608–1615.

14 Chow F, Ozols E, Nikolic-Paterson DJ, Atkins RC, Tesch GH: Macrophages in mouse type 2 diabetic nephropathy: correlation with diabetic state and progressive renal injury. Kidney Int 2004;65:116–128.

15 Gilbert RE, Cooper ME: The tubulointerstitium in progressive diabetic kidney disease: more than an aftermath of glomerular injury? Kidney Int 1999;56:1627–1637.

16 Sanai T, Sobka T, Johnson T, et al: Expression of cytoskeletal proteins during the course of experimental diabetic nephropathy. Diabetologia 2000;43:91–100.

17 Gugliucci A, Bendayan M: Renal fate of circulating advanced glycated end products (AGE): evidence for reabsorption and catabolism of AGE-peptides by renal proximal tubular cells. Diabetologia 1996;39:149–160.

18 Youssef S, Nguyen DT, Soulis T, Panagiotopoulos S, Jerums G, Cooper ME: Effect of diabetes and aminoguanidine therapy on renal advanced glycation end-product binding. Kidney Int 1999;55:907–916.

19 Tang SC, Leung JC, Chan LY, Tsang AW, Lai KN: Activation of tubular epithelial cells in diabetic nephropathy and the role of the peroxisome proliferator-activated receptor-gamma agonist. J Am Soc Nephrol 2006;17:1633–1643.

20 Tashiro K, Koyanagi I, Saitoh A, et al: Urinary levels of monocyte chemoattractant protein-1 (MCP-1) and interleukin-8 (IL-8), and renal injuries in patients with type 2 diabetic nephropathy. J Clin Lab Anal 2002;16:1–4.

21 Wada T, Tomosugi N, Naito T, et al: Prevention of proteinuria by the administration of anti-interleukin 8 antibody in experimental acute immune complex-induced glomerulonephritis. J Exp Med 1994;180:1135–1140.

22 Okada S, Shikata K, Matsuda M, et al: Intercellular adhesion molecule-1-deficient mice are resistant against renal injury after induction of diabetes. Diabetes 2003;52:2586–2593.

23 Chow FY, Nikolic-Paterson DJ, Ozols E, Atkins RC, Tesch GH: Intercellular adhesion molecule-1 deficiency is protective against nephropathy in type 2 diabetic db/db mice. J Am Soc Nephrol 2005;16:1711–1722.

24 Tang SC, Leung JC, Chan LY, Cheng AS, Lan HY, Lai KN: Renoprotection by rosiglitazone in accelerated type 2 diabetic nephropathy: role of STAT1 inhibition and nephrin restoration. Am J Nephrol 2010;32:145–155.

25 Chow FY, Nikolic-Paterson DJ, Ozols E, Atkins RC, Rollin BJ, Tesch GH: Monocyte chemoattractant protein-1 promotes the development of diabetic renal injury in streptozotocin-treated mice. Kidney Int 2006;69:73–80.

26 Morcos M, Sayed AA, Bierhaus A, et al: Activation of tubular epithelial cells in diabetic nephropathy. Diabetes 2002;51:3532–3544.

27 Stenvinkel P, Ketteler M, Johnson RJ, et al: IL-10, IL-6, and TNF-alpha: central factors in the altered cytokine network of uremia – the good, the bad, and the ugly. Kidney Int 2005;67:1216–1233.

28 Tang SC, Chan LY, Leung JC, et al: Differential effects of AGEs on renal tubular inflammation. Nephrology 2010, Epub ahead of print.

29 Tang SC, Chan LY, Leung JC, et al: Bradykinin and high glucose promote renal tubular inflammation. Nephrol Dial Transplant 2010;25:698–710.

30 Dasu MR, Devaraj S, Park S, Jialal I: Increased toll-like receptor (TLR) activation and TLR ligands in recently diagnosed type 2 diabetic subjects. Diabetes Care 2010;33:861–868.

31 Lin M, Au WS, Chan LY, et al: Toll-like receptor 4 mediates high glucose-induced tubular inflammation in diabetic nephropathy. J Am Soc Nephrol 2010;21:204A.

32 Li F, Yang N, Zhang L, et al: Increased expression of toll-like receptor 2 in rat diabetic nephropathy. Am J Nephrol 2010;32:179–186.

33 Qi W, Chen X, Gilbert RE, et al: High glucose-induced thioredoxin-interacting protein in renal proximal tubule cells is independent of transforming growth factor-beta1. Am J Pathol 2007;171:744–754.

34 Holian J, Qi W, Kelly DJ, et al: The Role of Kruppel-like factor 6 in transforming growth factor-beta-1-induced epithelial-mesenchymal transition of proximal tubule cells. Am J Physiol Renal Physiol 2008;295:F1388–F1396.

35 Qi W, Chen X, Holian J, Tan CY, Kelly DJ, Pollock CA. Transcription factors Kruppel-like factor 6 and peroxisome proliferator-activated receptor-gamma mediate high glucose-induced thioredoxin-interacting protein. Am J Pathol 2009;175:1858–1867.

36 Qi W, Holian J, Tan CY, Kelly DJ, Chen XM, Pollock CA: The roles of Kruppel-like factor 6 and peroxisome proliferator-activated receptor-gamma in the regulation of macrophage inflammatory protein-3alpha at early onset of diabetes. Int J Biochem Cell Biol 2011;43:383–392.

Prof. Sydney C.W. Tang, MD, PhD
Department of Medicine, University of Hong Kong
Hong Kong SAR (China)
Tel. +852 2255 4777, Fax +852 2816 2863, E-Mail scwtang@hkucc.hku.hk

 Tang · Leung · Lai

Lai KN, Tang SCW (eds): Diabetes and the Kidney.
Contrib Nephrol. Basel, Karger, 2011, vol 170, pp 135–144

The Renin-Angiotensin System

Kar Neng Lai · Joseph C.K. Leung · Sydney C.W. Tang

Department of Medicine, University of Hong Kong, Queen Mary Hospital,
Hong Kong SAR, China

Abstract

Diabetic nephropathy (DN) is a leading cause of end-stage renal disease in developed countries where type 2 diabetes mellitus has reached epidemic proportions. Although the exact pathogenesis of DN is not fully understood and is likely diverse in nature, there are convincing data that the renin-angiotensin system (RAS) is a major mediator of renal injury. Angiotensin II (Ang II), traditionally playing a central role as a mediator of glomerular hemodynamic adaptation and injury, is now recognized to exert proinflammatory action leading to upregulation of chemokines, adhesion molecules, and other fibrogenic growth factors that culminate in a decline of renal function. Hyperglycemia and mechanical stress deriving from glomerular hypertension are the key factors underlying pathogenesis of DN. The common signaling pathways stimulated by high glucose and mechanical insult may act synergistically, thereby accelerating the cell damage. Podocytes are subjected not only to the load of filtered glucose but also to diverse mechanical forces. Both high glucose and mechanical stress may impair the protein systems anchoring the podocyte foot processes in the glomerular basement membrane, therefore blunting resistance of these cells to mechanical forces in addition to the inflammatory insults. Loss of the podocytes is irreversible due to their inability to proliferate and to replenish damaged cells. Podocytes are injured early in the course of DN, which, most likely, underlies further glomerular and renal damage in diabetes. Under normal physiological conditions, podocytes play a specific role in the maintenance of intraglomerular RAS balance with enzymatic activities that predominantly lead to ANG1–7 and ANG1–9 formation, as well as Ang II degradation. ANG1–7 counteracts the proinflammatory actions of Ang II. These enzymatic activities are altered in a nonphysiological environment such as hyperglycemia that mimics diabetic kidney disease. An understanding of the local intraglomerular RAS will provide a novel approach for early stages of DN.

The renin-angiotensin system (RAS) has been implicated in the development of progressive glomerulosclerosis in diabetic and nondiabetic nephropathies.

Traditionally, angiotensin II (Ang II) plays a central role as a mediator of glomerular hemodynamic adaptation and injury. Ang II also plays a pivotal role in glomerulosclerosis through induction of transforming growth factor-β (TGF-β) expression in mesangial cells [1]. TGF-β stimulates extracellular matrix protein synthesis, increases matrix protein receptors, and alters protease/protease-inhibitor balances, thereby inhibiting matrix degradation. In vitro studies show TGF-β induces tubular epithelial-myofibroblast transdifferentiation supporting its involvement in the formation and evolution of glomerular and tubular fibrosis. The beneficial effect of either angiotensin-converting enzyme inhibitor (ACEI) or Ang II subtype-1 receptor (ATR1) antagonist on proteinuria and creatinine clearance (independent of blood pressure reduction) further raises the importance of this system in diabetic nephropathy (DN).

Renin-Angiotensin System in the Kidney

Systemic RAS activation has long been shown to be related to hypertension and cardiovascular risk. The classical view of the RAS as a circulating endocrine system has recently evolved to organ- and tissue-based systems that perform paracrine/autocrine functions. Local RAS exists in different organs including the kidney, heart, pancreas and bone marrow. In the kidney, all of the RAS components are present in resident kidney cells [2]. Intrarenal Ang II is formed by independent multiple mechanisms.

In situ hybridization studies have demonstrated that the angiotensinogen gene is specifically present in the proximal tubules of the kidneys [3]. Angiotensinogen mRNA is expressed largely in proximal convoluted tubules and proximal straight tubules, and only small amounts are present in glomeruli and vasa recta [4]. In the kidney, angiotensinogen protein is specifically located in the proximal convoluted tubules by immunohistochemistry. A strong positive immunostaining for angiotensinogen protein is present in proximal convoluted tubules and proximal straight tubules, and there is weak positive staining in glomeruli and vasa recta; however, there is no staining in distal tubules or collecting ducts [5]. The synthesized angiotensinogen in the kidney is secreted into the lumen, leading to angiotensin I generation and subsequent formation of Ang II. Renin mRNA and renin-like activity are also present in cultured proximal tubular cells [6]. In addition, low but measurable renin concentrations in proximal tubule fluid have been reported in rats. Abundant expression of angiotensin-converting enzyme (ACE) mRNA and protein have also been shown to be present in brush borders of proximal tubules of human kidneys. ACE has also been measured in proximal and distal tubular fluid but is more abundant in proximal tubule fluid. Thus, conditions are present in proximal tubules for Ang II generation. In addition to the classic RAS pathways, (pro) renin receptors and chymase are also involved in local Ang II formation in the kidney. Moreover, circulating Ang II is actively internalized into proximal tubular cells by ATR1-dependent mechanisms. Consequently, Ang II is compartmentalized in the renal interstitial fluid and the proximal tubular compartments with much higher local concentrations than those existing in the circulation. Recent evidence has also revealed that inappropriate activation of the intrarenal RAS is an important contributor to the pathogenesis of hypertension and renal injury. Details of the intrarenal RAS and the differential regulation of Ang II are well summarized in a review by Kobori et al. [7]. Ang II also stimulates the production of aldosterone. Add-on therapy of selective aldosterone antagonist provides an additive antiproteinuric effect of ACEI in a randomized trial of type 2 diabetics with macroalbuminuria [8].

Finally, the recently cloned (pro)renin receptor is an exciting new addition to the RAS [9]. (Pro)renin binding to the (pro)renin receptor not only causes a nonproteolytic activation of (pro)renin leading to the activation of the RAS, but also stimulates the receptor's own intracellular signaling pathways independent of the RAS. Within the kidney, the (pro)renin receptor is present in the glomerular mesangium and podocytes, which play an important role in the maintenance of the glomerular filtration barrier. Therefore, (pro)renin-receptor blockers, which competitively bind to the receptor as a decoy peptide, have superior benefits with regard to proteinuria and glomerulosclerosis in experimental animal models with elevated plasma (pro)renin levels such as diabetes and hypertension compared with conventional RAS inhibitors, possibly by inhibiting both the nonproteolytic activation of (pro)renin and RAS-independent intracellular signals.

Angiotensin II Receptors in the Kidney

There are two major types of Ang II receptor ATR1 and ATR2. However, there is much less ATR2 expression in adult kidneys. It has been reported that ATR1 mRNA has been localized to proximal convoluted and straight tubules, thick ascending limbs of the loop of Henle, cortical and medullary collecting duct cells, glomeruli, arterial vasculature, vasa recta, and juxtaglomerular cells. Studies using polyclonal and monoclonal antibodies to the ATR1 demonstrated that ATR1 protein is present on vascular smooth muscle cells throughout the vasculature, including the afferent and efferent arterioles and mesangial cells. In addition, ATR1 is present on proximal tubule brush border and basolateral membranes, thick ascending limb epithelia, distal tubules, collecting ducts, glomerular podocytes, and macula densa cells. These findings suggest that the intrarenal RAS works independently of the systemic RAS.

Saturation analysis of specific binding of Ang II to different renal cells revealed unique percentage of ATR1 and ATR2 in mesangial cells, podocytes and proximal tubular epithelial cells (fig. 1) [10–12]. The predominance of ATR1

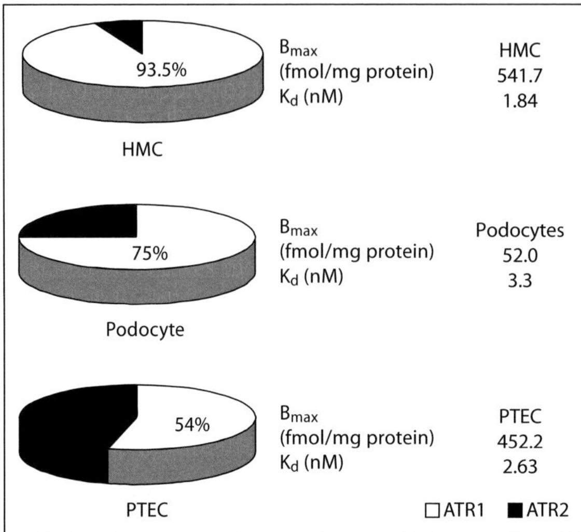

Fig. 1. Percentage of ATR1 and ATR2 and saturation analysis of specific binding of Ang II to different renal cells. B_{max} = Maximal density of specific binding sites; K_d = dissociation constant; HMC = human mesangial cells; PTEC = proximal tubular epithelial cells. Data adopted from Wang et al. [11], Chan et al [12], and unpublished data from Lai.

over ATR2 in mesangial cells and podocytes explains the previous observation that there is much less ATR2 expression in adult kidneys. With functional difference between these receptors, their differential distribution has important bearing in the pathophysiology and treatment by RAS blockade in various kidney disorders.

Renin-Angiotensin System Data from Murine Animals

Murine model has provided enormous data in the understanding of the importance of RAS in the development of salt-sensitive and genetic hypertension, and other renal diseases (summarized in a review by Kobori et al. [7]). However, murine data may not always be extrapolated to human being due to species difference.

In rodents, two subtypes of ATR1 mRNA (AT_{1a} receptor and AT_{1b} receptor) have been demonstrated in the vasculature and glomerulus and in all nephron segments [13]. The AT_{1a} receptor mRNA is the predominant subtype in nephron segments, whereas the AT_{1b} receptor is more abundant than AT_{1a} receptor in the glomerulus. Luminal AT_{1a} receptor mediates a biphasic regulation of bicarbonate absorption in proximal tubules by luminal Ang II in rodent, while no evidence is obtained for a role of ATR2. In contrast, Ang II induces apoptosis in

   Lai · Leung · Tang

proximal tubules (human and rodent) via ATR2 but not via ATR1. The human mast cell-specific protease 'chymase' efficiently converts angiotensin I (Ang I) to Ang II independent of ACE, whereas rat chymase and some mouse mast cell protease degrades Ang I to inactive fragments instead of Ang II.

Exogenous administration of Ang II elicits dose-dependent decreases in renal blood flow and glomerular filtration rate [14]. Although there is agreement that Ang II exerts substantial direct effects on the renal microvasculature and glomerular mesangium, there remains controversy regarding the intensity of actions at various sites and the relative contribution of systemically and intrarenally formed Ang II to the overall regulation of renal hemodynamics. The observation that Ang II increases the filtration fraction has frequently been used to support the notion that Ang II predominantly constricts the postglomerular arterioles. The plasma concentration of Ang II in rats receiving continuous administration of Ang II is 100-fold higher than that in anesthetized rats with values between 50 and 100 pm (10^{-10} M). Such concentration of Ang II is able to induce apoptosis in renal tubular epithelial cells. Hence, the pathophysiological findings in rats receiving super-pharmacological dosage of Ang II must be interpreted cautiously.

Intrarenal Expression of Renin-Angiotensin System in Diabetic Nephropathy

The RAS has been implicated in the progression of DN. Therapeutic blockade of the RAS slows the disease progression. For many years, ACEI and ATR1 antagonist were thought to improve the clinical outcome in diabetic patients via their blood pressure-lowering effects, acting, in part, to mitigate hyperfiltration-enhanced glomerular capillary pressure. Considerable evidence suggests RAS blockade also exerts antiproteinuric and anti-inflammatory effects in DN. In vitro studies show glucose increases the Ang II synthesis and induces the upregulation of (pro)renin receptor and its ligands (pro)renin and renin in mesangial cells [15], and enhances the angiotensinogen gene expression in tubular epithelial cells [16, 17]. Renal ACE activity and bradykinin levels are raised in nonobese diabetic mice [18]. There is also increased tubular ACE expression. Glucose also stimulates the TGF-β production in resident renal cells either directly or via Ang II which favors renal fibrosis [19]. Interestingly, complete blockade of the AT_{1a} receptor-mediated pathway has a minimal effect on the enhanced TGF-β/Smad signaling in the early stage of DN in knockout animal [19].

1,25-dihydroxyvitamin D_3 negatively regulates the RAS in mesangial cells and podocytes with renin and angiotensinogen as the main targets. The renoprotective mechanism of vitamin D analog in DN acts through suppressing hyperglycemia-induced angiotensinogen expression by blocking the NF-κB-mediated pathway [20].

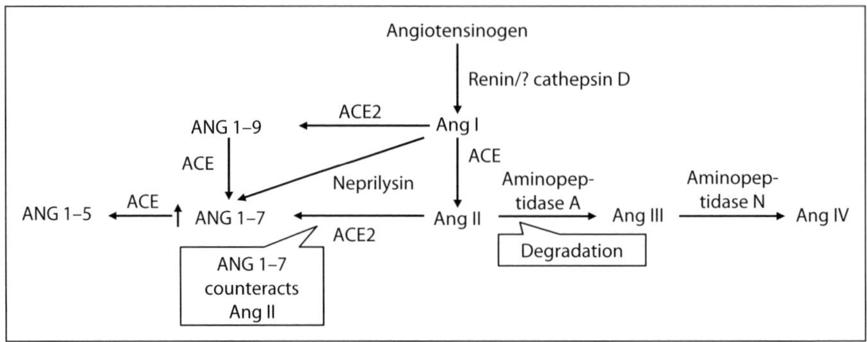

Fig. 2. Metabolism and enzymatic pathways of the RAS in podocytes.

Podocyte Expression of Renin-Angiotensin System in Diabetic Nephropathy

While glomerular hypertrophy, mesangial matrix expansion, and glomerular basement membrane thickening are classical hallmarks of DN, increasing evidence reveals that the onset of albuminuria is closely associated with podocytopathies, such as foot process effacement, podocyte hypertrophy, detachment, apoptosis and epithelial-myofibroblast transdifferentiation. Interestingly, detailed studies of the RAS and its enzymatic activities within podocytes only emerged recently [21]. Under normal physiological conditions, podocytes express a functional intrinsic RAS characterized by enzymatic activities that predominantly lead to ANG1–7 and ANG1–9 formation, as well as Ang II degradation (fig. 2). ANG1–7 counteracts the proinflammatory actions of Ang II. ACE2 is a novel monocarboxypeptidase member of the RAS family. As a homologue of ACE, it catalyzes the conversion of Ang II to ANG1–7. It appears that ACE2 activity buffers the deleterious actions of Ang II by limiting its levels and via the signaling of ANG1–7 through its cell surface Mas receptor [22]. It has been suggested that, in normal circumstances, podocytes play a specific role in the maintenance of intraglomerular RAS balance. Interestingly, glomerular ACE2 levels appear to be reduced in human [23] and rodent models of DN [24], which may allow for enhanced ATR1 signaling in podocytes.

Nonetheless, these enzymatic activities are altered in an in vitro nonphysiological environment such as hyperglycemia that mimics diabetic kidney disease (fig. 3). Indeed, podocytes are susceptible to the mechanical forces brought about by elevated glomerular capillary pressure. In vitro mechanical stretch, a mimic of elevated glomerular capillary pressure in vivo, renders podocytes more susceptible to apoptosis and actin cytoskeleton reorganization [24]. The local RAS in podocytes under a diabetic milieu (high glucose and mechanical stretch) induces apoptosis and TGF-β synthesis [25] with enhanced expression of

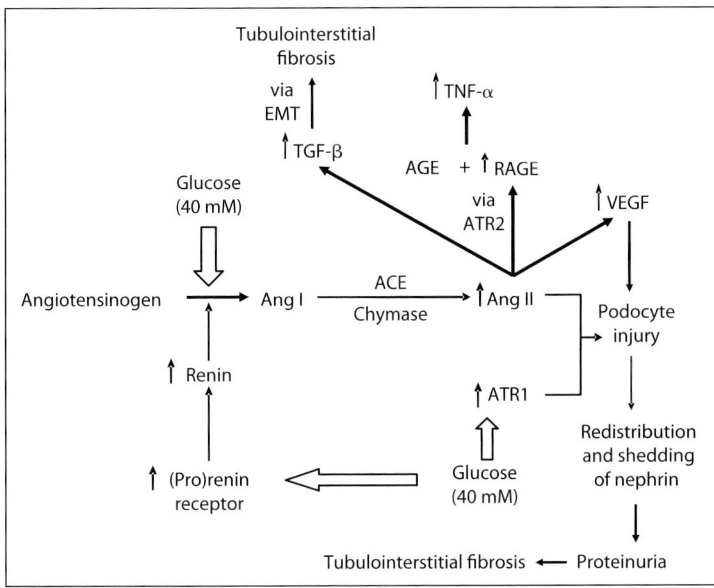

Fig. 3. Effect of hyperglycemia on the RAS in podocytes leading to tubulointerstitial fibrosis. EMT = Epithelial-myofibroblast transdifferentiation. Broad arrows represent the stimulatory effect of high glucose.

angiotensinogen gene as well as both (pro)renin and Ang II subtype-1 receptors [25–27]. Furthermore, high glucose enhances angiotensin I and II synthesis possibly through renin [19, 26, 27]. In vitro study shows that Ang II induces apoptosis in rat podocytes with upregulation of the expression of Fas, FasL, and Bax and downregulation of Bcl-2 expression [28]. Podocyte-derived vascular endothelial growth factor, a permeability and angiogenic factor whose expression is increased in diabetic kidney disease, is perhaps a major mediator of the increased protein filtration [29]. Proteinuria also occurs with reduced nephrin expression induced by Ang II. Proteinuria can then induce in tubular cells a genetic program leading to tubulointerstitial inflammation, fibrosis and tubular atrophy. Besides direct effects of albuminuria on tubular cells, pathophysiological changes in the ultrafiltration barrier lead to an increased tubular filtration of various growth factors (such as TGF-β and insulin-like growth factor I) that may further alter the function of tubular cells.

Both ATR1 and ATR2 are expressed in podocytes. Binding of Ang II to ATR2 induces the receptor for advanced glycation end product (AGE) expression which enhances tumor necrosis factor-α (TNF-α) production [30]. In addition, glycated albumin and Ang II reduce nephrin expression. Glycated albumin inhibits nephrin synthesis through the engagement of the receptor for AGEs (RAGE), whereas Ang II acts on cytoskeleton redistribution, inducing the shedding of nephrin [31]. Since intraglomerular Ang II levels are increased in DN, the interaction between the RAS and AGE/RAGE axis in podocytes is pivotal in the development of DN.

Gene Polymorphism of Renin-Angiotensin System in Diabetic Nephropathy

Despite the huge amount of studies looking for candidate genes, the ACE gene remains the unique, well-characterized locus clearly associated with progression of DN and with response to treatment with drugs that directly interfere with the RAS, such as ACEI and ATR1 antagonists [32]. The II genotype is protective against development and progression of type 1 and type 2 nephropathy. ACE inhibitors are particularly effective at the stage of normoalbuminuria or microalbuminuria in both type 1 and type 2 diabetics with the II genotype, whereas the DD genotype is associated with a better response to ATR1 antagonists in overt nephropathy of type 2 diabetes. The role of other RAS or non-RAS polymorphisms and their possible interactions with different ACE I/D genotypes are less clearly defined. Thus, evaluating the ACE I/D polymorphism is a reliable tool to identify patients at risk and those who may benefit the most from renoprotective therapy with ACEI or ATR1 antagonists.

Conclusion

Activation of a local RAS by high glucose promotes progressive podocyte injury and loss in DN. New antifibrotic strategies aiming at the (pro)renin receptor, Ang II, ATR1 and AGE/receptor for AGE signal pathways provide a promising approach for early stages of DN.

Acknowledgment

The study was supported by the Research Grant Council of Hong Kong (HKU 7689/10M), and J.C.K.L. was supported by the L&T Charitable Foundation and the House of INDOCAFE.

References

1 Kagami S, Border WA, Miller DE, Noble NA: Angiotensin II stimulates extracellular matrix protein synthesis through induction of transforming growth factor-beta expression in rat glomerular mesangial cells. J Clin Invest 1994;93:2431–2437.

2 Lai KN, Leung JC, Lai KB, To WY, Yeung VT, Lai FM: Gene expression of the renin-angiotensin system in human kidney. J Hypertens 1998;16:91–102.

3 Ingelfinger JR, Zuo WM, Fon EA, Ellison KE, Dzau VJ: In situ hybridization evidence for angiotensinogen messenger RNA in the rat proximal tubule. A hypothesis for the intrarenal renin angiotensin system. J Clin Invest 1990;85:417–423.

4 Terada Y, Tomita K, Nonoguchi H, Marumo F: PCR localization of angiotensin II receptor and angiotensinogen mRNAs in rat kidney. Kidney Int 1993;43:1251–1259.

5 Kobori H, Harrison-Bernard LM, Navar LG: Expression of angiotensinogen mRNA and protein in angiotensin II-dependent hypertension. J Am Soc Nephrol 2001;12:431–439.

6 Henrich WL, McAllister EA, Eskue A, Miller T, Moe OW: Renin regulation in cultured proximal tubular cells. Hypertension 1996; 27:1337–1340.

7 Kobori H, Nangaku M, Navar LG, Nishiyama A: The intrarenal renin-angiotensin system: from physiology to pathobiology of hypertension and kidney disease. Pharmacol Rev 2007;59:251–287.

8 Epstein M, Williams GH, Weinberger M, Lewin A, Krause S, Mukherjee R, Patni R, Beckerman B: Selective aldosterone blockade with eplerenone reduces albuminuria in patients with type 2 diabetes. Clin Am J Soc Nephrol 2006;1:940–951.

9 Ischihara A, Kaneshiro Y, Takemitsu T, Sakoda M, Itoh H: The (pro)renin receptor and the kidney. Semin Nephrol 2007;27: 524–528.

10 Lai KN, Chan LY, Tang SC, Tsang AW, Li FF, Lam MF, Lui SL, Leung JC: Mesangial expression of angiotensin II receptor in IgA nephropathy and its regulation by polymeric IgA1. Kidney Int 2004;66:1403–1416.

11 Wang L, Flannery PJ, Spurney RF: Characterization of angiotensin II-receptor subtypes in podocytes. J Lab Clin Med 2003; 142:313–321.

12 Chan LY, Leung JC, Tang SC, Choy BY, Lai KN: Tubular expression of angiotensin II receptors and their regulation in IgA nephropathy. J Am Soc Nephrol 2005;16: 2306–2317.

13 Miyata N, Park F, Li XF, Cowley AW Jr: Distribution of angiotensin AT_1 and AT_2 receptor subtypes in the rat kidney. Am J Physiol 1999;277:F437–F446.

14 Paul M, Mehr AP, Kreutz R: Physiology of local renin-angiotensin systems. Physiol Rev 2006;86:747–803.

15 Huang J, Siragy HM: Glucose promotes the production of interleukin-1beta and cyclooxygenase-2 in mesangial cells via enhanced (Pro)renin receptor expression. Endocrinology 2009;150:5557–5565.

16 Leehey DJ, Singh AK, Alavi N, Singh R: Role of angiotensin II in diabetic nephropathy. Kidney Int 2000;(suppl 77):S93–S98.

17 Zhang SL, To C, Chen X, Filep JG, Tang SS, Ingelfinger JR, Carriere S, Chan JS: Effect of renin-angiotensin system blockade on the expression of the angiotensinogen gene and induction of hypertrophy in rat kidney proximal tubular cells. Exp Nephrol 2001;9: 109–117.

18 Colucci JA, Arita DY, Cunha TS, Di Marco GS, Vio CP, Pacheco-Solva A, Casarini DE: Renin-angiotensin system may trigger kidney damage in NOD mice. J Renin Angiotensin Aldosterone Syst 2010, Epub ahead of print.

19 Okazaki Y, Yamasaki Y, Uchida HA, Okamoto K, Satoh M, Maruyama K, Maeshima Y, Sugiyama H, Sugaya T, Kashihara N, Makino H: Enhanced TGF-beta/Smad signaling in the early stage of diabetic nephropathy is independent of the AT_{1a} receptor. Clin Exp Nephrol 2007;11:77–87.

20 Deb DK, Chen Y, Zhang Z, Zhang Y, Szeto FL, Wong KE, Kong J, Li YC: 1,25-Dihydroxyvitamin D3 suppresses high glucose-induced angiotensinogen expression in kidney cells by blocking the NF-kappaB pathway. Am J Physiol Renal Physiol 2009; 296:F1212–F1218.

21 Velez JCQ, Bland AM, Arthur JM, Raymond JR, Janech MG: Characterization of renin-angiotensin system enzyme activities in cultured mouse podocytes. Am J Physiol Renal Physiol 2007;293:F398–F407.

22 Dilauro M, Burns KD: Angiotensin-(1–7) and its effects in the kidney. Sci World J 2009;9:522–535.

23 Reich HN, Oudit GY, Penninger JM, Scholey JW, Herzenberg AM: Decreased glomerular and tubular expression of ACE2 in patients with type 2 diabetic and kidney disease. Kidney Int 2008;74:1610–1616.

24 Ye M, Wysocki J, William J, Soler MJ, Cokic I, Batlle D: Glomerular localization and expression of angiotensin-converting enzyme 2 and angiotensin-converting enzyme: implication for albuminuria in diabetes. J Am Soc Nephrol 2006;17:3067–3075.

25 Durvasula RV, Petermann AT, Hiromura K, Blonski M, Pippin J, Mundel P, Pichler R, Griffin S, Couser WG, Shankland SJ: Activation of a local tissue angiotensin system in podocytes by mechanical strain. Kidney Int 2004;65:30–39.

26 Durvasula RV, Shankland SJ: Activation of a local renin angiotensin system in podocytes by glucose. Am J Physiol Renal Physiol 2008; 294:F830–F839.

27 Yoo TH, Li JJ, Kim JJ, Jung DS, Kwak SJ, Ryu DR, Choi HY, Kim JS, Kim HJ, Han SH, Lee JE, Han DS, Kang SW: Activation of the renin-angiotensin system within podocytes in diabetes. Kidney Int 2007;71:1019–1027.

28 Ding G, Reddy K, Kapasi AA, Franki N, Gibbons N, Kasinath BS, Singhal PC: Angiotensin II induces apoptosis in rat glomerular epithelial cells. Am J Physiol Renal Physiol 2002;283:F173–F180.

29 Ziyadeh FN, Wolf G: Pathogenesis of the podocytopathy and proteinuria in diabetic glomerulopathy. Curr Diabetes Rev 2008;4:39–45.

30 Ruster C, Bondeva T, Franke S, Tanaka N, Yamamoto H, Wolf G: Angiotensin II upregulates RAGE expression on podocytes: role of AT2 receptors. Am J Nephrol 2009;29: 538–550.

31 Doublier S, Salvidio G, Lupia E, Ruotsalainen V, Verzola D, Deferrari G, Camussi G: Nephrin expression is reduced in human diabetic nephropathy: evidence for a distinct role for glycated albumin and angiotensin II. Diabetes 2003;52:1023–1030.

32 Ruggenenti P, Bettinaglio P, Pinares F, Remuzzi G: Angiotensin converting enzyme insertion/deletion polymorphism and renoprotection in diabetic and nondiabetic nephropathies. Clin J Am Soc Nephrol 2008;3: 1511–1525.

Kar Neng Lai, MD, DSc
Department of Medicine, University of Hong Kong, Queen Mary Hospital
Hong Kong SAR (China)
Tel. +852 2255 4251, Fax +852 2816 2863, E-Mail knlai@hkucc.hku.hk

Lai·Leung·Tang

Lai KN, Tang SCW (eds): Diabetes and the Kidney.
Contrib Nephrol. Basel, Karger, 2011, vol 170, pp 145–155

The Kallikrein-Kinin System

Sydney C.W. Tang · Joseph C.K. Leung · Kar Neng Lai

Department of Medicine, University of Hong Kong, Queen Mary Hospital,
Hong Kong SAR, China

Abstract

Emerging evidence suggests a role of the kallikrein-kinin system (KKS) in the pathogenesis of diabetic nephropathy (DN). Tissue kallikrein 1 is a member of the tissue kallikrein family that is mainly responsible for the generation of kinins, and bradykinin (BK) is the principal kinin responsible for the biologic actions of the KKS that acts through the ubiquitous BK 2 receptor (B2R) and the inducible B1R. In the kidney, all KKS components are expressed. In particular, kallikrein 1 that is traditionally thought to be solely confined to the distal nephron has recently been identified in the proximal tubule of the human diabetic kidney. Current evidence suggests conflicting roles of the KKS in DN. For a renoprotective role of the KKS, BK reduces mesangial cell proliferation under the diabetic milieu; Akita B2R–/– or STZ-induced KLK–/– mice (T1DM) have more severe albuminuria and glomerulosclerosis, while antagonizing the B2R with icatibant attenuates the antiproteinuric effect of ramipril in *db/db* mice (T2DM). For a detrimental role of the KKS, BK upregulates tubular cell IL-6, CCL-2, and TGF-β expression via ERK1/2 activation; the B2R–/– status protects against the development of DN lesions in STZ-injected mice, while blocking B2R with icatibant allevi- ates biochemical and histologic injuries in uninephrectomized *db/db* mice. These opposite findings may arise from multiple factors and call for further evaluation to clarify the role of the KKS in DN and diabetic tubulopathy.

The Kallikrein-Kinin System

Since kallikrein was discovered as a vasodilatory substance in human urine, the kallikrein-kinin system (KKS) has been considered to play a physiological role in controlling blood pressure. A growing body of evidence has implicated the KKS in the pathogenesis of at least two microvascular complications of diabetes mellitus – nephropathy and retinopathy. Before discussing its involvement in diabetic nephropathy (DN), it will be worthwhile to have a brief overview of this system.

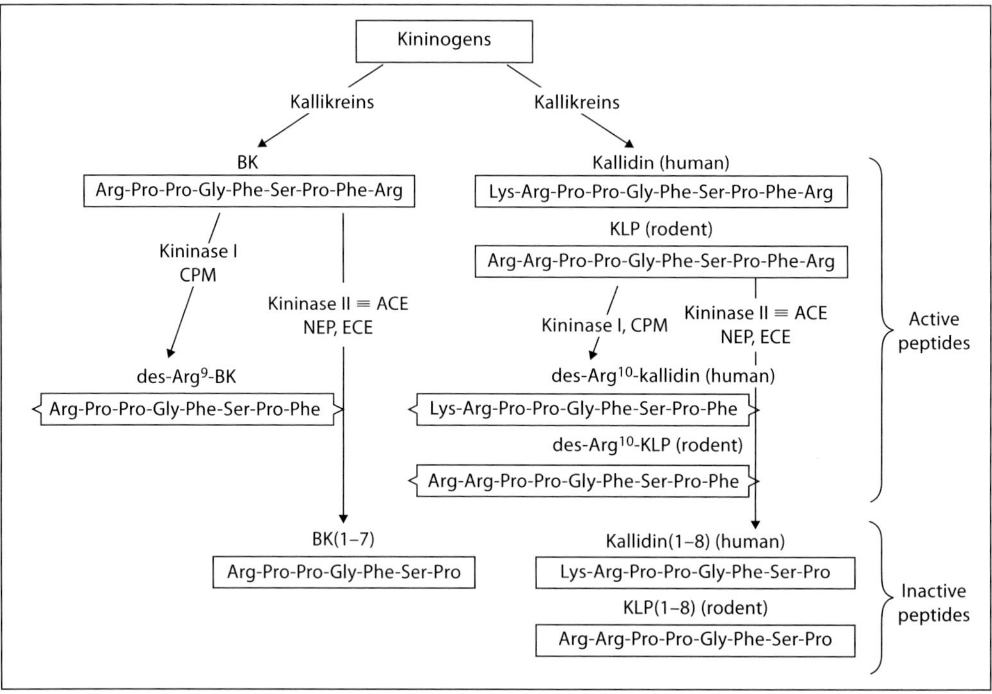

Fig. 1. Components of the KKS. Schema shows the biosynthetic and degradation pathways of kinins in human and rodents. CPM = Carboxypeptidase-M; NEP = neprilysin (endopeptidase 24.11); ECE = endothelin-converting enzyme; KLP = kallidin-like peptide. Reproduced with permission from Kakoki and Smithies [7].

The kinins, bradykinin (BK) and kallidin in humans or BK and the kallidin-like peptide in rodents, are generated from kininogens by kallikreins (fig. 1). Humans have one kininogen gene, whereas rodents have two closely linked kininogen genes [1]. Kallidin can be converted into BK by a plasma aminopeptidase. All the kinins are strong agonists of the BK 2 receptor (B2R), and to a lesser extent, the B1R. Kininase I (carboxypeptidase-N) and carboxypeptidase-M remove arginine from the carboxyl terminus of the kinins to generate their des-Arg derivatives, which are agonists mainly of the B1R. Kininase II (a synonym for angiotensin-I-converting enzyme, ACE), neprilysin (endopeptidase 24.11), and endothelin-converting enzyme all remove two amino acids (Phe and Arg) from the carboxyl terminus of the kinins to inactivate them. ACE is also a carboxydipeptidase that removes two amino acids from the carboxyl terminus of angiotensin I (an inactive peptide) and converts it into angiotensin II (an active vasopressor peptide). ACE also converts the active blood pressure-lowering kinins, BK(1–9) and kallidin(1–10), into their inactive forms, BK(1–7) and kallidin(1–8), respectively.

The actions of the KKS are primarily mediated by BK, a peptide that triggers vasodilation, vascular leakage, and secretion of proinflammatory mediators

Tang · Leung · Lai

[2, 3]. BK has a short plasma half-life of 27 ± 10 s [4], and its actions are regulated locally through its production and proteolytic processing within tissues. The generation of BK is mediated by both plasma and tissue kallikreins, which are serine protease enzymes. Tissue kallikrein 1 (KLK1) is a member of the tissue kallikrein family that has been referred to as 'true tissue kallikrein' because it is the only tissue KLK family member with significant kinin-generating activity [5]. Tissue KLK1 acts through 2 cell surface G-protein-coupled receptors: the widely expressed B2R and the inducible B1R [3]. The B2R protein is constitutively expressed in most tissues. In particular, vascular endothelial cells express B2R abundantly, where it is functionally linked to activation of endothelial nitric oxide synthase. B1R is constitutively absent, but is inducible by inflammation and toxins, and also by the absence of B2R when it will assume some of the hemodynamic functions of B2R [6]. The intracellular signals that follow stimulation of B1R and B2R are reviewed elsewhere [7].

Scientific Basis for Implicating the Kallikrein-Kinin System in Diabetic Nephropathy

All components of the KKS are expressed in the kidney and are regulators of intrarenal hemodynamics and tubular function. In the kidney, B2R mRNA is expressed in all segments under physiological conditions, whereas B1R is almost absent [7]. Disruption of the KLK1 gene dramatically reduces renal kinin production, suggesting that tissue kallikrein is the main BK-generating enzyme in the kidney. Little is known about the potential role of plasma kallikrein and the contact system on renal function and DN. In diabetes, renal kallikrein levels and excretion of urinary kallikrein are decreased [8].

In STZ-induced diabetic rats with moderate hyperglycemia, there is increased urinary excretion of active KLK kinins with reduced vascular resistance [8]. In contrast, severe STZ-induced hypofiltrating rats have downregulated KLK activity, indicating a direct correlation of KLK activity and renal plasma flow [9]. In addition to hemodynamic actions, kinins may influence inflammation and cell proliferation under diabetic conditions [10]. However, the available data, detailed below, show a discrepancy as to whether the renal KKS reduces or enhances renal injury in DN.

First Demonstration of Tissue Kallikrein in the Renal Proximal Tubule

In the kidney, kallikrein is traditionally thought to be almost exclusively localized to the connecting tubule at the distal part of the nephron. Emerging evidence suggests that the proximal tubule cell that contributes significantly to the pathogenesis of DN as detailed in chapter 14 [see chapter by Tang et al., this vol., pp. 124–134], also expresses KLK1. Using in situ hybridization, Chen et al. [11] reported tissue kallikrein mRNA expression in the proximal tubules at the urinary pole of human kidney sections, although the predominant localization was still at the distal tubules, collecting duct, and loop of Henle. In addition, kallikrein has also been localized in a rat model of cisplatin toxicity and in proximal tubule cells of the kidney in African women with hypertension of pregnancy [12].

More recently, our laboratory confirmed proximal tubular expression of KLK1 in human diabetic kidney tissue by immunohistochemical staining by using an anti-KLK1 Ab that specifically recognizes KLK1 and does not cross-react with other human kallikreins (fig. 2a). The specificity of the anti-KLK1 antibody for the immunohistochemical staining was confirmed by showing the loss of proximal tubular staining in diabetic kidney tissue when the antibody was preabsorbed with KLK1 peptide. Such findings open the door for further exploring the role of the KKS in diabetic tubulopathy.

Kallikrein-Kinin System Activation Plays a Renoprotective Role in Diabetic Nephropathy

It is well known that ACE inhibition is protective against DN. As this action increases the biologically active BK(1–9) (fig. 1), it may be inferred that augmentation of BK through inhibition of endogenous degradation may confer renoprotection during diabetes. In vitro, BK has been shown to reduce mesangial cell proliferation under the diabetic milieu [13]. Fibronectin secretion induced by macrophage-conditioned medium is reduced by perindoprilat or BK treatment [14].

In vivo, Akita type 1 diabetic mice with absent B2R have more severe albuminuria and glomerulosclerosis [15], and senescence-associated phenotypes [16] than wild-type (WT) animals. In STZ-induced rats, the B2R antagonist, icatibant (HOE-140) significantly attenuates the antiproteinuric effect of ramipril [17], signifying that the antiproteinuric action of the ACE inhibitor is mediated at least partially via the B2R. More recently, Bodin et al. [18] showed that KLK-knockout STZ-induced mice had more severe albuminuria than WT. However, this model is limited by the slow development of DN; thus, the effect of KLK deficiency on established DN cannot be adequately assessed. The direct effect of KLK deficiency in a respectable model of T2DM is as yet undefined, although Buleon et al. [19] recently showed indirectly that blocking B2R reduced the renoprotective effect of ramipril in *db/db* mice. The problem with this model is twofold: B1R is upregulated and able to take over the function of B2R during its inhibition [6, 20, 21], making interpretations based on single receptor blockade difficult; *db/db* mice do not develop DN lesions readily [22].

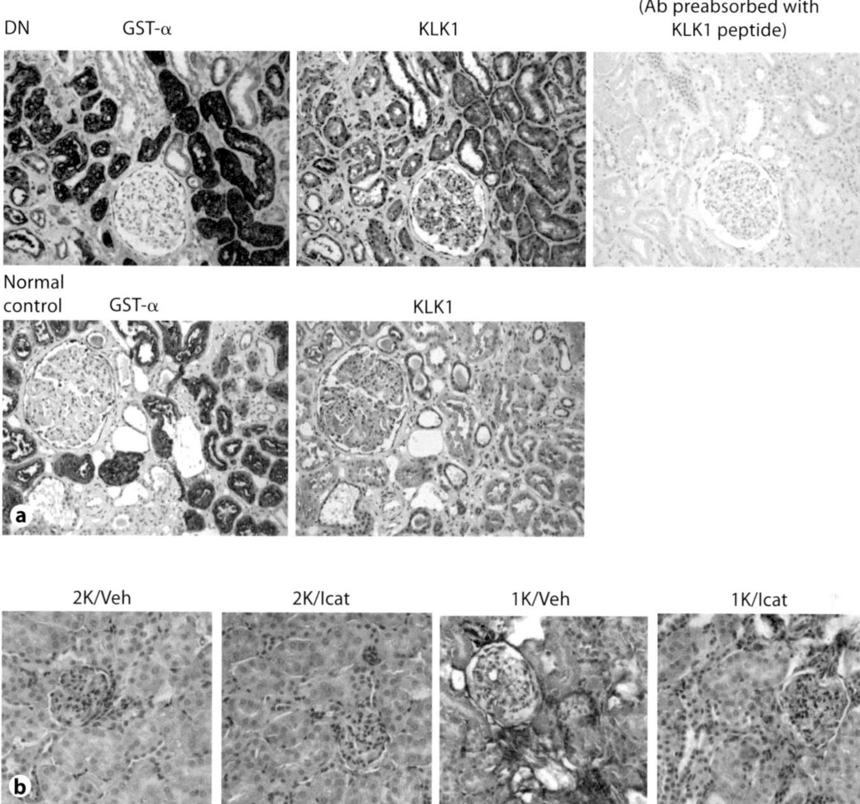

Fig. 2. Participation of the KKS in diabetic nephropathy. **a** Demonstration of KLK1 in the proximal tubules of the diabetic kidney [28, 30]. Immunostaining for the proximal tubular cell marker GST-α and tissue kallikrein in serial/contiguous sections of a representative human diabetic kidney biopsy and a normal control kidney section. There was positive staining for KLK1 in proximal tubules (in addition to connecting tubules) of the diabetic kidney, and the staining specificity was confirmed by preabsorption of the KLK1 Ab with 5 µg KLK1 peptide. ×400, all sections counterstained with hematoxylin. **b** Antifibrotic effect of blocking the B2R. Immunostaining of α-smooth muscle actin in db/db mice after sham operation or uninephrectomy and treatment with vehicle (Veh) or icatibant (Icat) for 8 weeks. 2K = Sham-operated mouse with two intact kidneys; 1K = uninephrectomized mouse with 1 kidney. ×400.

Kallikrein-Kinin System Activation Plays a Detrimental Role in Diabetic Nephropathy

Contrary to the aforementioned findings, emerging evidence suggest that the KKS may facilitate renal injury during diabetes. In vitro, BK induces proliferation of quiescent renal cells [23] and upregulates CTGF, TGF-β, and collagen I mRNA via extracellular signal-regulated kinase (ERK)1/2 activation in MC [24]. In vivo, renal B2R and TGF-β mRNA expression are significantly increased in

Table 1. Physical and biochemical parameters of *db/m* and *db/db* animals at baseline and after 8 weeks of treatment [30]

| | *db/m* mice | | | | | | *db/db* mice | |
| | Sham | Unx | | | | | Sham | |
	Veh	Veh	Met	Icat	Ros	Icat + Ros	Veh	Met
Body weight, g*								
Baseline	25.6±0.8	25.9±0.8	25.5±0.8	25.2±1.0	25.8±0.8	25.2±1.2	41.7±1.1	42.2±0.8
8 weeks	28.5±0.7	28.2±0.8	28.8±0.8	28.7±0.7	28.6±1.0	28.5±1.0	50.7±1.2	47.1±0.9[†]
Blood glucose, mg/dl*								
Baseline	120±6	124±5	123±6	123±4	120±6	122±5	427±11	427±8
8 weeks	124±4	124±4	122±7	123±5	126±4	126±5	632±13	248±14[†]
Serum creatinine, mg/dl*								
Baseline	0.14±0.03	0.15±0.02	0.15±0.03	0.15±0.02	0.15±0.02	0.14±0.03	0.17±0.02	0.17±0.02
8 weeks	0.16±0.03	0.19±0.02	0.17±0.02	0.17±0.02	0.17±0.02	0.17±0.02	0.25±0.03	0.25±0.02
UACR, µg/mg*								
Baseline	65±10	66±8	67±7	65±3	65±6	66±5	1,468±145	1,453±126
8 weeks	73±7	133±5	132±8	133±9	94±5	97±5	1,716±147	1,709±158

Sham = Sham operation; Unx = uninephrectomy; Veh = vehicle; Met = metformin; Icat = icatibant; Ros = rosiglitazone; UACR = urine albumin to creatinine ratio.
* p < 0.05 for comparisons between the corresponding *db/db* and *db/m* groups; [†] p < 0.05 vs. all other *db/db* groups at 8 weeks; [‡] p < 0.05 vs. sham-operated *db/db* animals; [§] p < 0.05 vs. Unx-Veh *db/db* animals; [#] p < 0.05 vs. Unx-Ros or Unx-Icat *db/db animals*.

STZ-induced animals [24]. Targeted deletion of B2R protects against the development of DN in which B2R–/– mice injected with STZ display less albuminuria and glomerular and tubular injuries than WT animals [25]. Deletion of B2R leads to a compensatory rise in renal B1R expression in B2R–/– mice. In STZ-induced diabetic hyperfiltrating rats, acute treatment with the nonspecific KLK inhibitor, aprotinin, nearly normalizes GFR and renal blood flow [8], and normalizes renal function [26], while blocking B1R or B2R markedly reduces proteinuria [27].

Factors contributing to these apparent differences in the role of B2R in DN include dissimilarities in background genetics of the animals studied, the method of induction of diabetes and the specific end points measured. In addition, the effect of BK and B2R blockers in mice may not be directly extrapolated

Unx							
Icat	Ros	Icat + Ros	Veh	Met	Icat	Ros	Icat + Ros
41.6±0.9	42.1±0.9	41.7±1.0	41.6±1.4	42.8±0.7	41.3±1.2	42.0±1.1	41.4±1.3
50.6±0.9	52.4±1.1	52.2±0.9	50.9±0.9	46.8±1.1[†]	50.8±0.9	52.7±1.1	52.3±1.3
424±8	425±12	427±7	424±12	421±9	427±12	423±11	425±11
628±10	235±8[†]	230±8[†]	633±10	246±13[†]	636±10	237±12[†]	230±11[†]
0.18±0.02	0.17±0.02	0.17±0.02	0.18±0.02	0.19±0.02	0.17±0.02	0.18±0.02	0.17±0.02
0.25±0.01	0.24±0.02	0.23±0.02	0.35±0.04[‡]	0.36±0.04[‡]	0.32±0.04[‡, §]	0.29±0.03[‡, §]	0.23±0.02[§, #]
1,450±64	1,451±89	1,452±137	1,434±149	1,373±86	1,442±167	1,461±149	1,450±145
1,696±115	1,678±88	1,661±90	2,178±154[‡]	2,000±76[‡]	2,016±129[‡, §]	1,839±133[§]	1,586±113[§, #]

to humans. Finally, most of the animal models used are T1DM models in either Akita mice or STZ-induced diabetes.

To circumvent the shortcomings in the aforementioned animal models, we recently employed a neat culture system of human proximal tubular epithelial cells (PTEC) and showed that high glucose (HG; 30 mM) induced B2R, KLK1 and low-molecular-weight kininogen expression in PTEC, and that HG-induced upregulation of IL-6, CCL-2 (MCP-1), TGF-β, and VEGF in PTEC was attenuated by the tissue kallikrein inhibitor aprotinin. In addition, BK upregulated tubular expression of IL-6, CCL-2, and TGF-β via ERK1/2. The B2R blocker, icatibant, downregulated BK- and HG-induced MAPK p42/p44 but not HG-induced PKC activation, and partially reduced both HG- and BK-induced IL-6, CCL-2, and TGF-β secretion [28]. As the PPAR-γ agonist in known to confer renoprotection in experimental DN [see chapter by Tang et al., this vol., pp. 124–134], via a mechanism (STAT1 inhibition) different from icatibant (ERK1/2 inhibition), we also explored whether combining the two agents may produce an additive effect. The PPAR-γ agonist, rosiglitazone, attenuated HG-induced PKC but not HG- or BK-induced ERK1/2 activation, and reduced HG-stimulated VEGF, along with IL-6, CCL-2, and TGF-β secretion. Rosiglitazone plus icatibant further reduced these effects of HG.

Extending the application of B2R blockade to an experimental model of T2DM, we employed the uninephrectomized *db/db* diabetic mice that have been reported to exhibit lesions resembling T2DM in human [29]. As with the in vitro studies, the effect of the PPAR-γ agonist is also explored. Biochemically, *db/db* mice had significantly higher blood glucose, serum creatinine and albuminuria at baseline compared with *db/m* mice (table 1). Uninephrectomy had no observable effect on blood glucose levels in nondiabetic *db/m* littermates or the diabetic *db/db* mice. Among *db/db* animals, blood glucose levels were similar among different groups at baseline and increased significantly after 8 weeks, except that mice treated with metformin or rosiglitazone had blood glucose lower than baseline levels after 8 weeks of treatment regardless of whether uninephrectomy or sham operation was performed. Notably, there was no difference in blood glucose level between metformin and rosiglitazone-treated animals. Among the *db/db* group with uninephrectomy, mice that received icatibant or rosiglitazone had significantly lower serum creatinine and albuminuria, which was further reduced when icatibant and rosiglitazone were given in combination. These beneficial effects were not observed in the metformin group that achieved similar glycemic control as rosiglitazone-treated animals. Likewise, the severity of reactive glomerular and proximal tubular hypertrophy, glomerulosclerosis, interstitial injury, cortical F4/80 and α-smooth muscle actin immunostaining, and CCL-2, ICAM-1, and TGF-β overexpression were all attenuated by icatibant and rosiglitazone, and these effects were enhanced when both agents were combined (fig. 2b). Immunohistochemical staining confirmed the proximal tubular expression of CCL-2 (inflammation) and TGF-β (fibrosis). Treatment with icatibant was associated with decreased B2R but increased B1R expression, which was exaggerated in uninephrectomized animals. At the signaling level, icatibant and rosiglitazone reduced ERK1/2 and STAT1 activation, respectively [30]. Collectively, these data suggest the presence of KLK1 in PTEC and the participation and possible deleterious effects of the KKS during DN (fig. 2).

Conclusion

Participation of the KKS in DN has become increasingly apparent. The novel observation of KLK1 expression in the renal proximal tubule suggests a role of

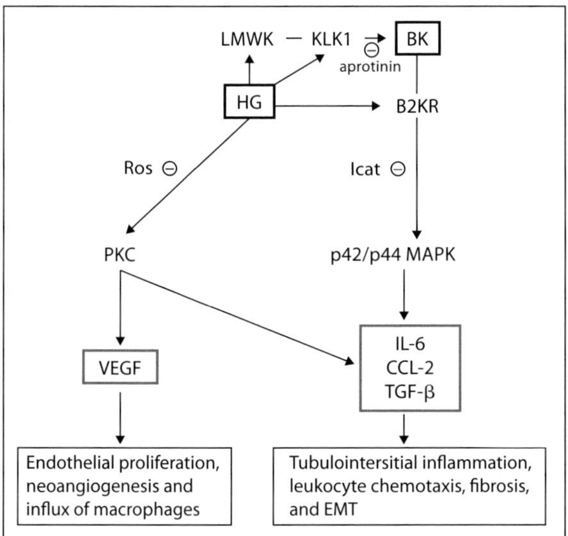

Fig. 3. Proposed schema for the KKS-inflammation-fibrosis axis [28]. Activation of renal tubular cells by HG and BK in promoting tubulointerstitial inflammation, fibrosis, and neovascularization is shown. Arrows denote stimulatory effects; encircled negative signs denote inhibitory effects. Ros = Rosiglitazone; Icat = icatibant; EMT = epithelial-mesenchymal transdifferentiation; LMWK = low-molecular-weight kininogen.

the KKS in diabetic tubulopathy. Based on the in vitro and animal data from our group, a schema is proposed for the role of the KKS-inflammation-fibrosis axis (fig. 3). The opposite results obtained by different investigators may be due to various reasons ranging from the differences in the strains of mice used, the method of induction of diabetes, and the cell type studied; but this only implies that many more studies on the involvement of the KKS in DN and diabetic tubulopathy are required to clarify our understanding of this currently underevaluated biologic system in DN.

Acknowledgments

This study was supported by a General Research Fund of the Research Grants Council (grant number: HKU 7760/08M) of Hong Kong, and J.C.K.L. was supported by the L & T Charitable Foundation and the House of INDOCAFE.

References

1 Shesely EG, Hu CB, Alhenc-Gelas F, Meneton P, Carretero OA: A second expressed kininogen gene in mice. Physiol Genomics 2006;26:152–157.

2 Joseph K, Kaplan AP: Formation of bradykinin: a major contributor to the innate inflammatory response. Adv Immunol 2005; 86:159–208.

3 Leeb-Lundberg LM, Marceau F, Muller-Esterl W, Pettibone DJ, Zuraw BL: International union of pharmacology. XLV. Classification of the kinin receptor family: from molecular mechanisms to pathophysiological consequences. Pharmacol Rev 2005;57:27–77.

4 Cyr M, Lepage Y, Blais C Jr, et al: Bradykinin and des-Arg(9)-bradykinin metabolic pathways and kinetics of activation of human plasma. Am J Physiol Heart Circ Physiol 2001;281:H275–H283.

5 Clements JA: Reflections on the tissue kallikrein and kallikrein-related peptidase family – from mice to men – what have we learnt in the last two decades? Biol Chem 2008;389: 1447–1454.

6 Duka I, Kintsurashvili E, Gavras I, Johns C, Bresnahan M, Gavras H: Vasoactive potential of the b(1) bradykinin receptor in normotension and hypertension. Circ Res 2001;88: 275–281.

7 Kakoki M, Smithies O: The kallikrein-kinin system in health and in diseases of the kidney. Kidney Int 2009;75:1019–1030.

8 Harvey JN, Jaffa AA, Margolius HS, Mayfield RK: Renal kallikrein and hemodynamic abnormalities of diabetic kidney. Diabetes 1990;39:299–304.

9 Tschope C, Reinecke A, Seidl U, et al: Functional, biochemical, and molecular investigations of renal kallikrein-kinin system in diabetic rats. Am J Physiol 1999;277: H2333–H2340.

10 Riad A, Zhuo JL, Schultheiss HP, Tschope C: The role of the renal kallikrein-kinin system in diabetic nephropathy. Curr Opin Nephrol Hypertens 2007;16:22–26.

11 Chen LM, Song Q, Chao L, Chao J: Cellular localization of tissue kallikrein and kallistatin mRNAs in human kidney. Kidney Int 1995; 48:690–697.

12 Naicker T, Khedun SM, Moodley J, Bhoola KD: Localisation of tissue kallikrein in the kidney of black African women with early onset pre-eclampsia: a pilot study. Immunopharmacology 1997;36:249–254.

13 Alric C, Pecher C, Cellier E, et al: Inhibition of IGF-I-induced Erk 1 and 2 activation and mitogenesis in mesangial cells by bradykinin. Kidney Int 2002;62:412–421.

14 Pawluczyk IZ, Patel SR, Harris KP: The role of bradykinin in the antifibrotic actions of perindoprilat on human mesangial cells. Kidney Int 2004;65:1240–1251.

15 Kakoki M, Takahashi N, Jennette JC, Smithies O: Diabetic nephropathy is markedly enhanced in mice lacking the bradykinin B2 receptor. Proc Natl Acad Sci U S A 2004;101:13302–13305.

16 Kakoki M, Kizer CM, Yi X, et al: Senescence-associated phenotypes in Akita diabetic mice are enhanced by absence of bradykinin B2 receptors. J Clin Invest 2006;116:1302–1309.

17 Tschope C, Seidl U, Reinecke A, et al: Kinins are involved in the antiproteinuric effect of angiotensin-converting enzyme inhibition in experimental diabetic nephropathy. Int Immunopharmacol 2003;3:335–344.

18 Bodin S, Chollet C, Goncalves-Mendes N, et al: Kallikrein protects against microalbuminuria in experimental type I diabetes. Kidney Int 2009;76:395–403.

19 Buleon M, Allard J, Jaafar A, et al: Pharmacological blockade of B2-kinin receptor reduces renal protective effect of angiotensin-converting enzyme inhibition in db/db mice model. Am J Physiol Renal Physiol 2008;294:F1249–F1256.

20 Duka I, Shenouda S, Johns C, Kintsurashvili E, Gavras I, Gavras H: Role of the B(2) receptor of bradykinin in insulin sensitivity. Hypertension 2001;38:1355–1360.

21 Griol-Charhbili V, Messadi-Laribi E, Bascands JL, et al: Role of tissue kallikrein in the cardioprotective effects of ischemic and pharmacological preconditioning in myocardial ischemia. FASEB J 2005;19:1172–1174.

22 Breyer MD, Bottinger E, Brosius FC, III, et al: Mouse models of diabetic nephropathy. J Am Soc Nephrol 2005;16:27–45.

23   Goldstein RH, Wall M: Activation of protein formation and cell division by bradykinin and des-Arg9-bradykinin. J Biol Chem 1984; 259:9263–9268.

24   Tan Y, Wang B, Keum JS, Jaffa AA: Mechanisms through which brady-kinin promotes glomerular injury in diabetes. Am J Physiol Renal Physiol 2005;288:F483–F492.

25   Tan Y, Keum JS, Wang B, McHenry MB, Lipsitz SR, Jaffa AA: Targeted deletion of B2-kinin receptors protects against the develop-ment of diabetic nephropathy. Am J Physiol Renal Physiol 2007;293:F1026–F1035.

26   Zuccollo A, Frontera M, Cueva F, Navarro M, Catanzaro OL: Effects of aprotinin on the kallikrein-kinin system in type I diabetes (insulitis). Immunopharmacology 1997;37: 251–256.

27   Zuccollo A, Navarro M, Catanzaro O: Effects of B1 and B2 kinin receptor antagonists in diabetic mice. Can J Physiol Pharmacol 1996;74:586–589.

28   Tang SC, Chan LY, Leung JC, et al: Bradykinin and high glucose promote renal tubular inflammation. Nephrol Dial Transplant 2010;25:698–710.

29   Tang SC, Leung JC, Chan LY, Cheng AS, Lan HY, Lai KN: Renoprotection by rosiglitazone in accelerated type 2 diabetic nephropathy: role of STAT1 inhibition and nephrin resto-ration. Am J Nephrol 2010;32:145–155.

30   Tang SC, Chan. LY, Leung JC, Cheng AS, Lan HY, Lai KN: Synergistic renoprotective effects of B2-kinin receptor blocker and PPAR-gamma agonist in uninephrectomized db/db mouse model. Lab Invest 2011, in press.

Sydney C.W. Tang, MD, PhD
Professor of Medicine, Department of Medicine, University of Hong Kong
Hong Kong SAR (China)
Tel. +852 2255 4777, E-Mail scwtang@hkucc.hku.hk

Lai KN, Tang SCW (eds): Diabetes and the Kidney.
Contrib Nephrol. Basel, Karger, 2011, vol 170, pp 156–164

Translating Experimental Diabetic Nephropathy Studies from Mice to Men

Matthew D. Breyer

Lead Generation Biology, Biotechnology Discovery Research, Lilly Research Laboratories, Indianapolis, Ind., USA

Abstract

The laboratory mouse is among the best characterized and flexible experimental platforms available for the study of diabetic nephropathy (DN). However, studies of progressive kidney disease in mice have underscored several important technical considerations for accurate phenotyping of renal function. Most mouse models of DN fail to exhibit progressive kidney disease. However, a few models have proved particularly useful. Despite the utility of these models, whether they can accurately predict clinical benefit of therapeutics in human DN remains to be established.

Diabetic nephropathy (DN) is the major single cause of end-stage renal disease in the industrialized world. The cost of dialyzing diabetic patients in the US alone is greater than USD 17 billion per year [1]. Given the increasing prevalence of type 2 diabetes mellitus, occurring in an ever younger demographic, the burden of DN on global health can only be expected to increase [2]. Despite this growing public health imperative, substantial hurdles exist to developing therapies that slow the relentless decline in renal function observed in the subset of diabetics who are destined to develop end-stage kidney failure. Among the impediments to developing novel therapies for DN is the lack of validated animal models that faithfully recapitulate the human disease and in which therapies prove equally efficacious in the animal model and the clinic. In the past decade, substantial effort has attempted to address this deficiency by utilizing the laboratory mouse to develop clinically relevant experimental models of DN. As will be discussed, assessing the severity of diabetic kidney disease in the mouse has proved challenging.

Modeling Diabetic Nephropathy in Mice

The laboratory mouse represents a unique experimental platform offering unprecedented genetic characterization and capability for genetic engineering [3, 4]. Transgenic and knockout technology provides experimental control of in vivo murine gene expression through selective suppression or stimulation of individual, predetermined genes [4] making mice an increasingly attractive experimental model. In addition, the availability of >450 strains of inbred mice provides diverse experimental platforms, each strain with a uniform genetic composition [3, 5]. Conversely, the genetic variation *between* strains may be used to model interindividual genetic diversity in human outbred populations. Importantly, only a minority of diabetic patients (10–30%) ever develop nephropathy [6, 7], and several studies suggest genetic risk factors predispose a subset of human diabetics to progressive kidney failure [8, 9]. Characterization of diabetes mellitus and its associated renal phenotype in different mouse strains offers hope that the right mouse model will develop clinically relevant disease and help to elucidate the pathophysiology of human DN.

The NIH sponsored Animal Models of Diabetic Complications Consortium (AMDCC) recently outlined three major phenotypes – histopathology, albuminuria, and glomerular filtration rate (GFR) – to benchmark the severity of DN in a mouse model [10, 11]. Unfortunately, GFR has been studied in only a few inbred strains of diabetic mice, and many of these strains, including C57BL6, A/J, and 129 mice, fail to develop the AMDCC requiste features of advanced DN: >10-fold increase in albuminuria; nodular glomerulosclerosis, and renal insufficiency [10, 12, 13]. The precise reasons for these deficiencies are unknown, but possibilities include insufficient duration of study (typically no longer than 24 weeks) and premature death of the diabetic mice. It is notable that many mice strains (e.g. DBA2/J, FVB, and BTBR) are particularly frail when diabetic, and succumb to infection, progressive weight loss, and general failure to thrive, thereby frustrating efforts at longer term studies. Perhaps it should come as no surprise that diabetic mice fail to survive long enough to develop DN given that DN was not recognized as an entity in humans until prolonged survival afforded by insulin therapy became widespread [14–16].

Even without these animal husbandry challenges, there remains a significant disconnect between the clinical versus laboratory experiment in DN. These two experimental paradigms typically focus on quite different endpoints (table 1). For a successful drug registration trial, regulatory agencies require that a therapy slow the rate of doubling of serum creatinine, impact the onset of dialysis or reduce the death rate. In contrast, exploration of potential renal therapeutics

Table 1. Experimental end points in human DN trials versus studies in diabetic mice

Experimental endpoint	Clinical trial	Preclinical animal models
Death/dialysis/	registration end point	not determined
Creatinine doubling/GFR	registration end point	not determined
Proteinuria/albuminuria	surrogate end point	end point
Histopathology	not determined	end point

in diabetic mice has never focused on reducing the rate of creatinine doubling. Instead, most laboratory studies focus on histopathologic changes and albuminuria with less attention to clinically relevant renal function outcomes. As discussed below, there also exist unique technical hurdles to phenotyping, renal histopathology, and renal function in mouse models of DN.

Renal Histology

It is a disconcerting general truth that most diabetic mouse models exhibit only mild to moderate histopathological changes of DN [11]. Many widely used diabetic mouse models including C57BL/6 (B6) streptozotocin (STZ)-treated, B6-Akita, B6-*db/db* mice fail to exhibit anything but the mildest mesangial expansion [12, 13, 17, 18]. Notably, clinical studies demonstrate that mild to moderate mesangial expansion do not predict progressive kidney disease in man. Only more advanced histopathological renal lesions including nodular glomerulosclerosis, arteriolar hyalinosis, and tubulointerstitial disease are associated with progressive kidney disease in humans [19–21], and these are uncommon enough in mice as to warrant publication. This is consistent with the observation that most mouse models fail to exhibit a progressive decrease in GFR or associated significant hypertension – two key features of human DN [15, 16]. Furthermore, while renal pathology is the most routinely assessed characteristic of mouse models of DN, it has the least direct relevance to clinical trial of therapeutic interventions because renal biopsy can rarely be ethically justified as a trial endpoint (or its costs afforded). Translation of a preclinical observation of a drug that improves histopathology into a clinical proof of concept depends on the hope that the drug will also rapidly and robustly impact measurable clinical endpoints such as blood pressure, albuminuria or changes in GFR.

In the absence of progressive renal dysfunction, the significance of the mild histopathological changes observed in most mouse models of DN remains uncertain. One might argue that these mild histopathological changes seen in diabetic mice mirror mild mesangial expansion observed in diabetic humans who do not exhibit nephropathy [22]. Because these relatively mild renal histopathological changes observed in mouse models are difficult to quantify, many publications utilize perfusion-fixed mouse kidneys to facilitate quantitative analysis of glomerulosclerosis [13, 17, 23]. In contrast, in those mouse models exhibiting more severe pathologic lesions, perfusion fixation may not be used [18, 24]. These methodological differences complicate comparisons between different models.

Albuminuria

Proteinuria is a major prognostic biomarker predictive of progressive kidney failure in human DN [25, 26]. Furthermore, the renal benefit of angiotensin-converting enzyme inhibitors (ACE) and angiotensin receptor type 1 blockers (ARB) is tightly linked to their capacity to reduce proteinuria in patients with overt proteinuria, and has also been observed in phase 2 trials of endothelin receptor antagonists [27, 28]. Similar linkage between efficacy of ACE/ARB and proteinuria has also been demonstrated in rat models of kidney disease.

In mice, albuminuria rather than proteinuria should be used to assess glomerular injury. This is because in mice tubules secrete large amounts of several proteins in the urine not found in humans including major urinary proteins that are excreted at rates between 50 and 20,000 µg/day [29] potentially overwhelming any contribution of glomerular proteinuria to total urinary protein. For this reason, specific measurements of urinary albumin should be obtained. Optimally, both a 24-hour urine collection and a urine albumin:creatinine ratio (ACR) should be obtained. In diabetic patients, there is significant diurnal variation of albumin excretion [30], and similar variation in mice could contribute to inconsistencies in spot urine ACR determinations, arguing for the 24-hour urine sample. Conversely, since hyperglycemic mice can be markedly polyuric with urine volumes in excess of 20 ml/day, dilutional errors in urine albumin may occur with a 24-hour sample. Quantification of albuminuria may be performed by semiquantitative gel electrophoresis or ELISA kits using a validated anti-mouse albumin antibody since there is relatively low protein homology between mouse and human or rat albumin. Significant differences in measured albuminuria are also seen using anti-mouse albumin antibodies from different vendors (for instance the Bethyl mouse albumin ELISA seems to yield higher values than Exocell ELISA). So comparisons between studies are best made when the same reagents are used. For this reason, an NIH consortium has adopted a single manufacturer (see www.AMDCC.org).

Although reductions in proteinuria have not yet been accepted as a surrogate registration endpoint by regulatory agencies [31], this endpoint is the only endpoint that is commonly reported in both mouse models of diabetes and human clinical trials.

Glomerular Filtration Rate

Repeated determinations of serum creatinine in individual patients during clinical trials are typically required to determine the doubling rate of serum creatinine in treated and untreated groups for drug registration. In contrast, serial determinations of serum creatinine are rarely obtained in preclinical animal studies, and when measured, serum creatinine is generally obtained only as a single study end point. Measurement of serum creatinine in mice is not trivial. Creatinine doubling studies in mice have not to the author's knowledge been performed and would be complicated by the extremely small blood volume in mice (1.5–2.5 ml) limiting the number of repeated bleeds [36, 37]. In addition, circulating levels of creatinine as determined by high pressure liquid chromatography and mass spectroscopy have shown that serum creatinine in the normal mouse is ≤0.1 mg/dl [38–41]. In contrast, colorimetric (picric acid-based) methods are in the range of 0.4–0.7 mg/dl, and so are largely artifactual and should be dropped from use [39–42]. Enzymatic creatinine assay methods may serve as a less resource-intensive alternative with significantly less artifact [42, 43].

However, even authentic serum creatinine measurements are not strictly determined by GFR since recent studies in normal mice show that creatinine clearance is nearly double inulin clearance and that approximately 35–50% of urinary creatinine can be blocked by PAH or cimetidine and therefore due to active tubular secretion [44]. Thus, the utility of serum creatinine as a biomarker of renal function in mice may be substantially compromised by conditions which alter secretion. It may be that addition of blood urea nitrogen will be required to substantiate alterations in renal function in those cases where inulin clearance cannot be determined. Nevertheless, these findings raise significant hurdles in terms of serial determination of creatinine and GFR in murine DN experiments.

Inulin clearance remains the gold standard for the measurement of GFR in mice [44–46]. GFR in the normal healthy mouse is approximately 250–400 µl/min (depending on the strain). Most diabetic mice only exhibit glomerular hyperfiltration without experiencing a fall in GFR [13]. Of the diabetic mouse models reported thus far, only the eNOS–/–;C57BLKS *db/db* has been documented to exhibit a decrease in GFR of greater than 50% (table 2) [23].

Choosing the Best Mouse Model

The choice of an optimal mouse model depends on the experimental goal. If one wishes to determine whether a mutation (e.g. knockout) exacerbates DN in mice, then it would be best to chose the nephropathy-resistant C57Bl6 mouse to look for the development of more robust nephropathy. Several different approaches to generating diabetes in this strain are available including the preferred low-dose STZ protocol [11, 13, 47] or the type I DM Akita

Table 2. Nephropathy phenotype in some selected diabetic mouse models

Mouse model	Pathology	Albuminuria	GFR, µl/min (DM vs. control)	Hypertension (DM vs. control)	Notes
C57BLKS db/db ± uninephrectomy [32–34]	++ mesangial expansion	262±24 µg/mg[a] 1,429±805 µg/mg[a]	366±39 vs. 331±25	~127 vs. 101[d] 146 vs. 133 mm Hg[c]	Good survival Infertile must breed heterogeneous × heterogeneous
C57BLKS db/db eNOS–/– [23, 25]	++++ nodular glomerulo-sclerosis	1,503±176 µg/mg[b] 7,614±3,246[a] µg/mg[a]	164 ± 27 vs. 331 ±25	158 vs. 110 mm Hg[c]	Difficult breeding, and high mortality
FVBove26 [24]	+++ some nodules	~15,000 µg/mg[b]	428.3±27 vs. 514±28 (9 months age)	~135 vs. 120 mm Hg[c]	Frail, difficult to maintain
BTBR ob/ob [18]	++ to +++ some nodules	809 ±134 µg/mg[a] 242 ±45 µg/24h[a]	No	~90 vs. 120 mm Hg (hypotensive)	Frail, difficult to maintain
129S6 Akita [17]	++ mesangial expansion	151±78 µg/24h[a]	~612 (est[e]) vs. 446 (est)	143 vs.130 mm Hg[c]	Good survival

[a] Exocell Albuwell M ELISA.
[b] Mouse Albumin ELISA Bethyl Laboratories.
[c] Tail cuff SBP.
[d] MAP radiotelemetry [32].
[e] Estimated from data in paper using an assumed body weight of 25 g.

mutation [17] or the type 2 DM *db/db* [22, 23]. Conversely, if one is in need of a severe form of nephropathy with reduced GFR, then the eNOS knockout C57BLKS *db/db* mouse would be a good choice [23, 35, 48] (table 2). This model may also be relevant to human disease since in some populations a nonsynonymous single nucleotide polymorphism in eNOS (NOSIII) changing glutamine 298 to asparagine is associated with reduced NOS activity and also with accelerated DN [49–51]. Importantly, and in contrast to most models of nephropathy studied to date, eNOS–/– C57BLKS *db/db* exhibit decreased GFRs to levels less than 50% of that in eNOS+/+ C57BLKS *db/db*, confirmed by increased serum creatinine and decreased inulin clearance. This model is also characterized by increased interstitial macrophage infiltration [52], a characteristic of human disease of DN [19, 53]. Unfortunately, this model is difficult to breed and maintain and is not amenable to high volume experiments. Other mouse strains including KK/HiJ [54, 55], BTBR *ob/ob* [18], and FVBove26 [24] have been proposed to exhibit features of more advanced DN but may be complicated by hydronephrosis [24, 56] or reduced survival and husbandry challenges. Indeed, the frailty and poor fecundity of many of these strains limit their utility. Because of its relative heartiness and respectable albuminuria, many remain attached to the C57BLKS *db/db* as the workhorse for DN studies. (For details of STZ-induced diabetic eNOS–/– mice, refer to chapter by Nakagawa et al., this vol., pp. 93–101)

References

1 Foley RN, Collins AJ: The growing economic burden of diabetic kidney disease. Curr Diab Rep 2009;9:460–465.
2 Gilbertson DT, Liu J, Xue JL, Louis TA, Solid CA, Ebben JP, Collins AJ: Projecting the number of patients with end-stage renal disease in the United States to the year 2015. J Am Soc Nephrol 2005;16:3736–3741.
3 Yang H, Bell TA, Churchill GA, Pardo-Manuel de Villena F: On the subspecific origin of the laboratory mouse. Nat Genet 2007; 39:1100–1107.
4 Kohan DE: Progress in gene targeting: using mutant mice to study renal function and disease. Kidney Int 2008;74:427–437.
5 Beck JA, Lloyd S, Hafezparast M, Lennon-Pierce M, Eppig JT, Festing MF, Fisher EM: Genealogies of mouse inbred strains. Nat Genet 2000;24:23–25.
6 Adler AI, Stevens RJ, Manley SE, BilousRW, Cull CA, Holman RR: Development and progression of nephropathy in type 2 diabetes: the United Kingdom Prospective Diabetes Study (UKPDS 64). Kidney Int 2003;63:225–232.
7 Finne P, Reunanen A, Stenman S, Groop PH, Gronhagen-Riska C: Incidence of end-stage renal disease in patients with type 1 diabetes. JAMA 2005;294:1782–1787.
8 Iyengar SK, Abboud, HE, Goddard KA, et al: Genome-wide scans for diabetic nephropathy and albuminuria in multiethnic populations: the family investigation of nephropathy and diabetes. Diabetes: 2007;56:1577–1585.
9 Rogus JJ, Poznik GD, Pezzolesi MG, et al: High-density single nucleotide polymorphism genome-wide linkage scan for susceptibility genes for diabetic nephropathy in type 1 diabetes: discordant sibpair approach. Diabetes 2008;57:2519–2526.
10 Breyer MD, Bottinger E, Brosius FC 3rd, Coffman TM, Harris RC, Heilig CW, Sharma K: Mouse models of diabetic nephropathy. J Am Soc Nephrol 2005;16:27–45.
11 Brosius FC 3rd, Alpers CE, Bottinger EP, et al: Mouse models of diabetic nephropathy. J Am Soc Nephrol 2009;20:2503–2512.
12 Gurley SB, Clare SE, Snow KP, Hu A, Meyer TW, Coffman TM: Impact of genetic background on nephropathy in diabetic mice. Am J Physiol Renal Physiol 2006;290:F214–F222.
13 Qi Z, Fujita H, Jin J, Davis LS, Wang Y, Fogo AB, Breyer MD: Characterization of susceptibility of inbred mouse strains to diabetic nephropathy. Diabetes 2005;54:2628–2637.
14 Kimmelstiel P, Wilson C: Intercapillary lesions in the glomeruli of kidney. Am J Pathol 1936;12:83–97.
15 Newburger, RA, Peters PJ: Intercapillary glomerulosclerosis: a syndrome of diabetes, hypertension, and albuminuria. Arch Int Med 1939;64:1252–1264.
16 Wilson JL, Root HF, Marble A: Diabetic nephropathy; a clinical syndrome. N Engl J Med 1951;245:513–517.
17 Gurley SB, Mach CL, Stegbauer J, Yang J, Snow KP, Hu A, Meyer TW, Coffman TM: Influence of genetic background on albuminuria and kidney injury in Ins2(+/C96Y) (Akita) mice. Am J Physiol Renal Physiol 2010;298:F788–F795.

18 Hudkins KL, Pichaiwong W, Wietecha T, et al: BTBR Ob/Ob mutant mice model progressive diabetic nephropathy. J Am Soc Nephrol 2010;21:1533–1542.

19 Bohle A, Wehrmann M, Bogenschutz O, Batz C, Muller CA, Muller GA: The pathogenesis of chronic renal failure in diabetic nephropathy. Investigation of 488 cases of diabetic glomerulosclerosis. Pathol Res Pract 1991; 187:251–259.

20 Lane P, Steffes MW, Mauer SM: Structural-functional relationships in type I insulin-dependent diabetes mellitus in humans. J Diabet Complications 1991;5:69–71.

21 White KE, Bilous RW: Type 2 diabetic patients with nephropathy show structural-functional relationships that are similar to type 1 disease. J Am Soc Nephrol 2000;11:1667–1673.

22 Mauer M, Zinman B, Gardiner R, et al: Renal and retinal effects of enalapril and losartan in type 1 diabetes. N Engl J Med 2009;361: 40–51.

23 Zhao HJ, Wang S, Cheng H, Zhang MZ, Takahashi T, Fogo AB, Breyer MD, Harris RC: Endothelial nitric oxide synthase deficiency produces accelerated nephropathy in diabetic mice. J Am Soc Nephrol 2006;17: 2664–2669.

24 Zheng S, Noonan WT, Metreveli NS, Coventry S, Kralik PM, Carlson EC, Epstein PN: Development of late-stage diabetic nephropathy in OVE26 diabetic mice. Diabetes 2004;53:3248–3257.

25 Ninomiya, T, Perkovic V, de Galan BE, et al: Albuminuria and kidney function independently predict cardiovascular and renal outcomes in diabetes. J Am Soc Nephrol 2009; 20:1813–1821.

26 Eijkelkamp WB, Zhang Z, Remuzzi G, et al: Albuminuria is a target for renoprotective therapy independent from blood pressure in patients with type 2 diabetic nephropathy: post hoc analysis from the Reduction of Endpoints in NIDDM with the Angiotensin II Antagonist Losartan (RENAAL) trial. J Am Soc Nephrol 2007;18:1540–1546.

27 Mann JF, Green D, Jamerson K, Ruilope, LM, Kuranoff SJ, Littke T, Viberti G: Avosentan for overt diabetic nephropathy. J Am Soc Nephrol 2010;21:527–535.

28 de Zeeuw D, Remuzzi G, Parving HH, Keane WF, Zhang Z, Shahinfar S, Snapinn S, Cooper ME, Mitch WE, Brenner BM: Albuminuria, a therapeutic target for cardiovascular protection in type 2 diabetic patients with nephropathy. Circulation 2004;110:921–927.

29 Szoka PR, Paigen K: Regulation of mouse major urinary protein production by the Mup-A gene. Genetics 1978;90:597–612.

30 Hansen HP, Hovind P, Jensen BR, Parving HH: Diurnal variations of glomerular filtration rate and albuminuria in diabetic nephropathy. Kidney Int 2002;61:163–168.

31 Levey, AS, Cattran, D, Friedman A, Miller WG, Sedor J, Tuttle K, Kasiske B, Hostetter T: Proteinuria as a surrogate outcome in CKD: report of a scientific workshop sponsored by the National Kidney Foundation and the US Food and Drug Administration. Am J Kidney Dis 2009;54:205–226.

32 Senador D, Kanakamedala K, Irigoyen MC, Morris M, Elased KM: Cardiovascular and autonomic phenotype of db/db diabetic mice. Exp Physiol 2009;94:648–658.

33 Sharma K, McCueP, Dunn SR: Diabetic kidney disease in the db/db mouse. Am J Physiol Renal Physiol 2003;284:F1138–F1144.

34 Bagi Z, Erdei N, Toth A, Li W, Hintze TH, Koller A, Kaley G: Type 2 diabetic mice have increased arteriolar tone and blood pressure: enhanced release of COX-2-derived constrictor prostaglandins. Arterioscler Thromb Vasc Biol 2005;25:1610–1616.

35 Mohan S, Reddick RL, Musi N, Horn DA, Yan B, Prihoda TJ, Natarajan M, Abboud-Werner SL: Diabetic eNOS knockout mice develop distinct macro- and microvascular complications. Lab Invest 2008;88:515–528.

36 Riches AC, Sharp JG, Thomas DB, Smith SV: Blood volume determination in the mouse. J Physiol 1973;228:279–284.

37 Sluiter W, Oomens LW, Brand A, Van Furth R: Determination of blood volume in the mouse with 51 chromium-labelled erythrocytes. J Immunol Methods 1984;73:221–225.

38 Takahashi N, Boysen G, Li F, Li Y, Swenberg JA: Tandem mass spectrometry measurements of creatinine in mouse plasma and urine for determining glomerular filtration rate. Kidney Int 2007;71:266–271.

39 Dunn SR, Qi Z, Bottinger EP, Breyer MD, Sharma K: Utility of endogenous creatinine clearance as a measure of renal function in mice. Kidney Int 2004;65:1959–1967.

40 Palm M, Lundblad A: Creatinine concentration in plasma from dog, rat, and mouse: a comparison of 3 different methods. Vet Clin Pathol 2005;34:232–236.

41 Yuen PS, Dunn SR, Miyaji T, Yasuda H, Sharma K, Star RA: A simplified method for HPLC determination of creatinine in mouse serum. Am J Physiol Renal Physiol 2004;286:F1116–F1119.

42 Keppler A, Gretz N, Schmidt R, Kloetzer HM, Groene HJ, Lelongt B, Meyer M, Sadick M, Pill J: Plasma creatinine determination in mice and rats: an enzymatic method compares favorably with a high-performance liquid chromatography assay. Kidney Int 2007;71:74–78.

43 Evans GO: The use of an enzymatic kit to measure plasma creatinine in the mouse and three other species. Comp Biochem Physiol B 1986;85:193–195.

44 Eisner C, Faulhaber-Walter R, Wang Y, Leelahavanichkul A, Yuen PS, Mizel D, Star RA, Briggs, JP, Levine M, Schnermann J: Major contribution of tubular secretion to creatinine clearance in mice. Kidney Int 2010;77:519–526.

45 Lorenz JN, Gruenstein E: A simple, non-radioactive method for evaluating single-nephron filtration rate using FITC-inulin. Am J Physiol 1999;276:F172–F177.

46 Qi Z, Whitt I, Mehta A, Jin J, Zhao M, Harris RC, Fogo AB, Breyer MD: Serial determination of glomerular filtration rate in conscious mice using FITC-inulin clearance. Am J Physiol Renal Physiol 2004;286:F590–F596.

47 Leiter EH: Multiple low-dose streptozotocin-induced hyperglycemia and insulitis in C57BL mice: influence of inbred background, sex, and thymus. Proc Natl Acad Sci USA 1982;79:630–634.

48 Nakagawa T, Sato W, Glushakova O, Heinig M, Clarke T, Campbell-Thompson M, Yuzawa Y, Atkinson MA, Johnson RJ, Croker B: Diabetic endothelial nitric oxide synthase knockout mice develop advanced diabetic nephropathy. J Am Soc Nephrol 2007;18:539–550.

49 Shimasaki Y, Yasue H, Yoshimura M, et al: Association of the missense Glu298Asp variant of the endothelial nitric oxide synthase gene with myocardial infarction. J Am Coll Cardiol 1998;31:1506–1510.

50 Zanchi A, Moczulski DK, Hanna, LS, Wantman M, Warram JH, Krolewski AS: Risk of advanced diabetic nephropathy in type 1 diabetes is associated with endothelial nitric oxide synthase gene polymorphism. Kidney Int 2000;57:405–413.

51 Shin Shin, Y, Baek SH, Chang KY, Park CW, Yang CW, Jin DC, Kim YS, Chang YS, Bang BK: Relations between eNOS Glu298Asp polymorphism and progression of diabetic nephropathy. Diabetes Res Clin Pract 2004;65:257–265.

52 Sato W, Kosugi T, Zhang L, et al: The pivotal role of VEGF on glomerular macrophage infiltration in advanced diabetic nephropathy. Lab Invest 2008;88:949–961.

53 Lewis A, Steadman R, Manley P, Craig K, de la Motte C, Hascall V, Phillips AO: Diabetic nephropathy, inflammation, hyaluronan and interstitial fibrosis. Histol Histopathol 2008;23:731–739.

54 Okazaki M, Saito Y, Udaka Y, Maruyama M, Murakami H, Ota S, Kikuchi T, Oguchi K: Diabetic nephropathy in KK and KK-Ay mice. Exp Anim 2002;51:191–196.

55 Ito T, Tanimoto M, Yamada K, et al: Glomerular changes in the KK-Ay/Ta mouse: a possible model for human type 2 diabetic nephropathy. Nephrology (Carlton) 2006;11:29–35.

56 Ninomiya H, Inomata T, Ogihara K: Microvasculature of hydronephrotic kidneys in KK-A(Y) mice. J Vet Med Sci 2000;62:1093–1098.

Matthew D. Breyer
Lead Generation Biology, Biotechnology Discovery Research, Lilly Research Laboratories
355 E. Merrill St.
Indianapolis, IN 46285 (USA)
Tel. +1 317 655 6783, Fax +1 317 277 2934, E-Mail breyerma@lilly.com

Lai KN, Tang SCW (eds): Diabetes and the Kidney.
Contrib Nephrol. Basel, Karger, 2011, vol 170, pp 165–171

Role of Triglyceride-Rich Lipoproteins in Renal Injury

Kit Fai Ng · Hnin Hnin Aung · John C. Rutledge

Division of Cardiovascular Medicine, University of California Davis School of Medicine,
 Davis, Calif., USA

Abstract

Dyslipidemia is implicated as a risk factor for the development of atherosclerosis. Specifically triglyceride-rich lipoproteins and their lipolysis products are shown to be proinflammatory and proapoptosis in both in vivo and in vitro studies with endothelium. However, the role of triglyceride-rich lipoproteins in the progression of kidney diseases is not clear. Epidemiology studies demonstrated a correlation between renal disease and blood lipids. Recent evidence suggests that the mechanism may involve cellular uptake of lipid and de novo lipogenesis. Further studies are needed to establish the relevance of these mechanistic studies in human pathophysiology.

Clinical and experimental studies have suggested that lipids and/or lipoproteins promote glomerular and tubulointerstitial injury through mediators such as reactive oxygen species, chemokines, cytokines, increased renal transforming growth factor-β1 (TGF-β1) gene expression, and macrophage extravasations into the glomeruli and tubules. In this review, we will examine the pathogenic mechanisms for lipid and lipoprotein-induced renal injury.

Epidemiology Linking Renal Disease with Elevated Blood Triglycerides

Several epidemiology studies have suggested a positive correlation between serum cholesterol, triglyceride (TG) levels and chronic kidney disease (CKD) [1, 2]. Dyslipidemia is commonly found among diabetic patients suffering from CKD, which includes elevated levels of TGs, atherogenic TG-rich apolipoprotein (apo) B-containing very-low-density lipoprotein cholesterol, low-density lipoprotein cholesterol, and intermediate-density lipoprotein cholesterol, but lower levels of high-density lipoprotein (HDL-C). The contribution of these TG-rich lipoproteins towards diabetic nephropathy (DN) has recently been reviewed [3]. Dyslipidemia has been suggested as a risk factor for the progression of renal insufficiency through development of glomerulosclerosis and tubulointerstitial lesions with concomitant accelerated atherosclerosis [4, 5], and is often found in patients with microalbuminuria and overt proteinuria. These patients also are noted for their elevated plasma level of circulating apoB [6], apoC-III, and apoA [7]. Cardiovascular risk factors such as elevated levels of circulating cell-adhesion molecules and systemic inflammatory markers can be found in the early stages of diabetic and nondiabetic CKD. Reduction in HDL-C level is also associated with an increased risk of DN as lower levels of HDL-C have been shown to predispose type 1 diabetics to suffer from renal injury more than those with higher levels of HDL-C [8]. This evidence suggests a parallel between the pattern of cardiovascular disease and renal diseases. However, the apoA5 $-1131T \rightarrow C$ polymorphism that affects TG levels has not been correlated with DN [9], although this may be the result of the differences between measurements of fasting and postprandial TG levels.

Clinical studies have also demonstrated lipid-lowering treatment can affect progression of renal diseases. Studies examining the effects of lipid-lowering drugs on diabetic kidney disease have recently been reviewed [3]. Of particular interest is the association between low estimated glomerular filtration rate (eGFR) and dyslipidemia in patients with CKD and cardiovascular disease. Among patients undergoing exercise therapy, improvements in eGFR are positively correlated with HDL-C level, and negatively correlated with TG level [10].

Most literature has suggested that the elevation of 3-hydroxy-3methylglutaryl coenzyme A (HMG-CoA) reductase activity and severity of albuminuria are markers of DN [11]. Large post-hoc data analyses show that HMG-CoA reductase inhibitors (statins) improved both lipid profile and renal function in patients with CKD [12, 13]. A meta-analysis study also showed that lipid and lipoprotein reduction decreases proteinuria in patients with CKD by preserving GFR [14]. However, the mechanism is still unknown, and it is possible that the renal protection conferred by statins is lipid independent. Fenofibrate, a TG-lowering drug that is also a peroxisome proliferator-activated receptor-α (PPAR-α) agonist, has been shown in the FIELD Helsinki substudy to decrease eGFR with a concomitant increase in plasma creatinine, complicating the analysis. Therefore, the benefit of this TG-lowering drug in improving renal function is still unclear [15]. Further studies will be needed to separate out the effect of lowering TGs and activating PPAR-α. Although the mechanism of lipid-induced renal injury remains to be determined in these human studies, they nonetheless showed benefit from lipid lowering and improvement in the prognosis of renal diseases.

Experimental Models of Renal Disease and Hypertriglyceridemia

Animal studies have corroborated the evidence from human studies and demonstrated benefit of lipid lowering therapy in experimental renal and metabolic diseases. For example, diabetic rats treated with a lipid-lowering statin drug, lovastatin, for 8 weeks showed significantly preserved GFR compared to untreated diabetic rats [16]. In addition, 12 months of lovastatin treatment suppressed glomerular TGF-β1 gene expression, albuminuria, and preserved renal mass and glomerular volume in diabetic rats [17]. Similar observations were made in statin-treated nondiabetic rats with kidney disease with the additional benefits of decreased glomerular and tubulointerstitial macrophage accumulation and reduced renal injury [18]. Zucker fatty rats treated with pioglitazone, an insulin sensitizer, have reduced serum TG levels and basement membrane thickening [19]. These studies suggest that the lipid profile improvement correlates with attenuation of classical renal disease symptoms.

The question remains, however, whether dyslipidemia is a cause or metabolic sequelae of renal diseases and the mechanism remains unclear. Cell culture studies have shown that plasma lipids are injurious to many renal cell types, leading to inflammation, cytokine production, increased lipid accumulation, and in some cases foam cell formation. The toxic effects of lipid accumulation are thought to be mediated through the increase in free fatty acid metabolites [see 35]. Very-low-density lipoprotein has been shown to stimulate cytokine production such as monocyte chemotactic protein-1 in human mesangial cell culture, resulting in increased monocyte adhesion [20]. Oxidized LDL also increases lectin-like oxidized low-density lipoprotein receptor 1 activity and inflammatory marker inter-cellular adhesion molecule 1 expression in the rat renal epithelial cell line NRK52E [21]. There is also evidence that saturated free fatty acids, a major class of lipolysis products, induce apoptosis in mesangial cells and podocytes [22, 23]. Low-density lipoprotein and scavenger receptors are also found on mesangial cells that can preferentially bind to atherogenic oxidized LDL and lead to cellular lipid accumulation. Recently, Berfield et al. [24] has demonstrated that insulin growth factor-1, a cytokine that is upregulated in DN, induces foam cell formation in mesangial cells by increasing caveola-mediated endocytosis with no effect on LDL and scavenger receptors. Mesangial cells are also shown to preferentially accumulate TG and downregulate PPAR-δ which decrease fatty acid oxidation [25]. These observations confirm in vivo observations of lipid accumulation in renal injuries.

However, it is also increasingly clear that increased lipid uptake is not solely responsible for lipid accumulation in mesangial cells. Studies have also pointed out increased cellular lipid content can be due to de novo lipogenesis. A series of experiments done by the Levi group have demonstrated the importance of sterol regulatory element-binding protein 1 (SREBP-1) in renal lipid accumulation. Using streptozotocin-treated rat as a type 1 diabetes model [26], FVB *db/db* mice as type 2 diabetes model [27], and diet-induced obese C57BL/6J mice as a metabolic syndrome model [28], they have demonstrated that increased renal expression of SREBP-1 leads to increased lipid accumulation. SREBP-1 expression increases fatty acid and TG synthesis through upregulating expression of enzymes such as fatty acid synthase and acetyl-CoA carboxylase. These animal models developed classical manifestations of DN such as lipid droplet formation in the mesangium and tubules, increased TGF-β synthesis, and mesangium expansion, further demonstrating the correlation between glomerulus foam cell formation and early signs of renal failure. Of particular interest is that these changes might not associate with changes in serum TG, except in the case of the type 2 diabetes FVB *db/db* mice model [29].

Dietary lipid studies have also demonstrated the link between renal SREBP-1 expression and high fat diet. In streptozotocin-treated rats, a diet high in saturated fatty acid induced SREBP-1 expression in the kidney with lipid accumulation in the glomerulus and tubular cells. These rats also had increased plasma TGs, albuminuria, and renal TG content [30]. In the same study, a polyunsaturated fatty acid (PUFA) diet seemed to normalize this SREBP-1 expression. The same SREBP-1 normalization can be seen in *db/db* mice fed with PUFA, resulting in reduced renal TG content and expansion of mesangial matrix [31]. These studies suggest that SREBP-1 expression and its impact on lipid accumulation in the kidney may be diet induced. Interestingly, SREBP-1 expression might also be related to age, as aging C57BL/6 mice have increased renal SREBP-1 expression that can be reversed by caloric restriction [28]. Modulating SREBP-1 expression using a farnesoid X receptor-activating ligand reverses deleterious effects in the kidney of Western diet-fed mice, including proteinuria, podocyte loss, mesangial expansion, and renal lipid accumulation [29]. In another study, feeding homozygous LDL receptor mice a Western diet with an apoA-1 mimetic peptide reduced renal SREBP-1c mRNA levels and TG levels, without changes in the serum level of lipids and lipoproteins levels [32]. Thus, targeting SREBP-1 may prove to be an effective treatment to lipid-induced renal injuries and warrants further clinical studies.

Another signaling pathway that plays an important role in lipid-induced renal disease is the peroxisome proliferator-activated receptor pathway. Clinical studies using fenofibrate, which is a PPAR-α agonist, have demonstrated some success in attenuating the progression of renal failure. A study comparing wild-type and heterozygous PPAR-γ mice, both fed a high-fat diet, has demonstrated that PPAR-γ insufficiency attenuates renal injury, lipid accumulation, and abnormal lipid metabolism [33]. On the other hand, in another study using Zucker fatty rats, treatment with a novel PPAR-γ ligand not only reduced serum TG concentration, but also reduced renal injuries such as glomerular expansion and proximal tubular damage as measured by urine N-acetyl-β-D-glucosaminidase [34]. Thus far, the results have been inconclusive about the PPAR-γ pathway, and further studies are needed to clarify its role.

Finally, lipids could damage the proximal tubules by codelivery with filtered albumin as described in the 'Trojan horse' hypothesis [35]. It is well known that albumin is capable of absorbing blood substances such as toxins and particularly free fatty acid. Elevated filtration of serum albumin is common in patients with renal diseases and can therefore increase exposure of albumin-bound free fatty acid to the proximal tubules which causes inflammation and injuries [36]. Oxidizable lipoproteins are also found in the urine of proteinuric subjects, and therefore it is possible for atherogenic lipid to injure proximal tubules in this way as well.

Conclusion

These epidemiological and animal studies suggest an association between concentration of plasma lipoproteins and lipids and development and progression of CKD. However, the renal injury could also involve de novo lipogenesis. Lipid control appears to be important in prevention and treatment of CKD, but further large-scale, prospective, randomized and controlled studies are still needed to confirm the benefits of lipid-lowering treatment in human as well as to clarify the mechanisms involved.

Acknowledgement

This work was supported by Nora Eccles Treadwell Foundation and NIH-HL55667.

References

1 Attman PO, Nyberg G, William-Olsson T, Knight-Gibson C, Alaupovic P: Dyslipoproteinemia in diabetic renal failure. Kidney Int 1992;42:1381–1389.

2 Rosario RF, Prabhakar S: Lipids and diabetic nephropathy. Curr Diab Rep 2006;6:455–462.

3 Rutledge JC, Ng KF, Aung HH, Wilson DW: Role of triglyceride-rich lipoproteins in diabetic nephropathy. Nat Rev Nephrol 2010;6:361–370.

4 Gerdes C, Fisher RM, Nicaud V, Boer J, Humphries SE, Talmud PJ, Faergeman O: Lipoprotein lipase variants D9N and N291S are associated with increased plasma triglyceride and lower high-density lipoprotein cholesterol concentrations: studies in the fasting and postprandial states: the European atherosclerosis research studies. Circulation 1997;96:733–740.

5 Dallongeville J, Lussier-Cacan S, Davignon J: Modulation of plasma triglyceride levels by apoE phenotype: a meta-analysis. J Lipid Res 1992;33:447–454.

6   Samuelsson O, Aurell M, Knight-Gibson C, Alaupovic P, Attman PO: Apolipoprotein-β-containing lipoproteins and the progression of renal insufficiency. Nephron 1993;63: 279–285.

7   Attman PO, Samuelsson O, Alaupovic P: Lipoprotein metabolism and renal failure. Am J Kidney Dis 1993;21:573–592.

8   Nelson CL, Karschimkus CS, Dragicevic G, Packham DK, Wilson AM, O'Neal D, Becker GJ, Best JD, Jenkins AJ: Systemic and vascular inflammation is elevated in early IgA and type 1 diabetic nephropathies and relates to vascular disease risk factors and renal function. Nephrol Dial Transplant 2005;20: 2420–2426.

9   Baum L, Ng MC, So WY, Poon E, Wang Y, Lam VK, Tomlinson B, Chan JC: A case-control study of apoA5 –1131t→c polymorphism that examines the role of triglyceride levels in diabetic nephropathy. J Diabetes Complications 2007;21:158–163.

10 Toyama K, Sugiyama S, Oka H, Sumida H, Ogawa H: Exercise therapy correlates with improving renal function through modifying lipid metabolism in patients with cardiovascular disease and chronic kidney disease. J Cardiol 2010;56:142–146.

11 McFarlane SI, Muniyappa R, Francisco R, Sowers JR: Clinical review 145: pleiotropic effects of statins: lipid reduction and beyond. J Clin Endocrinol Metab 2002;87:1451–1458.

12 Steinmetz OM, Panzer U, Stahl RA, Wenzel UO: Statin therapy in patients with chronic kidney disease: to use or not to use. Eur J Clin Invest 2006;36:519–527.

13 Tonelli M, Isles C, Curhan GC, Tonkin A, Pfeffer MA, Shepherd J, Sacks FM, Furberg C, Cobbe SM, Simes J, Craven T, West M: Effect of pravastatin on cardiovascular events in people with chronic kidney disease. Circulation 2004;110:1557–1563.

14 Fried LF, Orchard TJ, Kasiske BL: Effect of lipid reduction on the progression of renal disease: a meta-analysis. Kidney Int 2001;59:260–269.

15 Forsblom C, Hiukka A, Leinonen ES, Sundvall J, Groop PH, Taskinen MR: Effects of long-term fenofibrate treatment on markers of renal function in type 2 diabetes: The field Helsinki substudy. Diabetes Care 2010;33:215–220.

16 Inman SR, Stowe NT, Cressman MD, Brouhard BH, Nally JV Jr, Satoh S, Satodate R, Vidt DG: Lovastatin preserves renal function in experimental diabetes. Am J Med Sci 1999;317:215–221.

17 Kim SI, Han DC, Lee HB: Lovastatin inhibits transforming growth factor-beta1 expression in diabetic rat glomeruli and cultured rat mesangial cells. J Am Soc Nephrol 2000;11:80–87.

18 Jandeleit-Dahm K, Cao Z, Cox AJ, Kelly DJ, Gilbert RE, Cooper ME: Role of hyperlipidemia in progressive renal disease: focus on diabetic nephropathy. Kidney Int Suppl 1999;71:S31–S36.

19 Hirasawa Y, Matsui Y, Yamane K, Yabuki SY, Kawasaki Y, Toyoshi T, Kyuki K, Ito M, Sakai T, Nagamatsu T: Pioglitazone improves obesity type diabetic nephropathy: relation to the mitigation of renal oxidative reaction. Exp Anim 2008;57:423–432.

20 Lynn EG, Siow YL, O K: Very low-density lipoprotein stimulates the expression of monocyte chemoattractant protein-1 in mesangial cells. Kidney Int 2000;57:1472–1483.

21 Kelly KJ, Wu P, Patterson CE, Temm C, Dominguez JH: LOX-1 and inflammation: a new mechanism for renal injury in obesity and diabetes. Am J Physiol Renal Physiol 2008;294:F1136–F1145.

22 Mishra R, Simonson MS: Saturated free fatty acids and apoptosis in microvascular mesangial cells: palmitate activates pro-apoptotic signaling involving caspase 9 and mitochondrial release of endonuclease G. Cardiovasc Diabetol 2005;4:2.

23 Sieber J, Lindenmeyer MT, Kampe K, Campbell KN, Cohen CD, Hopfer H, Mundel P, Jehle AW: Regulation of podocyte survival and endoplasmic reticulum stress by fatty acids. Am J Physiol Renal Physiol 2010;299:F821–F829.

24 Berfield AK, Abrass CK: IGF-1 induces foam cell formation in rat glomerular mesangial cells. J Histochem Cytochem 2002;50:395–403.

25 Berfield AK, Chait A, Oram JF, Zager RA, Johnson AC, Abrass CK: IGF-1 induces rat glomerular mesangial cells to accumulate triglyceride. Am J Physiol Renal Physiol 2006;290:F138–F147.

26 Sun L, Halaihel N, Zhang W, Rogers T, Levi M: Role of sterol regulatory element-binding protein 1 in regulation of renal lipid metabolism and glomerulosclerosis in diabetes mellitus. J Biol Chem 2002;277:18919–18927.

27 Wang Z, Jiang T, Li J, Proctor G, McManaman JL, Lucia S, Chua S, Levi M: Regulation of renal lipid metabolism, lipid accumulation, and glomerulosclerosis in FVBdb/db mice with type 2 diabetes. Diabetes 2005;54:2328–2335.

28 Jiang T, Wang Z, Proctor G, Moskowitz S, Liebman SE, Rogers T, Lucia MS, Li J, Levi M: Diet-induced obesity in C57BL/6J mice causes increased renal lipid accumulation and glomerulosclerosis via a sterol regulatory element-binding protein-1c-dependent pathway. J Biol Chem 2005;280:32317–32325.

29 Wang XX, Jiang T, Shen Y, Adorini L, Pruzanski M, Gonzalez FJ, Scherzer P, Lewis L, Miyazaki-Anzai S, Levi M: The farnesoid X receptor modulates renal lipid metabolism and diet-induced renal inflammation, fibrosis, and proteinuria. Am J Physiol Renal Physiol 2009;297:F1587–F1596.

30 Yokoyama M, Tanigawa K, Murata T, Kobayashi Y, Tada E, Suzuki I, Nakabou Y, Kuwahata M, Kido Y: Dietary polyunsaturated fatty acids slow the progression of diabetic nephropathy in streptozotocin-induced diabetic rats. Nutr Res 2010;30:217–225.

31 Chin HJ, Fu YY, Ahn JM, Na KY, Kim YS, Kim S, Chae DW: Omacor, n-3 polyunsaturated fatty acid, attenuated albuminuria and renal dysfunction with decrease of SREBP-1 expression and triglyceride amount in the kidney of type II diabetic animals. Nephrol Dial Transplant 2010;25:1450–1457.

32 Buga GM, Frank JS, Mottino GA, Hakhamian A, Narasimha A, Watson AD, Yekta B, Navab M, Reddy ST, Anantharamaiah GM, Fogelman AM: D-4F reduces EO6 immunoreactivity, SREBP-1c mRNA levels, and renal inflammation in LDL receptor-null mice fed a Western diet. J Lipid Res 2008;49:192–205.

33 Kume S, Uzu T, Araki S, Sugimoto T, Isshiki K, Chin-Kanasaki M, Sakaguchi M, Kubota N, Terauchi Y, Kadowaki T, Haneda M, Kashiwagi A, Koya D: Role of altered renal lipid metabolism in the development of renal injury induced by a high-fat diet. J Am Soc Nephrol 2007;18:2715–2723.

34 Nakano R, Kurosaki E, Shimaya A, Kajikawa S, Shibasaki M: Ym440, a novel hypoglycemic agent, protects against nephropathy in Zucker fatty rats via plasma triglyceride reduction. Eur J Pharmacol 2006;549:185–191.

35 Bobulescu IA: Renal lipid metabolism and lipotoxicity. Curr Opin Nephrol Hypertens 2010;19:393–402.

36 Arici M, Chana R, Lewington A, Brown J, Brunskill NJ: Stimulation of proximal tubular cell apoptosis by albumin-bound fatty acids mediated by peroxisome proliferator activated receptor-gamma. J Am Soc Nephrol 2003;14:17–27.

John C. Rutledge, MD
Division of Cardiovascular Medicine, University of California Davis School of Medicine
GBSF 5404, 451 E. Health Sciences Drive
Davis, CA 95616 (USA)
Tel. +1 530 752 4713, Fax +1 530 752 3470, E-Mail jcrutledge@ucdavis.edu

Lai KN, Tang SCW (eds): Diabetes and the Kidney.
Contrib Nephrol. Basel, Karger, 2011, vol 170, pp 172–183

Study of Diabetic Nephropathy in the Proteomic Era

Visith Thongboonkerd

Medical Proteomics Unit, Office for Research and Development, Faculty of Medicine
Siriraj Hospital, and Center for Research in Complex Systems Science, Mahidol University,
Bangkok, Thailand

Abstract

Diabetic nephropathy (DN) remains a major complication of diabetes leading to end-stage renal disease (ESRD). The number of diabetic patients with ESRD who require renal replacement therapy has been increasing, implicating unsuccessful prevention of diabetic renal complication. This unfavorable outcome reflects insufficient knowledge on pathogenic mechanisms of DN and its detection at late stage. Currently, microalbuminuria is used for diagnosis of DN. However, some patients with microalbuminuria have advanced renal pathological changes indicating that microalbuminuria is not the perfect marker for early detection of DN and a better biomarker is urgently needed. Recently, particularly after the completion of the Human Genome Project, proteomics (systematic analysis of proteins for their identity, quantity and function) has been recognized as an emerging subdiscipline of modern sciences. During the past decade, proteomics has been widely applied to several areas of biomedical research, including the investigation of DN. This chapter summarizes recent progress of proteomics applied to DN with ultimate goals to better understand its pathogenic mechanisms and to search for novel biomarkers for earlier diagnosis.

Despite appropriate treatment, diabetic nephropathy (DN) remains a major complication in diabetic patients and is the leading cause of end-stage renal disease. This unfavorable outcome is most likely due to two main factors. First, the complexity of pathogenic mechanisms of DN is not well understood. Although several previous studies have been conducted to understand this diabetic renal complication, many pieces of the jigsaw of the entire image of its pathogenic mechanisms are still missing. It is not surprising that the prevention of this important diabetic complication is not so successful when this image remains hazy. Therefore, better understanding of its pathogenesis and pathophysiology is crucial. Second, detection of DN is problematic. Currently, microalbuminuria is the best noninvasive marker available for the diagnosis of DN [1]. However, many patients, who were normoalbuminuric and just recently had positive test for microalbuminuria, have advanced renal histopathological changes [2–4], indicating that microalbuminuria is not the perfect marker for early detection of DN. This may be due to a limitation to measure urinary albumin by conventional laboratory method, which is based on an immunoassay that can detect only the immunoreactive forms of albumin, whereas other forms of albumin remain undetectable. Recently, Candiano et al. [5] reported that there are at least 63 forms or fragments of albumin observed in the urine. To be able to detect all forms of albumin and to search for better biomarkers for earlier diagnosis of DN, an unbiased method that can simultaneously detect all forms of proteins is crucially required.

In the postgenomic era, with the completion of the Human Genome Project, several biotechnologies have been developed to utilize the flooding genomic information to explain biology, (patho)physiology, and pathogenic mechanisms of diseases. One of these postgenomic biotechnologies is proteomics (systematic analysis of proteins for their identity, quantity and function [6]), which emerged after the term 'proteome' (set of proteins encoded by the genome) was coined by Marc Wilkins in 1994 [7]. With the success of protein separation science and mass spectrometry, proteomics can be used for simultaneous analyses of a large number of proteins on the genomic scale. Because proteomics can be used for high-throughput screening of proteins that may be involved in biological phenomena and disease pathogenic mechanisms without any need for their specific antibodies, proteomics is considered as an ideal method to examine both immunoreactive and immunounreactive forms of albumin, as well as other proteins in the urine. In addition, this unbiased method is also useful for biomarker discovery and identification of new therapeutic targets. During the past decade, proteomics has been widely applied to several areas of biomedical research, including the investigation of DN. This chapter summarizes recent progress of proteomics applied to DN with ultimate goals to better understand its pathogenic mechanisms and to search for novel biomarkers for earlier diagnosis.

Brief Overview of Proteomic Technologies: Advantages/Disadvantages of Each Technology

The challenge in proteome analysis is that it involves the complex mixture of proteins, of which physicochemical properties vary and some components are low-abundant proteins. An ideal analytical technique used for proteome analysis should have high resolution, high sensitivity, and high throughput capabilities of protein separation and identification. In the postgenomic biomedical

Table 1. Advantages and disadvantages of various proteomic methodologies

Method	Advantages	Disadvantages
2-DE	available in most of proteomic labs, simple to perform, practical for quantitative analysis of multiple samples	time-consuming, no automation, not applicable for polypeptides/proteins with molecular masses <10 or >200 kDa, limited use for highly hydrophobic proteins
LC-MS/MS	automation, multidimensional, high sensitivity	time-consuming, quantitative analysis is not simple, too sensitive towards interfering compounds
SELDI-TOF MS	easy to use, high throughput, automation, low sample volume required	restricted to selected proteins, low-resolution mass spectrometry, low information content, reproducibility is still problematic, difficult to identify proteins/peptides from SELDI spectra
CE-MS	automation, high sensitivity, multidimensional, low sample volume required	not applicable for proteins with molecular masses >20 kDa, difficult to identify proteins/peptides from CE-MS data
Microarrays	ideal approach, high-throughput, capable of examining low molecular weight and low abundant proteins/peptides	currently does not cover all urinary proteins, conservation of protein functionality during immobilization
Microfluidic technology on a chip	easy to use, automation, high sensitivity, low sample volume required	subsequent identification of the resolved proteins is still limited

research, commonly used proteomic technologies include two-dimensional gel electrophoresis (2-DE), liquid chromatography coupled with tandem mass spectrometry (LC-MS/MS), surface-enhanced laser desorption/ionization time-of-flight mass spectrometry (SELDI-TOF MS), capillary electrophoresis coupled with mass spectrometry (CE-MS), protein microarrays, and microfluidic technology on a chip. Each of these technologies has its own advantages and disadvantages (table 1). Two-dimensional gel electrophoresis is available in most of proteomic laboratories, is simple to perform, and is practical for quantitative analysis of multiple samples. However, it is time-consuming, not applicable for proteins/peptides with molecular masses <10 or >200 kDa, and its utility for highly hydrophobic proteins is still limited [8, 9]. Liquid chromatography coupled with tandem mass spectrometry can be automated with a high sensitivity. However, it is time-consuming and too sensitive for interfering compounds [8, 9]. Moreover, quantitative analysis using LC-MS/MS is not simple and may not be applicable in a large sample size. Surface-enhanced laser desorption/ionization time-of-flight mass spectrometry is an easy-to-use system with an automation and high-throughput manner. Only a small volume, such as <10 µl, of urine sample is analyzable by SELDI-TOF MS. However, the information obtained from SELDI-TOF MS is restricted only to a particular group of proteins and its reproducibility is still an issue of concern [8, 9]. Moreover, identification of proteins/peptides from SELDI spectra is not easy. Capillary electrophoresis coupled with mass spectrometry is another sensitive system with a high-throughput mode that requires a small volume of urine sample. However, it is not suitable for analysis of proteins with molecular masses >20 kDa [8, 9]. Similar to SELDI-TOF MS, identification of proteins/peptides from CE-MS spectra is not easy. Microarray technology is one of the ideal methods for urinary proteome profiling [10]. This approach allows the high-throughput profiling of multiple proteins in urine samples and offers an opportunity to discover low molecular weight and low abundant biomarkers that may be missed by other (more commonly used) techniques [11]. However, its applications are handicapped by the difficulty to make the arrays to cover all urinary proteins and by the limited availability of antibodies or other ligands [11]. Other limitations include the conservation of protein functionality during immobilization as well as the provision of the required absolute and relative sensitivity [11]. Microfluidic technology on a chip is the latest technology applied to urinary proteome profiling [12]. It is also an easy-to-use system with automation and high throughput. Only 4 µl of urine sample is required for analysis by microfluidic technology on a chip. However, the identification of the resolved proteins is still problematic, as in the case of SELDI-TOF MS and CE-MS (table 1).

Proteomic Analysis of Diabetic Nephropathy

A PubMed search using the keywords 'proteomics' or 'proteomic' or 'proteome' together with 'diabetic nephropathy' retrieved 86 articles published between 2002 and 2010 (fig. 1). This number underscores the great potential of proteomics in the investigation of DN. The following sections summarize some of these recent studies.

Better Understanding of Pathophysiology/Pathogenic Mechanisms of Diabetic Nephropathy by Proteomics Approach
The most commonly used proteomic technique applied to DN is 2-DE, which separates proteins by differential pH in the first dimension and by differential molecular masses in the second dimension. After spot matching and comparative intensity analysis, spots of interest (differentially expressed spots) are excised and subjected to in-gel tryptic digestion and identification by mass spectrometry (mainly matrix-assisted laser desorption/ionization time-of-flight mass

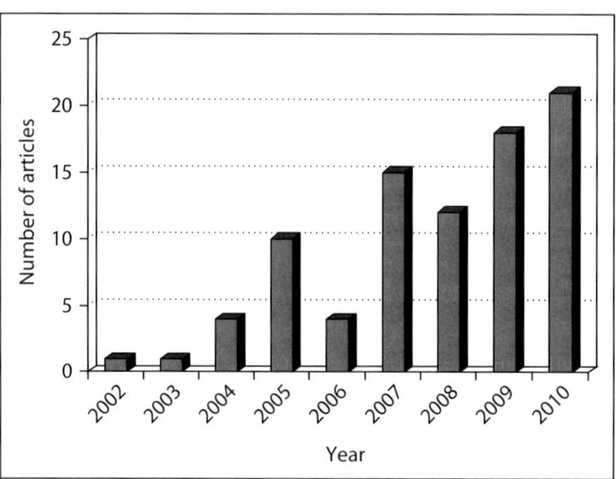

Fig. 1. Number of published articles related to proteomic analysis of DN.

spectrometry or MALDI-TOF MS followed by peptide mass fingerprinting for gel-based proteomic analysis). Thongboonkerd et al. [13] applied a gel-based proteomics approach to examine changes in the renal proteome of OVE26 murine model of type 1 diabetes and observed changes in renal tissue levels of 41 proteins. From these expression data, they focused their attention on the coordinated changes in elastin-elastase pathway. Elastase inhibitor was increased, whereas elastase was decreased. These investigators therefore hypothesized that elastin, which is highly hydrophobic extracellular matrix protein and could not be identified by 2-DE (because of its high hydrophobicity), should be accumulated in diabetic kidney [13]. They then performed immunohistochemical study confirming the hypothesis in both murine and human diabetic kidneys. Elastin was increased in the juxtaglomerular apparatus and renal tubules in the diabetic kidney. This is a good example to demonstrate that proteomics can be used to generate a new hypothesis, which can be confirmed by a conventional method [13].

The same laboratory also identified an increased level of renal calbindin-D28k, which is a vitamin-D-dependent calcium-binding protein, in OVE26 diabetic kidney [14]. The increased level of renal calbindin-D28k was confirmed by Western blot analysis. Immunohistochemistry revealed the increase in renal calbindin-D28k predominately in the distal nephron, implicating its role in tubular reabsorption of calcium during DN. These investigators also extended their study to type 2 diabetes using the *db/db* murine model, and found 39 altered renal proteins in this model [15]. Comparing changes identified in both type 1 and type 2 models, it was observed that some proteins had similar changes in both models, which might simply reflect hypoglycemic effects, whereas some other changes were unique in either type 1 or type 2 model, which might be the effects of hypoinsulinemia or hyperinsulinemia (insulin resistance). Some proteins had contradictory changes, which might be due to several factors (table 2).

Recently, several other groups also applied gel-based proteomics to DN. Tilton et al. [16] examined changes in renal cortical proteome in *db/db* mice and found alterations in the levels of 147 proteins. These altered proteins play significant roles in several biological processes, particularly in metabolic pathways. The investigators also created an interaction map of all the altered proteins, demonstrating that peroxisome proliferator-activated receptor-α was the common node of these protein-protein interactions. Another study by Ramachandra et al. [17] examined changes in membrane and cytosolic subproteomes in human mesangial cells under high-glucose condition. Four proteins were identified as altered proteins in the membrane fraction, whereas five altered proteins were identified in the cytosolic fraction. Among these altered proteins, calmodulin was identified as one of the increased proteins in the cytosolic fraction. Immunostaining confirmed the increase in calmodulin in mesangial cells exposed to high-glucose environment. Moreover, Western blot analysis confirmed the increase in calmodulin not only in high-glucose-exposed mesangial cells but also in rat and mouse diabetic kidney tissues. To address functional significance of calmodulin in diabetic kidney, these investigators examined effect of calmodulin inhibition on glucose uptake in mesangial cells. Functional analysis revealed that inhibition of calmodulin activity by two compounds, W7 and trifluoperazine, completely abolished transforming growth factor-β-induced glucose uptake in mesangial cells. These data indicate that calmodulin plays significant role in glucose transport in mesangial cells [17]. More recently, a similar study was performed by Li et al. [18] to identify changes in cellular proteome of rat mesangial cells upon exposure to high-glucose culture medium. Using 2-DE followed by MALDI-TOF MS, the investigators identified 28 proteins with altered levels in high-glucose condition. Among these, the reduction in proteasome subunit-α type 6 was confirmed not only in rat mesangial cells under the high-glucose condition, but also in streptozotocin-induced diabetic rats and spontaneous OLETF diabetic rat model.

Biomarker Discovery in Diabetic Nephropathy Using Proteomics Approach
An attempt to find urinary biomarkers for DN (and other glomerular diseases such as focal segmental glomerulosclerosis, lupus nephritis class IV, and membranous nephropathy) was made by Varghese et al. [19]. They found no single biomarker could differentiate type or cause of glomerular diseases. However, they used an artificial neural network to create a prediction algorithm and to define disease-specific urinary proteome profile or molecular signature containing multiple markers that could differentiate types of the diseases. They successfully defined such molecular signature containing 21 proteins, which provided a sensitivity of 75–86% and a specificity of 67–92% for the differentiation in a validation set of samples.

Table 2. Comparisons of changes in renal proteome between type 1 and type 2 diabetes

Altered renal proteins	Changes in renal tissue		Hypothesis
	type 1	type 2	
Antithrombin-III	increased	increased	hyperglycemic effects
Calbindin	increased	increased	
Calmodulin	increased	increased	
Complement component 1, q subcomponent-binding protein	increased	increased	
Myosin, light chain, smooth muscle	increased	increased	
Elastase inhibitor	increased	increased	
T-kininogen	increased	increased	
Apolipoprotein A-IV precursor	increased	unchanged	hypoinsulinemic effects
Cellular retinol-binding protein	increased	unchanged	
Contraception-associated protein 1	increased	unchanged	
Crocalbin-like protein	increased	unchanged	
Cytoskeletal tropomyosin	increased	unchanged	
Deoxyribonuclease I	increased	unchanged	
Elastase	decreased	unchanged	
Ferritin heavy chain	increased	unchanged	
Ferritin heavy chain	increased	unchanged	
Hippocampal cholinergic neurostimulating peptide precursor	increased	unchanged	
Histone H3.2	increased	unchanged	
HSPC207	increased	unchanged	
Na-H exchanger, isoform 3 regulator 1	increased	unchanged	
Skeletal muscle tropomyosin	increased	unchanged	
Vimentin	increased	unchanged	
A1AT	unchanged	decreased	hyperinsulinemic effects (or effects from insulin resistance)
Antithrombin-III	unchanged	increased	
Antithrombin-III	unchanged	increased	
Antithrombin-III	unchanged	increased	
Antithrombin-III	unchanged	increased	
ATP synthase delta chain, mitochondrial precursor	unchanged	increased	
Nucleobindin	unchanged	decreased	
Serine protease inhibitor 1-1	unchanged	decreased	
Ferritin light chain 1	increased	decreased	multifactorial
Phosphatidylethanolamine-binding protein	decreased	increased	

Biomarkers were also searched in serum using gel-based proteomics approach. Kim et al. [20] compared serum proteome profile of type 2 diabetic patients with microalbuminuria or end-stage renal disease to normoalbuminuric patients. They found changes in serum levels of 26 proteins in the two groups with renal involvement compared to normoalbuminuric patients. Some of the changes were confirmed by Western blot analysis, e.g. the decrease in extracellular glutathione peroxidase and apolipoprotein E. They also verified their results in a larger, independent set of samples, which clearly confirmed the decrease in glutathione peroxidase in diabetic patients with renal involvement.

Sharma et al. [21] applied gel-based proteomics to DN, but with another technique, the so-called difference in-gel electrophoresis, to find differentially excreted proteins in the urine. Basically, two groups of samples labeled with different dyes were run together within the same gel to avoid gel-to-gel variability. Using this technique, they identified 63 differentially excreted proteins in the urine of patients with long-standing diabetes, impaired renal function and overt proteinuria compared to the controls. One of these differentially excreted proteins was identified by mass spectrometry as α_1-antitrypsin (A1AT). ELISA clearly confirmed the marked increase in A1AT in the urine obtained from diabetic patients. Moreover, immunohistochemical study also revealed the increased tissue expression of A1AT in the diabetic kidney.

Subsequent difference in-gel electrophoresis study by Rao et al. [22] compared the normal urine with that obtained from type 2 diabetic patients with normoalbuminuria, microalbuminuria and macroalbuminuria. Using this technique, they successfully identified several proteins, which had differential levels among the four groups. Interestingly, they found a group of proteins with progressive changes in association with the degree of proteinuria. Surprisingly, they observed the decrease in urinary A1AT in diabetic patients, in contrast to the findings reported by Sharma et al. [21] as discussed above. As A1AT has multiple forms or fragments in the urine, these disparate results might reflect differential forms detected in these two studies.

Capillary electrophoresis coupled with mass spectrometry is another powerful tool for proteome profiling. The system contains capillary loops and electrospray ionization-TOF mass spectrometer. The specific software is used for analyzing the peptide signals and profiles. CE-MS was utilized by Meier et al. [23] to differentiate urinary polypeptide profiles of patients with type 1 diabetes from those of age-matched healthy controls. The analysis revealed that urinary polypeptide pattern of patients with diabetes significantly differed from that of the normal controls. They also performed subgroup analysis and found a pattern of potential biomarkers that could differentiate patients with DN from those without nephropathy. The same group also employed CE-MS to differentiate urinary polypeptide profiles of patients with type 2 diabetes from those of age-matched healthy controls [24]. In the latter study, they also observed specific urinary polypeptide profile for each subgroup based upon the degree of albuminuria.

Surface-enhanced laser desorption/ionization time-of-flight mass spectrometry or ProteinChip technology is widely used for examination of human body fluids. Protein sample is applied onto a chip surface carrying a functional group. After incubation, proteins that do not bind to the surface are removed, and the bound polypeptides or proteins are analyzed by TOF mass spectrometer. Recently, Cho et al. [25] utilized SELDI-TOF MS to search for potential biomarkers in the sera of streptozotocin-induced diabetic rats compared to control animals. They observed eight potential biomarkers in the serum, one of which was identified as C-reactive protein. The increased serum level of C-reactive protein was confirmed by ELISA.

In another study, Dihazi et al. [26] performed SELDI proteome profiling of urine samples obtained from type 2 diabetic patients with nephropathy, compared to those without nephropathy, proteinuric patients from nondiabetic causes, and healthy controls. They identified three protein spectra with mass/charge (or m/z) of 6,188, 11,744, and 14,766 that were differentially excreted among groups. Interestingly, these three spectra could differentiate diabetic patients with nephropathy from those without nephropathy. Subsequently, these SELDI spectra were identified by mass spectrometry as ubiquitin, β_2-microglobulin, and UbA52 ubiquitin ribosomal fusion protein, respectively. Therefore, these proteins may serve as potential biomarkers for the diagnosis of DN. These data were consistent with the findings reported in a subsequent similar study by Papale et al. [27], in which ubiquitin and β_2-microglobulin were identified as potential biomarkers of DN after initial SELDI profiling in a larger number of type 2 diabetic patients.

Additionally, Otu et al. [28] conducted a SELDI study of baseline urine samples collected from Pima Indians with type 2 diabetes 10 years before microalbuminuria occurred. A comparison with those who remained normoalbuminuric after 10 years in the first sample collection, revealed a molecular signature containing 12 SELDI spectra that could differentiate patients who subsequently developed DN from those who remained normoalbuminuric. This is a good work that yielded significant data to predict DN in a long-term follow-up study.

The recent advanced technology applied to urinary proteomics is a microfluidic technology on a chip [29]. Only 4 μl of the untreated urine could be analyzed by this system. Thongboonkerd et al. [29] employed this system to define disease-specific urinary biomarkers for DN and IgA nephropathy. By multiple comparisons, there were significant differences between the normal and diseased urine, and between DN and IgA nephropathy. There were three protein spectra or bands at approximately 12–15, 27–28 and 34–35 kDa that could distinguish DN from IgA nephropathy. Therefore, these differences may serve as the disease-specific biomarkers.

More recently, Merchant et al. [30] differentiated peptidome profiles of patients with type 1 diabetes with microalbuminuria who had normal renal function from those of patients who had declined renal function. Using LC-

MALDI-TOF MS profiling, they identified three peptides that were decreased [including fragments of collagen α_1(IV) and α_1(V) and tenascin-X] and three peptides that were increased (fragments of inositol polyphosphate 2-kinase, zona occludens 3, and FAT tumor suppressor 2) in the urine of patients with declined renal function. Immunohistochemical study also revealed the increased expression levels of tissue polyphosphate 2-kinase and zona occludens 3, consistent with the urinary peptidome data. These peptides may serve as the predictors of DN progression.

Conclusion

Since 2002, proteomics has played a significant role in the study of DN to address its pathophysiology/pathogenic mechanisms and to search for novel biomarkers for its earlier diagnosis. From these proteomic studies, several datasets have been reported to better understand pathophysiology/pathogenic mechanisms of DN, and a large number of biomarker candidates for diagnosis and prediction of DN have been identified. However, the information obtained from these studies remains on bench and is currently not closed to clinical application. To achieve the ultimate goal of clinical application, validation of the proteomic data in a larger sample size with appropriate control groups (i.e. other kidney diseases with similar clinical presentations and laboratory findings that mimic DN) is crucial. Although an attempt has been made in Europe to validate proteomic biomarkers in the urine of patients with DN, this multicenter study [31] limited its scope only to CE-MS data. There should be broader application of the data generated by other proteomic technologies as well. More importantly, translation of all these data to the ready and easy-to-use diagnostic kit(s) is needed. Even with some degree of progression in proteomics applied to DN, the total number of 86 articles published during 2002–2010 is much fewer than anticipated. Urgent attention is required to promote proteomic investigations of DN. It is envisaged that within the next 5–10 years proteomics will be more extensively applied to DN to unravel the complexity of its pathophysiology/pathogenic mechanisms and to define novel biomarkers for more accurate and earlier diagnosis as well as prognosis of DN. Thereafter, bedside application of proteomic data to diabetic patients will be achievable.

Acknowledgements

This work was supported by Office of the Higher Education Commission and Mahidol University under the National Research Universities Initiative, and The Thailand Research Fund (RTA5380005).

References

1 Caramori ML, Fioretto P, Mauer M: The need for early predictors of diabetic nephropathy risk: is albumin excretion rate sufficient? Diabetes 2000;49:1399–1408.

2 Chavers BM, Bilous RW, Ellis EN, Steffes MW, Mauer SM: Glomerular lesions and urinary albumin excretion in type I diabetes without overt proteinuria. N Engl J Med 1989;320:966–970.

3 Fioretto P, Steffes MW, Mauer M: Glomerular structure in nonproteinuric IDDM patients with various levels of albuminuria. Diabetes 1994;43:1358–1364.

4 Bangstad HJ, Osterby R, Dahl-Jorgensen K, Berg KJ, Hartmann A, Nyberg G, Frahm BS, Hanssen KF: Early glomerulopathy is present in young, type 1 (insulin-dependent) diabetic patients with microalbuminuria. Diabetologia 1993;36:523–529.

5 Candiano G, Musante L, Bruschi M, Petretto A, Santucci L, Del Boccio P, Pavone B, Perfumo F, Urbani A, Scolari F, Ghiggeri GM: Repetitive fragmentation products of albumin and alpha1-antitrypsin in glomerular diseases associated with nephrotic syndrome. J Am Soc Nephrol 2006;17:3139–3148.

6 Peng J, Gygi SP: Proteomics: the move to mixtures. J Mass Spectrom 2001;36: 1083–1091.

7 Wilkins MR, Sanchez JC, Gooley AA, Appel RD, Humphery-Smith I, Hochstrasser DF, Williams KL: Progress with proteome projects: why all proteins expressed by a genome should be identified and how to do it. Biotechnol Genet Eng Rev 1996;13:19–50.

8 Fliser D, Novak J, Thongboonkerd V, Argiles A, Jankowski V, Girolami MA, Jankowski J, Mischak H: Advances in urinary proteome analysis and biomarker discovery. J Am Soc Nephrol 2007;18:1057–1071.

9 Thongboonkerd V: Recent progress in urinary proteomics. Proteomics Clin Appl 2007;1:780–791.

10 Liu BC, Zhang L, Lv LL, Wang YL, Liu DG, Zhang XL: Application of antibody array technology in the analysis of urinary cytokine profiles in patients with chronic kidney disease. Am J Nephrol 2006;26:483–490.

11 Angenendt P: Progress in protein and antibody microarray technology. Drug Discov Today 2005;10:503–511.

12 Thongboonkerd V, Songtawee N, Sritippayawan S: Urinary proteome profiling using microfluidic technology on a chip. J Proteome Res 2007;6:2011–2018.

13 Thongboonkerd V, Barati MT, McLeish KR, Benarafa C, Remold-O'Donnell E, Zheng S, Rovin BH, Pierce WM, Epstein PN, Klein JB: Alterations in the renal elastin-elastase system in type 1 diabetic nephropathy identified by proteomic analysis. J Am Soc Nephrol 2004;15:650–662.

14 Thongboonkerd V, Zheng S, McLeish KR, Epstein PN, Klein JB: Proteomic identification and immunolocalization of increased renal calbindin-D28k expression in OVE26 diabetic mice. Rev Diab Stud 2005;2:17–24.

15 Thongboonkerd V, Barati MT, McLeish KR, Rovin BH, Pierce WM, Epstein PN, Klein JB: Altered elastase inhibitor and elastin expression in type 2 diabetic kidneys defined by proteomic analysis. J Am Soc Nephrol 2003;14 (suppl):600A.

16 Tilton RG, Haidacher SJ, Lejeune WS, Zhang X, Zhao Y, Kurosky A, Brasier AR, Denner L: Diabetes-induced changes in the renal cortical proteome assessed with two-dimensional gel electrophoresis and mass spectrometry. Proteomics 2007;7:1729–1742.

17 Ramachandra Rao SP, Wassell R, Shaw MA, Sharma K: Profiling of human mesangial cell subproteomes reveals a role for calmodulin in glucose uptake. Am J Physiol Renal Physiol 2007;292:F1182–F1189.

18 Li Z, Zhang H, Dong X, Burczynski FJ, Choy P, Yang F, Liu H, Li P, Gong Y: Proteomic profile of primary isolated rat mesangial cells in high-glucose culture condition and decreased expression of PSMA6 in renal cortex of diabetic rats. Biochem Cell Biol 2010;88:635–648.

19 Varghese SA, Powell TB, Budisavljevic MN, Oates JC, Raymond JR, Almeida JS, Arthur JM: Urine biomarkers predict the cause of glomerular disease. J Am Soc Nephrol 2007;18:913–922.

20 Kim HJ, Cho EH, Yoo JH, Kim PK, Shin JS, Kim MR, Kim CW: Proteome analysis of serum from type 2 diabetics with nephropathy. J Proteome Res 2007;6:735–743.

21 Sharma K, Lee S, Han S, Lee S, Francos B, McCue P, Wassell R, Shaw MA, RamachandraRao SP: Two-dimensional fluorescence difference gel electrophoresis analysis of the urine proteome in human diabetic nephropathy. Proteomics 2005;5:2648–2655.

22 Rao PV, Lu X, Standley M, Pattee P, Neelima G, Girisesh G, Dakshinamurthy KV, Roberts CT Jr, Nagalla SR: Proteomic identification of urinary biomarkers of diabetic nephropathy. Diabetes Care 2007;30:629–637.

23 Meier M, Kaiser T, Herrmann A, Knueppel S, Hillmann M, Koester P, Danne T, Haller H, Fliser D, Mischak H: Identification of urinary protein pattern in type 1 diabetic adolescents with early diabetic nephropathy by a novel combined proteome analysis. J Diabetes Complications 2005;19:223–232.

24 Mischak H, Kaiser T, Walden M, Hillmann M, Wittke S, Herrmann A, Knueppel S, Haller H, Fliser D: Proteomic analysis for the assessment of diabetic renal damage in humans. Clin Sci (Lond) 2004;107:485–495.

25 Cho WC, Yip TT, Chung WS, Leung AW, Cheng CH, Yue KK: Differential expression of proteins in kidney, eye, aorta, and serum of diabetic and non-diabetic rats. J Cell Biochem 2006;99:256–268.

26 Dihazi H, Muller GA, Lindner S, Meyer M, Asif AR, Oellerich M, Strutz F: Characterization of diabetic nephropathy by urinary proteomic analysis: identification of a processed ubiquitin form as a differentially excreted protein in diabetic nephropathy patients. Clin Chem 2007;53:1636–1645.

27 Papale M, Di Paolo S, Magistroni R, Lamacchia O, Di Palma AM, De Mattia A, Rocchetti MT, Furci L, Pasquali S, De Cosmo S, Cignarelli M, Gesualdo L: Urine proteome analysis may allow noninvasive differential diagnosis of diabetic nephropathy. Diabetes Care 2010;33:2409–2415.

28 Otu HH, Can H, Spentzos D, Nelson RG, Hanson RL, Looker HC, Knowler WC, Monroy M, Libermann TA, Karumanchi SA, Thadhani R: Prediction of diabetic nephropathy using urine proteomic profiling 10 years prior to development of nephropathy. Diabetes Care 2007;30:638–643.

29 Thongboonkerd V, Songtawee N, Sritippayawan S: Urinary proteome profiling using microfluidic technology on a chip. J Proteome Res 2007;6:2011–2018.

30 Merchant ML, Perkins BA, Boratyn GM, Ficociello LH, Wilkey DW, Barati MT, Bertram CC, Page GP, Rovin BH, Warram JH, Krolewski AS, Klein JB: Urinary peptidome may predict renal function decline in type 1 diabetes and microalbuminuria. J Am Soc Nephrol 2009;20:2065–2074.

31 Alkhalaf A, Zurbig P, Bakker SJ, Bilo HJ, Cerna M, Fischer C, Fuchs S, Janssen B, Medek K, Mischak H, Roob JM, Rossing K, Rossing P, Rychlik I, Sourij H, Tiran B, Winklhofer-Roob BM, Navis GJ: Multicentric validation of proteomic biomarkers in urine specific for diabetic nephropathy. PLoS One 2010;5:e13421.

Visith Thongboonkerd, MD, FRCPT
Medical Proteomics Unit, Office for Research and Development, Faculty of Medicine Siriraj Hospital, and
Center for Research in Complex Systems Science, Mahidol University
12th Floor Adulyadej Vikrom Building, 2 Prannok Road
Bangkoknoi, Bangkok 10700 (Thailand)
Tel./Fax +66 2 4184793, E-Mail thongboonkerd@dr.com

Lai KN, Tang SCW (eds): Diabetes and the Kidney.
Contrib Nephrol. Basel, Karger, 2011, vol 170, pp 184–195

Treatment and Landmark Clinical Trials for Renoprotection

Rose Z.W. Ting[a] · Andrea O.Y. Luk[a] · Juliana C.N. Chan[a,b]

[a]Department of Medicine and Therapeutics, and [b]Hong Kong Institute of Diabetes and Obesity,
The Chinese University of Hong Kong, The Prince of Wales Hospital, Hong Kong SAR, China

Abstract

During the last two decades, many large-scale randomized clinical trials have confirmed the importance of lowering blood pressure and inhibiting the renin-angiotensin-aldosterone system (RAAS) to preserve renal function in patients with chronic kidney disease due to different causes. With growing epidemic of type 2 diabetes, the burden of diabetic nephropathy (DN) will continue to grow. Based on a large body of epidemiological, experimental and interventional studies, strict glycemic control, blood pressure lowering and RAAS blockade are now the recommended strategies in the prevention and control of DN. Both angiotensin-converting enzyme inhibitor (ACEI) and angiotensin receptor blocker (ARB) have been shown unequivocally to reduce proteinuria and preserve renal function in affected patients. Importantly, therapeutic responses are dose dependent, and these drugs should be given in maximal dosages as tolerated. Combined use of ACEI and ARB reduced proteinuria and rate of decline of renal function, although it did not appear to confer extra renoprotection over the use of either agent alone. Spironolactone and aliskiren are alternative drugs that block the RAAS and have been demonstrated to further reduce proteinuria when added to ACEI or ARB. In the absence of long-term data about their safety and efficacy on renal function, combined use of these agents must be administered with caution after careful consideration of risk-benefit ratio with close monitoring for adverse effects, notably hyperkalemia. Pending further evidence, the use of a team approach to attain multiple treatment goals will improve renal outcomes in these high risk subjects.

Copyright © 2011 S. Karger AG, Basel

In this global epidemic of diabetes, diabetic nephropathy (DN) has caused profound impact on public health and society. Diabetes is the most common cause of end-stage renal disease (ESRD) and the major factor driving the increasing number of patients requiring renal replacement therapy worldwide.

Albuminuria and chronic kidney disease are also strong determinants of cardiovascular diseases. Thus, early identification and prompt treatment to halt or slow down the progression of DN are important strategies to reduce cardiovascular events and ESRD [1, 2]. In this chapter, we review the evidence for conventional therapies recommended for renoprotection especially in type 2 diabetes.

Glycemic Control Optimization

Diabetes is characterized by chronic hyperglycemia which causes proteinuria and renal damage through a number of mechanisms including cytokines, intracellular, metabolic, hemodynamic and growth factors [3]. Several landmark studies have confirmed that strict glycemic control can prevent the development of DN. The UKPDS (United Kingdom Prospective Diabetes Study) [4] recruited patients with type 2 diabetes to receive intensive blood glucose lowering or conventional therapy. At 10 years of follow-up, the intensive arm achieved a median glycated hemoglobin (HbA1c) of 7.0%, while the conventional arm had a median HbA1c of 7.9%. At close to 1% difference in HbA1c, the risk of incident microalbuminuria was 33% lower in the intensive arm (p < 0.001). Similar findings were reported in patients with type 1 diabetes in the DCCT (Diabetes Control and Complications Trial) [5]. In the recent ADVANCE (Action in Diabetes and Vascular Disease: Preterax and Diamicron Modified Release Controlled Evaluation) Study, intensive blood glucose control in type 2 diabetic patients with an achieved HbA1c of 6.5% after a mean duration of 5 years reduced the incidence of major microvascular events by 14%, mainly due to reduction in the incidence of DN (4.1 vs. 5.2%) compared to the conventional treatment group with an HbA1c of 7.3% [6]. However, due to the narrow risk-benefit ratio for intensive blood glucose lowering and the increased risk of coronary heart disease with hypoglycemia [7], aiming at HbA1c <7% is generally recommended by most international guidelines [8].

Blood Pressure Control Optimization

Hypertension is a strong predictor of development of diabetic vascular complications including nephropathy. Hypertension is highly prevalent in patients with type 2 diabetes and often co-occurs with other components of the metabolic syndrome such as central obesity and dyslipidemia [9]. Tight blood pressure control has been shown to reduce the risk of renal end points. In a prospective observational study using the UKPDS data [10], for every 10 mm Hg decrease in systolic blood pressure, the risk of microvascular complications including renal failure was lowered by 13% (p < 0.0001). Analysis of a large number of randomized controlled trials showed linear relationship between systolic blood pressure and rate of decline in glomerular filtration rate [11].

Most treatment guidelines recommend achieving a systolic blood pressure less than 130 mm Hg and diastolic blood pressure less than 80 mm Hg to preserve renal function [8]. However, reducing blood pressure to levels lower than these recommended values may not confer additional clinical benefits. The post-hoc analysis of the IDNT (Irbesartan Diabetic Nephropathy Trial) data revealed that while progressive lowering of systolic blood pressure to 120 mm Hg was associated with improved renal outcome, further reduction to less than 120 mm Hg increased the risk of all cause mortality [12]. In the ACCORD (Action to Control Cardiovascular Risk in Diabetes) blood pressure trial, intensive blood pressure control (systolic BP <120 mm Hg) was associated with an increased risk of elevated serum creatinine to more than 133 μmol/L in men or 115 μmol/L in women [13]. Based on this evidence, it is reasonable to aim to achieve a systolic blood pressure of 120–130 mm Hg in type 2 diabetic patients.

There is strong evidence suggesting that inhibition of the renin-angiotensin-aldosterone system (RAAS) confers antiproteinuric and renoprotective effects independent of blood pressure lowering. Angiotensin II receptor blockers (ARB) and angiotensin-converting enzyme inhibitors (ACEI) are now the preferred first-line antihypertensive drugs for diabetic patients especially those with nephropathy unless contraindicated. However, most studies have indicated that two or more drugs are needed to control blood pressure in these subjects [14]. Some studies have shown that dihydropyridine calcium channel blockers, such as nifedipine and amlodipine, might not confer renoprotection probably due to abolishment of renal autoregulation [15]. However, given its potent blood pressure-lowering effects, combined use of this drug class with inhibitors of the RAAS has been shown to confer additional cardiovascular benefits compared to monotherapy with either drug [16]. In the randomized BENEDICT (Bergamo Nephrologic Diabetes Complications Trial), 1,204 type 2 diabetic subjects with normoalbuminuria were randomly assigned to receive at least 3 years of treatment with trandolapril (an ACEI) plus verapamil (a non-dihydroperidine calcium channel blocker), trandolapril alone, verapamil alone or placebo aiming at a blood pressure of 120/80 mm Hg. After adjusting for baseline variables, treatment with trandolapril plus verapamil and trandolapril alone delayed the onset of microalbuminuria by factors of 2.6 and 2.1 compared to placebo, while the effect of verapamil was similar to that of placebo [17].

Blockade of the Renin-Angiotensin-Aldosterone System

The RAAS plays a key role in the pathogenesis of DN. Besides causing systemic and local hypertension, angiotensin II causes vasoconstriction of the glomerular

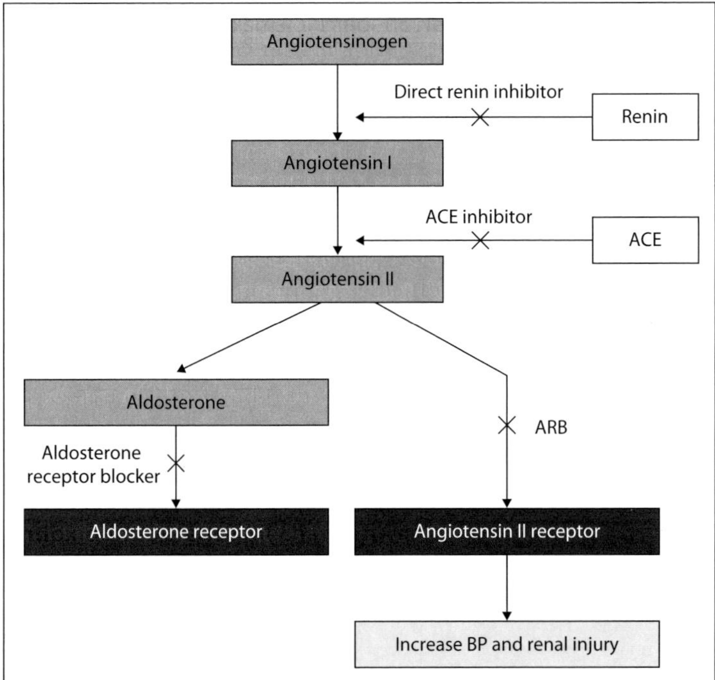

Fig. 1. Drugs that block the RAAS in DN.

efferent arteriole to increase filtration pressure and proteinuria. Besides, tissue activation of the RAAS promotes local inflammation and fibrosis resulting in loss of structure and function [18]. Thus, RAAS blockade confers renoprotection by attenuating maladaptive hemodynamic, functional and structural changes. Four drug classes are known to antagonize the action of angiotensin II and aldosterone through direct or indirect mechanisms: (a) ACEI, (b) ARB, (c) direct renin inhibitor, and (d) aldosterone receptor antagonist (fig. 1).

Angiotensin-Converting Enzyme Inhibitors
The effect of ACEI on DN was investigated mainly in patients with type 1 diabetes. In a 3-year study of 409 type 1 diabetic patients with established nephropathy, Lewis et al. [19] observed a significant 48% risk reduction in doubling of serum creatinine and 50% risk reduction in reaching combined end point of dialysis, renal transplantation and death in patients assigned to receive captopril compared to those receiving placebo. The rate of increase in serum creatinine was lower in the captopril than the placebo group (p = 0.004). Treatment with captopril was also associated with sustained reduction in urinary protein excretion (p = 0.001). Renal outcome was independent of blood pressure. Based on current evidence, ACEI should be used in all type 1 diabetic patients with nephropathy.

Angiotensin Receptor Blocker

The benefit of angiotensin receptor blocker (ARB) in renal preservation has been confirmed in several large randomized trials of patients with type 2 diabetes. In the IRMA-2 (Irbesartan Microalbuminuria in Hypertensive Patients with Type 2 Diabetes) study [20], 590 patients with microalbuminuria and normal renal function were randomized to receive irbesartan or placebo. After 2 years, the risk of developing macroalbuminuria was reduced by up to 68% in the irbesartan arm (p < 0.001). Moreover, treatment with irbesartan resulted in significant reduction in urinary albumin excretion (p < 0.001).

The IDNT investigated 1,715 type 2 diabetic patients with overt nephropathy and compared treatment with irbesartan 300 mg daily, amlodipine 10 mg daily or placebo [21]. Over a mean follow-up period of 2.6 years, the progression to primary composite end point of doubling of serum creatinine, ESRD and death of any cause was 20% (p = 0.02) and 23% (p = 0.006) lower in patients assigned to irbesartan compared to placebo and amlodipine, respectively. Treatment with irbesartan attenuated the rate of increase in serum creatinine and prolonged the time to doubling of serum creatinine. Amongst patients with doubling of serum creatinine, 50% were predicted to develop ESRD requiring dialysis in 12 months.

The RENAAL (Reduction of Endpoints in NIDDM with the Angiotensin II Antagonist Losartan) study enrolled 1,513 type 2 diabetic patients with overt nephropathy followed up for an average of 3.4 years [22]. Compared to placebo, treatment with losartan was associated with 25% risk reduction in doubling of serum creatinine (p = 0.006) and 28% risk reduction in the incidence of ESRD (p = 0.002). Reduction in proteinuria at 6 months was predictive of long-term renoprotective effects [23]. In a subgroup analysis of the RENAAL study with 13% of its participants coming from Asia, Asians benefited more from losartan treatment than their white counterparts [24]. The composite primary end point of doubling of serum creatinine, ESRD and death, was reduced by 35% (p = 0.02) in Asian subjects treated with losartan [24], compared to 16% (p = 0.02) in the entire cohort [23]. In all these studies, the observed antiproteinuric effect and stabilization of renal function with the use of ARB were independent of blood pressure. These findings concur with the known pluripotent properties of this drug class including local anti-inflammatory and antifibrotic effects.

Importantly, the renoprotective response was dose dependent, and therefore the maximum tolerated dose of ARB should be used [25]. In an open-label randomized study, 360 non-diabetic patients with renal insufficiency were randomly assigned to four groups: conventional dosage of benazepril (10 mg daily), individual uptitration of benazepril (median 20 mg daily; range 10–40 mg daily), conventional dosage of losartan (50 mg daily), or individual uptitration of losartan (median 100 mg daily; range 50–200 mg daily). After a median follow-up period of 3.7 years, optimal antiproteinuric dosages of benazepril and losartan achieved through uptitration were associated with 51 and 53% risk reduction for primary end point of dialysis, doubling of serum creatinine and all-cause death (p = 0.028 and 0.022, respectively). These effects were independent of blood pressure control and not associated with increased risk of major adverse events [26].

There is limited information regarding the comparative effectiveness of ACEI and ARB in DN, although the general consensus is that the two drug classes are of equivalent efficacy and are interchangeable. In a 5-year study involving 250 type 2 diabetic subjects with early nephropathy, treatments with telmisartan (80 mg daily), an ARB, and enalapril (20 mg daily), an ACEI, were associated with similar changes in glomerular filtration rate (–17.5 vs. –15.0 ml/min/1.73 m^2) suggesting clinical equivalence of these two classes of drugs at least in subjects with early nephropathy [27].

About 10% of patients treated with ACEI develop dry cough [28] with certain ethnic groups including Chinese being more frequently affected [29]. This is considered a class effect due to increased levels of kinins associated with angiotensin-converting enzyme (ACE) inhibition. In patients intolerant of ACEI due to intractable cough, an ARB can be substituted. Both ACEI and ARB can cause hyperkalemia requiring monitoring of serum potassium. This is especially important in patients with renal insufficiency and tendency to develop hyperkalemia such as type 4 renal tubular acidosis (hyporeninemic hypoaldosteronism). Hyperkalemia can be managed by restricting dietary intake of potassium, addition of sodium bicarbonate and/or thiazide diuretics [30].

Combination Therapy with Angiotensin-Converting Enzyme Inhibitor and Angiotensin Receptor Blocker
Whether the combined use of ACEI and ARB provides greater renal benefit than use of single agent remains inconclusive. An antiproteinuric response is not universal to all patients treated with ACEI or ARB. In the RENAAL study, up to 30% of patients treated with losartan progressed to ESRD or had doubling of serum creatinine despite receiving optimal care in a clinical trial setting [22]. This may be due to incomplete and nonsustained blockade of the RAAS due to generation of angiotensin II via non-ACE pathway or rebound in angiotensin II and aldosterone levels after initial treatment with ACEI.

Meta-analysis has suggested that dual therapy of ARB and ACEI may be more effective than either drug alone in reducing proteinuria [31], and that the extent of albuminuria reduction was closely correlated with risk reduction in renal end point [32]. In the ONTARGET (Ongoing Telmisartan Alone and in Combination with Ramipril Global Endpoint Trial) [33] that enrolled over 25,600 patients of high cardiovascular risk with either diabetes or established atherosclerotic vascular disease, combination therapy of telmisartan and ramipril slowed the progression of proteinuria. However, the risk of composite end point of doubling of serum creatinine, dialysis, renal transplantation and death was not superior to treatment with either drug alone. Furthermore, combination therapy was associated with significantly greater risk of systemic hypotension resulting in higher rate of treatment withdrawal. Some of these inconsistent findings may be due to the heterogeneity of the study population with the majority of patients not having diabetes.

In the ORIENT (Olmesartan Reducing Incidence of ESRD in DN Trial) which enrolled 577 Asian type 2 diabetic patients with proteinuria and renal insufficiency, 75% of whom were treated with ACEI, treatment with olmesartan, an ARB, was more effective in reducing albuminuria and rate of change of reciprocal serum creatinine (–0.071 vs. –0.089 dl/mg/year, p = 0.04), independent of blood pressure lowering, although the incidence of primary renal end point was similar between the two groups [34].

Direct Renin Inhibitor
Direct renin inhibitor blocks the conversion from angiotensinogen to angiotensin I and provides alternative mechanism for downregulation of the RAAS. Aliskiren is the only direct renin inhibitor that has been evaluated for its effect on renoprotection in DN. In the AVOID (Aliskiren in the Evaluation of Proteinuria in Diabetes) study [35], 599 type 2 diabetic patients with hypertension and overt nephropathy on losartan at baseline were randomized to receive aliskiren or placebo for 6 months. Treatment with aliskiren reduced mean urinary albumin-creatinine ratio by 20% compared to placebo (p < 0.001) after adjustment for blood pressure. The incidence of severe hyperkalemia, defined as serum potassium greater than 6.0 mmol/L, tended to be higher in the aliskiren than the placebo group (4.7 vs. 1.7%, p = 0.06). Otherwise, aliskiren was well tolerated with similar incidence of adverse events compared with placebo. Thus, aliskiren on top of maximal conventional renoprotective treatment reduced diabetic albuminuria, at least in the short-term, although its long-term effects on renoprotection remain to be confirmed.

Aldosterone Receptor Antagonist
Aldosterone receptor antagonist, such as spironolactone, has been shown to further reduce proteinuria when added to an ACEI or ARB. Mehdi et al. [36] compared the antiproteinuric effect of 48-week treatment with spironolactone, losartan and placebo in type 2 diabetic patients with macroalbuminuria. For the same degree of blood pressure lowering, the group that received spironolactone had the largest reduction in proteinuria. In another study by Schjoedt et al. [37], spironolactone as an add-on therapy to ACEI or ARB resulted in additional reduction in proteinuria in patients with DN. Nevertheless, since most studies that evaluated the renal effect of spironolactone were of small scale and short duration, it is premature to draw a conclusion regarding its long-term clinical benefit. Furthermore, the risk

of hyperkalemia is expected to increase significantly with concurrent use of spironolactone and ACEI or ARB.

Multifactorial Treatment to Prevent Diabetic Nephropathy

The pathogenesis of DN is multifactorial and involves hemodynamic, metabolic, inflammatory and growth factors [3]. In a prospective cohort of 5,829 Chinese type 2 diabetic patients with a median follow-up period of 5 years, 12% developed incident chronic kidney disease defined as estimated glomerular filtration rate less than 60 ml/min/1.73 m^2. The major predictors included age, male sex, smoking, disease duration, HbA1c, low body mass index, albuminuria, glomerular filtrate rate, LDL cholesterol and components of metabolic syndrome including waist circumference, blood pressure and high triglyceride [38]. There are experimental studies supporting the pathogenetic roles of lipotoxicity and inflammation in DN [2]. Although epidemiological and meta-analyses have suggested possible benefits of lipid lowering on reducing proteinuria and onset of chronic kidney disease, randomized clinical studies are needed to confirm these observational studies [39].

In keeping with the multicausality of chronic kidney disease, attainment of multiple treatment goals in blood pressure, glucose and lipid control and blockade of the RAAS was associated with increased probability of remission, and reduced probability of progression of diabetic albuminuria [40, 41]. In a multicentre randomized study involving 205 Chinese type 2 diabetic patients with renal insufficiency, patients treated with structured care delivered by a doctor-nurse team were 3-fold more likely to attain multiple treatment goals compared to usual care (61 vs. 28%). Patients who attained multiple treatment goals had 57% risk reduction in reaching composite renal end point including dialysis, serum creatinine ≥500 μmol/L and all-cause death [42]. As in the Steno 2 study [43], these findings highlight the importance of using a multifactorial strategy to optimize multiple risk factors to preserve renal function in type 2 diabetic patients (fig. 2).

Conclusion

The conventional strategies for renoprotection especially in type 2 diabetic patients include optimizing blood pressure and glycemic control, and the use of therapeutic agents (mainly ACEI or ARB) to block the RAAS. The development of a system which promotes the use of a team approach to attain multiple treatment goals and ensure the use of these renoprotective drugs is important in translating these results of efficacy to clinical effectiveness. These treatment strategies should be combined with an awareness and detection program to

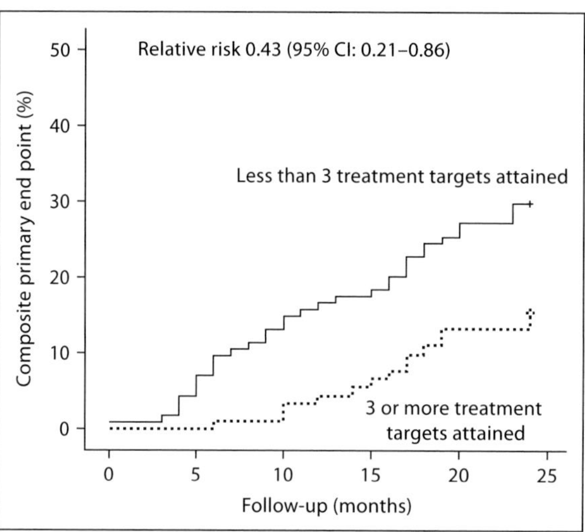

Fig. 2. Beneficial effects of attaining multiple treatment targets on development of ESRD and all-cause death. Kaplan-Meier analysis showing the cumulative incidence of the primary composite end point of death or ESRD defined as dialysis or the need for dialysis or plasma creatinine level ≥500 μmol/L in type 2 diabetic patients with renal insufficiency stratified by attainment of ≥3 prespecified treatment targets. Treatment targets: (1) blood pressure <130/80 mm Hg, (2) AIC <7%, (3) LDL cholesterol <2.6 mmol/L, (4) fasting plasma triglyceride <2 mmol/L, and (5) treatment with ACE inhibitors and/or ARBs. From Chan et al. [42] with permission.

reduce the morbidity and mortality associated with chronic kidney disease, especially in patients with diabetes.

References

1 Rossing P: Diabetic nephropathy: worldwide epidemic and effects of current treatment on natural history. Curr Diab Rep 2006;6: 479–483.

2 Luk A, Chan JC: Diabetic nephropathy – what are the unmet needs? Diabetes Res Clin Pract 2008;82(suppl 1):S15–S20.

3 Schrijvers BF, De Vriese AS, Flyvbjerg A: From hyperglycemia to diabetic kidney disease: the role of metabolic, hemodynamic, intracellular factors and growth factors/cytokines. Endocr Rev 2004;25:971–1010.

4 UK Prospective Diabetes Study (UKPDS) Group: Intensive blood glucose control with sulphonylureas or insulin compared with conventional treatment and risk of complications in patients with type 2 diabetes (UKPDS 33). Lancet 1998;352:837–853.

5 The Diabetes Control and Complications Trial Research Group: The effect of intensive treatment of diabetes on the development and progression of long-term complications in insulin-dependent diabetes mellitus. N Engl J Med 1993;329:977–986.

6   Patel A, MacMahon S, Chalmers J, Neal B, Billot L, Woodward M, Marre M, Cooper M, Glasziou P, Grobbee D, et al: Intensive blood glucose control and vascular outcomes in patients with type 2 diabetes. N Engl J Med 2008;358:2560–2572.

7   Desouza CV, Bolli GB, Fonseca V: Hypoglycemia, diabetes, and cardiovascular events. Diabetes Care 2010;33:1389–1394.

8   American Diabetes Association: Standards of medical care in diabetes. Diabetes Care 2010;33(suppl 1):S11–S61.

9   Alberti KG, Eckel RH, Grundy SM, Zimmet PZ, Cleeman JI, Donato KA, Fruchart JC, James WP, Loria CM, Smith SC Jr: Harmonizing the metabolic syndrome: a joint interim statement of the International Diabetes Federation Task Force on Epidemiology and Prevention; National Heart, Lung, and Blood Institute; American Heart Association; World Heart Federation; International Atherosclerosis Society; and International Association for the Study of Obesity. Circulation 2009;120:1640–1645.

10 Adler AI, Stratton IM, Neil HA, Yudkin JS, Matthews DR, Cull CA, Wright AD, Turner RC, Holman RR: Association of systolic blood pressure with macrovascular and microvascular complications of type 2 diabetes (UKPDS 36): prospective observational study. BMJ 2000;321:412–419.

11 Bakris GL, Williams M, Dworkin L, Elliott WJ, Epstein M, Toto R, Tuttle K, Douglas J, Hsueh W, Sowers J: Preserving renal function in adults with hypertension and diabetes: a consensus approach. National Kidney Foundation Hypertension and Diabetes Executive Committees Working Group. Am J Kidney Dis 2000;36:646–661.

12 Pohl MA, Blumenthal S, Cordonnier DJ, De Alvaro F, Deferrari G, Eisner G, Esmatjes E, Gilbert RE, Hunsicker LG, de Faria JB, et al: Independent and additive impact of blood pressure control and angiotensin II receptor blockade on renal outcomes in the irbesartan diabetic nephropathy trial: clinical implications and limitations. J Am Soc Nephrol 2005;16:3027–3037.

13 Cushman WC, Evans GW, Byington RP, Goff DC Jr, Grimm RH Jr, Cutler JA, Simons-Morton DG, Basile JN, Corson MA, Probstfield JL, et al: Effects of intensive blood-pressure control in type 2 diabetes mellitus. N Engl J Med 2010;362:1575–1585.

14 Vijan S, Hayward RA: Treatment of hypertension in type 2 diabetes mellitus: blood pressure goals, choice of agents, and setting priorities in diabetes care. Ann Int Med 2003;138:593–602.

15 Bakris GL, Weir MR, Shanifar S, Zhang Z, Douglas J, van Dijk DJ, Brenner BM: Effects of blood pressure level on progression of diabetic nephropathy: results from the RENAAL study. Arch Int Med 2003;163:1555–1565.

16 Tatti P, Pahor M, Byington R, Mauro PD, Guarisco R, Strolol G, Strollo F: Outcome results of the fosinopril versus amlodipine cardiovascular events randomised trial (FACET) in patients with hypertension and NIDDM. Diabetes Care 1998;21:597–603.

17 Ruggenenti P, Fassi A, Ilieva AP, Bruno S, Iliev IP, Brusegan V, Rubis N, Gherardi G, Arnoldi F, Ganeva M, et al: Preventing microalbuminuria in type 2 diabetes. N Engl J Med 2004;351:1941–1951.

18 Remuzzi G, Perico N, Macia M, Ruggenenti P: The role of renin-angiotensin-aldosterone system in the progression of chronic kidney disease. Kidney Int Suppl 2005;99:S57–S65.

19 Lewis EJ, Hunsicker LG, Bain RP, Rohde RD: The effect of angiotensin-converting enzyme inhibition on diabetic nephropathy: The Collaborative Study Group. N Engl J Med 1993;329:1456–1462.

20 Parving HH, Lehnert H, Brochner-Mortensen J, Gomis R, Anderesen S, Arner P, for the Irbesartan in Patients with Type 2 Diabetes and Microalbuminuria Study Group: The effect of irbesartan on the development of diabetic nephropathy in patients with type 2 diabetes. N Engl J Med 2001;345:870–878.

21 Lewis EJ, Hunsicker LG, Clarke WR, Berl T, Pohl MA, Lewis JB, Ritz E, Atkins RC, Rohde R, Raz I: Renoprotective effect of the angiotensin-receptor antagonist irbesartan in patients with nephropathy due to type 2 diabetes. N Engl J Med 2001;345:851–860.

22 Brenner BM, Cooper ME, de Zeeuw D, Keane WF, Mitch WE, Parving HH, Remuzzi G, Snapinn SM, Zhang Z, Shahinfar S: Effects of losartan on renal and cardiovascular outcomes in patients with type 2 diabetes and nephropathy. N Engl J Med 2001;345:861–869.

23 de Zeeuw D, Remuzzi G, Parving HH, Keane WF, Zhang Z, Shahinfar S, Snapinn S, Cooper ME, Mitch WE, Brenner BM: Proteinuria, a target for renoprotection in patients with type 2 diabetic nephropathy: lessons from RENAAL. Kidney Int 2004;65:2309–2320.

24 Chan JC, Wat NM, So WY, Lam KS, Chua CT, Wong KS, Morad Z, Dickson TZ, Hille D, Zhang Z, et al: Renin angiotensin aldosterone system blockade and renal disease in patients with type 2 diabetes. An Asian perspective from the RENAAL Study. Diabetes Care 2004;27:874–879.

25 Gansevoort RT, de Zeeuw D, de Jong PE: Is the antiproteinuric effect of ACE inhibition mediated by interference in the renin-angiotensin system? Kidney Int 1994;45:861–867.

26 Hou FF, Xie D, Zhang X, Chen PY, Zhang WR, Liang M, Guo ZJ, Jiang JP: Renoprotection of Optimal Antiproteinuric Doses (ROAD) Study: a randomized controlled study of benazepril and losartan in chronic renal insufficiency. J Am Soc Nephrol 2007;18:1889–1898.

27 Barnett AH, Bain SC, Bouter P, Karlberg B, Madsbad S, Jervell J, Mustonen J: Angiotensin-receptor blockade versus converting-enzyme inhibition in type 2 diabetes and nephropathy. N Engl J Med 2004;351:1952–1961.

28 Matchar DB, McCrory DC, Orlando LA, Patel MR, Patel UD, Patwardhan MB, Powers B, Samsa GP, Gray RN: Systematic review: comparative effectiveness of angiotensin-converting enzyme inhibitors and angiotensin II receptor blockers for treating essential hypertension. Ann Int Med 2008;148:16–29.

29 Woo KS, Nicholls MG: High prevalence of persistent cough with angiotensin converting enzyme inhibitors in Chinese. Br J Clin Pharmacol 1995;40:141–144.

30 Williams GH: Hyporeninemic hypoaldosteronism. N Engl J Med 1986;314:1041–1042.

31 Kunz R, Friedrich C, Wolbers M, Mann JF: Meta-analysis: effect of monotherapy and combination therapy with inhibitors of the renin angiotensin system on proteinuria in renal disease. Ann Int Med 2008;148:30–48.

32 Lea J, Greene T, Hebert L, Lipkowitz M, Massry S, Middleton J, Rostand SG, Miller E, Smith W, Bakris GL: The relationship between magnitude of proteinuria reduction and risk of end-stage renal disease: results of the African American study of kidney disease and hypertension. Arch Int Med 2005;165:947–953.

33 Mann JF, Schmieder RE, McQueen M, Dyal L, Schumacher H, Pogue J, Wang X, Maggioni A, Budaj A, Chaithiraphan S, et al: Renal outcomes with telmisartan, ramipril, or both, in people at high vascular risk (the ONTARGET study): a multicentre, randomised, double-blind, controlled trial. Lancet 2008;372:547–553.

34 Imai E, Chan JC, Ito S, Haneda M, Makino H: Effects of olmesartan on renal and cardiovascular outcomes in type 2 diabetes with overt nephropathy; a randomized study, in press.

35 Parving HH, Persson F, Lewis JB, Lewis EJ, Hollenberg NK: Aliskiren combined with losartan in type 2 diabetes and nephropathy. N Engl J Med 2008;358:2433–2446.

36 Mehdi UF, Adams-Huet B, Raskin P, Vega GL, Toto RD: Addition of angiotensin receptor blockade or mineralocorticoid antagonism to maximal angiotensin-converting enzyme inhibition in diabetic nephropathy. J Am Soc Nephrol 2009;20:2641–2650.

37 Schjoedt KJ, Jacobsen P, Rossing K, Boomsma F, Parving HH: Dual blockade of the renin-angiotensin-aldosterone system in diabetic nephropathy: the role of aldosterone. Horm Metab Res 2005;37(suppl 1):4–8.

38 Luk AO, So WY, Ma RC, Kong AP, Ozaki R, Ng VS, Yu LW, Lau WW, Yang X, Chow FC, et al: Metabolic syndrome predicts new onset of chronic kidney disease in 5,829 patients with type 2 diabetes: a 5-year prospective analysis of the Hong Kong Diabetes Registry. Diabetes Care 2008;31:2357–2361.

39 Luk AO, Yang X, Ma RC, Ng VW, Yu LW, Lau WW, Ozaki R, Chow FC, Kong AP, Tong PC, et al: Association of statin use and development of renal dysfunction in type 2 diabetes – the Hong Kong Diabetes Registry. Diabetes Res Clin Pract 2010;88:227–233.

40 Araki S, Haneda M, Sugimoto T, Isono M, Isshiki K, Kashiwagi A, Koya D: Factors associated with frequent remission of microalbuminuria in patients with type 2 diabetes. Diabetes 2005;54:2983–2987.

41 Tu ST, Chang SJ, Chen JF, Tien KJ, Hsiao JY, Chen HC, Hsieh MC: Prevention of diabetic nephropathy by tight target control in an Asian population with type 2 diabetes mellitus: a 4-year prospective analysis. Arch Int Med 2010;170:155–161.

42 Chan JC, So WY, Yeung CY, Ko GT, Lau IT, Tsang MW, Lau KP, Siu SC, Li JK, Yeung VT, et al: Effects of structured versus usual care on renal endpoint in type 2 diabetes: the SURE study: a randomized multicenter translational study. Diabetes Care 2009;32:977–982.

43 Gaede P, Lund-Andersen H, Parving HH, Pedersen O: Effect of a multifactorial intervention on mortality in type 2 diabetes. N Engl J Med 2008;358:580–591.

Juliana C.N. Chan, MD
Department of Medicine and Therapeutics and Hong Kong Institute of Diabetes and Obesity
The Chinese University of Hong Kong, The Prince of Wales Hospital
Hong Kong SAR (China)
Tel. +852 26323138, Fax +852 26323108, E-Mail jchan@cuhk.edu.hk

Lai KN, Tang SCW (eds): Diabetes and the Kidney.
Contrib Nephrol. Basel, Karger, 2011, vol 170, pp 196–208

Intensive Glycemic Control and Renal Outcome

Min Jun · Vlado Perkovic · Alan Cass

Renal and Metabolic Division, The George Institute for Global Health, Sydney Medical School, University of Sydney, Camperdown, N.S.W., Australia

Abstract

Diabetes is the leading cause of chronic kidney disease (CKD) in many countries around the world, accounting for up to 50% of people who develop end-stage renal disease. The risk of morbidity and mortality is particularly high in people with both conditions, highlighting the importance of preventing the development and progression of earlier stage CKD in people with diabetes. Inadequate glycemic control has been associated with poor outcomes in diabetes, with early observational studies suggesting that tighter glycemic control may improve microvascular and macrovascular outcomes in patients with diabetes. Since then, large trials assessing the effect of intensive glycemic control in the general diabetic population have provided strong evidence that intensive therapy reduces the incidence and progression of microvascular outcomes including microalbuminuria, a key early marker of diabetic nephropathy (DN). These results have been consistent across both type 1 and type 2 diabetic populations demonstrating that intensive glycemic therapy provides clear benefits, delays the onset or progression of DN, particularly in its early stages. Whilst the benefits of intensive glycemic therapy for people with diabetes and early stage CKD have been well established, controversy remains as to whether intensive therapy slows the progression of established DN, particularly among individuals who have a reduced glomerular filtration rate. In addition, severe hypoglycemia has been associated with intensive glycemic therapy, raising safety concerns that may be of particular relevance for patients with decreased kidney function (CKD stages 3–5). Until further data are available, an individualized approach to glucose management is recommended in people with reduced glomerular filtration rate.

Glucose levels have been linked to the risk of incident and progressive nephropathy in patients with diabetes for many years. Observational studies have shown that hyperglycemia is strongly associated with the risk of developing microvascular complications (including new onset microalbuminuria) and the strong relationship between declining glycemic control and new onset nephropathy, retinopathy, and neuropathy [1, 2]. As microalbuminuria is an independent marker for the development of chronic kidney disease (CKD) and loss of glomerular filtration rate (GFR), its prevention is of critical importance in avoiding or delaying CKD in diabetic patients. Other observational data have demonstrated direct log-linear relationships between elevated blood glucose levels and end-stage renal disease (ESRD) or renal death [3]. In the individual participant data meta-analysis reported by the APCSC (Asia Pacific Cohort Studies Collaboration), each standard deviation increase in fasting blood glucose was associated with a 60% increase in the risk of renal death (95% confidence interval, CI: 42–81%) [3]. In light of these relationships, there has been great interest in assessing the role of intensive glycemic control in preventing or delaying the onset and progression of microvascular disease including diabetic nephropathy (DN).

When considering the relationship between intensive glycemic control and kidney disease in patients with diabetes, this chapter summarizes and critically reviews currently available evidence to address the following issues:

1 Does intensive glycemic therapy reduce the incidence of DN in patients with diabetes?
2 Does intensive glycemic therapy slow the progression of DN?
3 Does intensive glycemic therapy provide benefits in patients with later stages of CKD (i.e. reduced GFR or ESRD)?

Does Intensive Glycemic Therapy Reduce the Incidence of Microvascular Disease in Patients with Diabetes?

Observational data from the United Kingdom Prospective Diabetes Study (UKPDS) showed that there was a direct relationship between the risk of developing microvascular disease and glycemic control over time (fig. 1) [4]. A linear relationship between the adjusted incidence rate of microvascular disease and glycated hemoglobin (HbA1c) levels was observed, with the lowest incidence in the group with the tightest glycemic control (HbA1c <6%). Interestingly, there was no evidence of a threshold at either the top or bottom of the observed range of HbA1c levels. Each 1% lower mean HbA1c was associated with a 37% reduction in the risk of developing microvascular disease (95% CI: 33–41%) [4]. Large trials to date have established that intensively lowering HbA1c levels (6–7% in intensive therapy versus 7–9% in standard therapy) significantly reduces the risk of microvascular outcomes including new-onset microalbuminuria and new or worsening nephropathy in both type 1 and 2 diabetes mellitus patients [5–7]. These results form the basis of the current recommendation of maintaining a target HbA1c of <7% in the general diabetic population [8]. It is important to

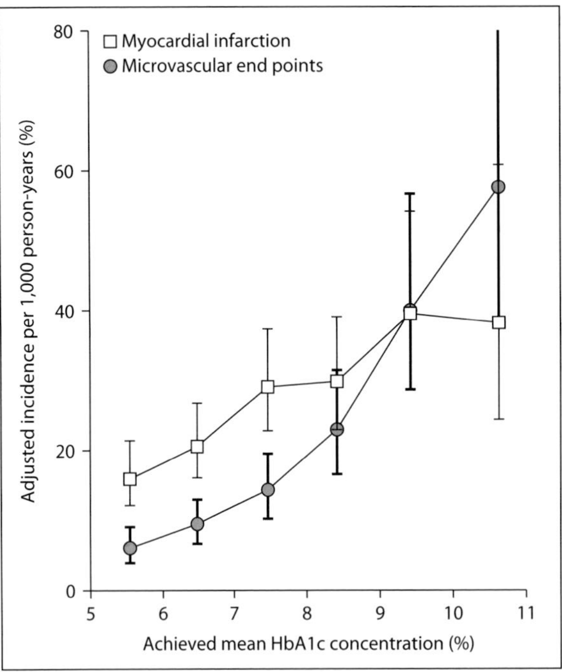

Fig. 1. Incidence rates of microvascular complications by category of achieved mean HbA1c concentration adjusted for age, sex, and ethnic group. Adapted from Stratton et al. [4].

note that this recommendation is made irrespective of the presence or absence of CKD.

A number of large randomized controlled trials have been undertaken in diabetic patients (table 1) aiming to assess the effects of intensive glycemic control including its role in reducing the risk of microvascular disease. The findings for type 1 and type 2 diabetes mellitus will be described separately.

Type 1 Diabetes
The Stockholm Diabetes Intervention Study (SDIS) randomized 102 patients with insulin-dependent diabetes mellitus to intensive glycemic therapy or standard glycemic control [9]. After 18 months, intensive therapy produced a significantly lower HbA1c compared to that of standard therapy (p < 0.0001; HbA1c values not provided). After 7.5 years of follow-up, in the intensive therapy group, HbA1c was reduced from a mean of 9.5% at baseline to 7.1% whilst in the standard group, HbA1c was reduced from 9.4 to 8.5% (p = 0.001) [10]. The study reported that after 18 months, 3 of the 34 patients in the intensive group who had normoalbuminuria developed microalbuminuria (defined as urinary albumin excretion rate of 15–300 μg/min), whilst 11 of 36 patients in the standard group developed microalbuminuria (p = 0.023).

A landmark study in the area of intensive glycemic control in diabetes, the Diabetes Control and Complications Trial (DCCT) randomized 1,441 patients with type 1 diabetes to either intensive or standard glycemic control therapies to assess the effect of intensive glycemic control on the development and progression of long-term complications of type 1 diabetes [5]. The aim of intensive therapy was to reach an HbA1c level of <6.05%. A significantly lower HbA1c level was achieved throughout the study for the intensive group (approximately 7 vs. 9% in standard) during a mean follow-up of 6.5 years. The study showed that intensive therapy reduced the incidence of microalbuminuria by 34% (95% CI: 2–56%) in the primary prevention cohort. The EDIC (Epidemiology of Diabetes Interventions and Complications) study was a long-term, observational follow-up study of the DCCT designed to assess the long-term effects of intensive glycemic control. A total of 1,349 patients from the original cohort were followed for an additional 8 years after the completion of the DCCT. The study, based on an intention-to-treat analysis, showed a 59% risk reduction for microalbuminuria (95% CI: 39–73%) [11]. There were significantly fewer patients in the intensive therapy arm that reached a serum creatinine level of 2 mg/dl or greater (5 vs. 19, p = 0.004).

Type 2 Diabetes
The UKPDS was, at the time, the largest trial involving people with diabetes. The study was designed to assess whether intensive glycemic control in patients with type 2 diabetes reduced the risk of macrovascular and microvascular outcomes [7]. The study randomly assigned 3,867 newly diagnosed patients with type 2 diabetes to intensive glycemic therapy using a sulphonylurea agent, insulin, or standard therapy using diet only. During its 10-year follow-up, the study achieved a median HbA1c of 7.0% in the intensive therapy group and 7.9% in the standard therapy group. The UKPDS showed a 24% risk reduction in the risk of microalbuminuria (95% CI: 9–38%). The effect size was not as large when compared to the DCCT, and this may have been related to the smaller difference in HbA1c achieved in this study. The 10-year postinterventional results of the UKPDS reported that despite the loss of HbA1c separation between the groups after 1 year posttrial, a 24% risk reduction for microvascular disease in the sulfonylurea-insulin group was seen (95% CI: 11–36%) [12]. These observational data showed that the benefits of intensive glycemic control persisted for up to 10 years.

The Veterans Affairs Cooperative Study on glycemic control and complications in type 2 diabetes was a study which randomized 153 patients, of whom 95 had no history of microalbuminuria, to either intensive glycemic control or standard glycemic control [13]. At 2 years, the study achieved a mean HbA1c level of 7.1% in the intensive group and 9.1% in the standard group. The study found that in patients with no previous history of microalbuminuria, 17% (7/42) developed microalbuminuria in the intensive group compared to 35% (16/46) in the standard group (p = 0.05).

Table 1. Characteristics of trials assessing the effect of intensive glycemic control in patients with type 1 or type 2 diabetes

	Study	Study size	Intensive therapy	Standard therapy	Mean follow-up
Type 1 diabetes	SDIS (1988) [9]	102	Individual education and continuous face-to-face and telephone tutoring	Routine diabetes care	7.5 years (10)
	DCCT (1993) [5]	1,441	≥3 daily insulin injections; HbA1c 7%	1–2 daily insulin injections; HbA1c 9%	6.5 years
Type 2 diabetes	Kumamoto (1995) [17]	110	≥3 daily insulin injections	1–2 daily insulin injections	6 years
	UKPDS (1998) [7]	3,867	FPG <6 mM and pre-meal glucose conc. 4–7 mM for insulin group, Int group1: sulphonylurea (chlorpropamide 100–500 mg, glibenclamide 2.5–20 mg, glipizide 2.5–40 mg), Int group 2: insulin	FPG <15 mM without symptoms of hyperglycemia using diet	10 years
	ACCORD (2008) [15]	10,251	Target HbA1c <6.0%	Target HbA1c 7.0–7.9%	3.5 years

Outcomes	Achieved HbA1c	Results
Normoalbuminuria (<15 µg/min), microalbuminuria (15–300 µg/min), nephropathy (>300 µg/min)	Int = 7.2%, Stan = 8.7% (10)	3 of the 34 patients in the intensive group who had normoalbuminuria developed microalbuminuria, whilst 11 of 36 patients in the standard group developed microalbuminuria (p = 0.023)
New-onset microalbuminuria (≥40 mg/24h); progression of albuminuria	Int = 7.3%, Stan = 9.0%	Primary prevention: 34% risk reduction for microalbuminuria (95% CI: 2–56%); secondary prevention: 43% risk reduction for microalbuminuria (95% CI: 21–58%); 54% risk reduction for progression of albuminuria; advanced nephropathy (UAE ≥300 mg/24 h and CCI <70 ml/min/1.73 m2): 2 in Int and 5 in Stan
Nephropathy defined as normoalbuminuria (<30 mg/24 h), microalbuminuria (30–300 mg/24 h), albuminuria (>300 mg/24 h) with changes from any stage to a higher stage defined as progression	Int = 7.0%, Stan = 7.9%	Primary prevention: cumulative % of the development and progression of nephropathy, Int = 7.7%, Stan = 28.0%, p = 0.032; secondary prevention: cumulative % of the development and progression of nephropathy, Int = 11.5%, Stan = 32.0%, p = 0.044
Microvascular outcomes	Int = 7.0%, Stan = 7.9%	25% risk reduction for microvascular endpoints (95% CI: 7–40%)
First composite outcome: renal failure (defined as initiation of dialysis or end-stage kidney disease, renal transplantation, rise of serum creatinine >291.72 µM), retinal photocoagulation, vitrectromy to treat retinopathy; second composite outcome: nephropathy (microalbuminuria, macroalbuminuria, end-stage kidney disease, doubling of serum creatinine or >20 U eGFR decrease), retinopathy, neuropathy	Int = 6.3%, Stan = 7.6%	First composite: HR: 1.00, 95% CI: 0.88–1.14, p=0.9969; second composite (nephropathy): 21% risk reduction for incident microalbuminuria (95% CI: 10–31%), 31% risk reduction for incident macroalbuminuria (95% CI: 15–45%), end-stage kidney disease HR: 0.95, 95% CI: 0.73–1.24), p = 0.7126, no effect of reducing doubling SCr or slowing GFR: HR: 1.07, 95% CI: 1.01–1.13, p = 0.016

	Study	Study size	Intensive therapy	Standard therapy	Mean follow-up
	ADVANCE (2008) [6]	1,1140	Target HbA1c ≤6%; gliclazide (30–120 mg daily)	Target HbA1c defined on basis of local guidelines; any glucose lowering medication	5 years
	VADT (2009) [18]	1,791	Maximum doses of either metformin and rosiglitazone or glimepiride and rosiglitazone; target HbA1c was an absolute reduction of 1.5% points compared with standard	Half of maximum doses of either metformin and rosiglitazaone or glimepiride and rosiglitazone	5.6 years

Outcomes described are those relevant to renal outcomes. Int = Intensive therapy; Stan = standard therapy; UAE = urinary albumin excretion; CCl = creatinine clearance; FPG = fasting plasma glucose.

The Action to Control Cardiovascular Risk in Diabetes (ACCORD) trial randomized 10,251 patients with type 2 diabetes to intensive therapy or standard therapy [14]. As part of a prespecified secondary analysis, the study assessed the effect of intensive glycemic control on the rate of microvascular complications in type 2 diabetes. As has been reported, the ACCORD glycemic control trial was stopped early when an increase in all-cause mortality was observed in the intensive therapy group [14]. At cessation of study-related intensive glycemic therapy, the mean HbA1c in the intensive group was 6.3% compared to 7.6% in the standard group [15]. At the end of the randomized treatment stage of the study, a 21% reduction in the risk of microalbuminuria was observed (95% CI: 10–31%). Regarding the incidence of macroalbuminuria, a 31% risk reduction was observed (95% CI: 15–45%).

Outcomes	Achieved HbA1c	Results
Microvascular outcomes: new or worsening nephropathy [defined as development of macroalbuminuria (UAC >300 µg albumin/mg creatinine) or doubling of serum creatinine to ≥200 µM or need for renal replacement therapy, or death due to renal disease] or retinopathy	Int = 6.5%, Stan = 7.3%	21% risk reduction for new or worsening nephropathy (95% CI: 7–34); 9% risk reduction for new-onset microalbuminuria (95% CI: 2–15%); 30% risk reduction for progression to macroalbuminuria (95% CI: 15–43%); trend towards reduction in the need for renal replacement therapy or death from renal disease (HR: 0.64, 95% CI: 0.38–1.08, p = 0.09); no effect on doubling of serum creatinine (HR: 1.15, 95% CI: 0.82–1.63)
Microvascular complications including severe nephropathy being defined as doubling of serum creatinine, creatinine level >3 mg/dl, or GFR <15 ml/min; progression of albuminuria defined as increase in albuminuria for at least 2 successive yearly visits without reversion to an improved level	Int = 6.9%, Stan = 8.4%	No effect on severe nephropathy, significant effect on the change from normal to microalbuminuria or macroalbuminuria (incidence in Stan = 14.7% vs. Int = 10.0%, p = 0.03), effect on incidence of any increase in albuminuria (Stan = 13.8% vs. Int = 9.1%, p = 0.01)

The Kumamoto study was designed to assess the effect of intensive insulin therapy on the incidence and progression of microvascular outcomes in Japanese patients with type 2 diabetes [16]. The trial randomized 110 patients with type 2 diabetes to either intensive glycemic therapy (multiple insulin injections) or standard therapy (conventional insulin injections once or twice daily). The study found that after 6 years of follow-up, the cumulative rates of incident nephropathy (defined as microalbuminuria: 30–300 mg/24 h) in the primary prevention cohort were 7.7% for the intensive group and 28.0% for the standard group (p = 0.032).

The largest trial assessing the effects of intensive glycemic control on microvascular outcomes reported to date is the ADVANCE (Action in Diabetes and Vascular Disease: Preterax and Diamicron Modified Release Controlled Evaluation) trial [6]. The ADVANCE randomized 11,140 type 2 diabetes patients to either intensive or standard glycemic control and assessed the effects on a range of microvascular and microvascular outcomes. Twenty-seven percent of the study population had a history of microalbuminuria at baseline, and an additional 3–4% had macroalbuminuria. At the end of follow-up (median follow-up of 5 years), a mean HbA1c of 6.5% was achieved in the intensive group compared to 7.3% in the standard group. The study found that intensive therapy produced a 9% risk reduction in the risk of new-onset microalbuminuria (95% CI: 2–15%).

The Veterans Affairs Diabetes Trial (VADT) randomized 1,791 military veterans with type 2 diabetes to either intensive or standard glycemic therapy [17]. The study achieved a median HbA1c of 6.9% in the intensive group and 8.4% in the standard group. In contrast to previous large trials, the reduction in the incidence of microalbuminuria did not reach significance in the VADT (13.2% in standard group versus 9.7% in intensive group; p = 0.12). However, intensive therapy significantly reduced the progression of nephropathy (defined as a change to any higher stage, i.e. normoalbuminuria, microalbuminuria, and macroalbuminuria; 13.8% in standard group vs. 9.1% in intensive group, p = 0.01).

Taken together, the data in type 1 and type 2 diabetes have clearly demonstrated that intensive glycemic control prevents the development of microalbuminuria relatively consistently across the completed trials. The effects on harder renal outcomes will be reviewed in the next section.

Does Intensive Glycemic Therapy Slow the Progression of Diabetic Nephropathy?

Numerous large trials involving people with diabetes and established nephropathy have shown that tight glucose control can reduce the progression of kidney disease.

Type 1 Diabetes Mellitus
The DCCT assessed the effects of intensive glycemic control on the progression of DN in a secondary prevention cohort. The study showed that the risk of progression of albuminuria (defined as any change from normoalbuminuria to microalbuminuria or macroalbuminuria, or microalbuminuria to macroalbuminuria) was reduced by 56% (95% CI: 18–76%). Regarding the long-term effects of intensive glycemic control on renal outcomes, EDIC, the extended follow-up study of DCCT, reported that new cases of albuminuria occurred in 9 (1.4%) of the patients in the original intensive treatment group compared to 59 (9.4%) of those in the original standard treatment group (84% reduction in odds, 95% CI: 67–92%) [11]. A total of 24 patients reached a serum creatinine >2 mg/dl, of which 5 were in the intensive group and 19 in the standard group (p = 0.004). Of those patients, 11 required dialysis or renal transplant (4 in the intensive group versus 7 in the standard group; p = 0.36) [11].

The 7.5-year follow-up results of the SDIS reported that nephropathy (defined as urinary albumin excretion rate of >200 µg/min) developed in 1 patient in the intensive therapy group compared to 9 patients in the standard therapy group (p = 0.01). No patients receiving intensive therapy developed nephropathy with subnormal GFR (normal range defined as 77–127 ml/min) compared to 6 patients in the standard group (p = 0.02) [10].

Type 2 Diabetes Mellitus
The ADVANCE trial showed that intensive glycemic therapy reduced the risk of renal events including new or worsening nephropathy (defined as development of macroalbuminuria, doubling of serum creatinine, need for renal replacement therapy, or death due to renal disease) by 21% (95% CI: 7–34). This risk reduction was primarily observed based on the reduction of development of macroalbuminuria from intensive glycemic therapy (hazard ratio, HR: 0.70, 95% CI: 0.57–0.85). A nonsignificant trend towards a reduction in the need for renal replacement therapy or death from renal causes was also observed (HR: 0.64, 95% CI: 0.38–1.08; p = 0.09); however, there was no effect on the doubling of serum creatinine (HR: 1.15, 95% CI: 0.82–1.63). The unclear effects on these harder renal outcomes are likely a product of the relative paucity of these events in ADVANCE, leading to low power for this outcome.

The Kumamoto study reported that intensive glycemic control reduced the progression of nephropathy [defined as a change from a lower stage to any higher stage; stages defined as: normoalbuminuria (30 mg/24 h), microalbuminuria (30–300 mg/24 h), or albuminuria (>300 mg/24 h)]. The cumulative percentages of the progression in nephropathy were 11.5% in the intensive group and 32% in the standard group (p = 0.044) [16]. The subsequent follow-up study reporting the 8-year results, showed that for the primary prevention group the cumulative percentage of patients who developed nephropathy after 8 years in the intensive group was 11.5% compared to that of 43.5% in the standard group (p = 0.029) [18]. In the long-term follow-up study of the Kumamoto study, overall benefit for the progression of nephropathy was observed for the secondary prevention group (those with albuminuria present at baseline), where 16% in the intensive group progressed compared to 40% in the standard group (p = 0.043) [18].

The ACCORD trial showed that the incidence of macroalbuminuria was significantly reduced in the intensive therapy group (HR: 0.69, 95% CI: 0.55–0.85) [14]. However, intensive glycemic therapy had no clear effect on the risk of ESRD (HR: 0.95, 95% CI: 0.73–1.24). Importantly, the study found a 3-fold increase in the frequency of severe hypoglycemia in the intensive therapy group, a finding also reported earlier by the SDIS in the type 1 diabetes population [9], which has been postulated to be a potential mechanism by which mortality may have been increased. This finding highlighted the risk of serious hypoglycemia involved with intensive glycemic therapy which may be particularly pertinent to patients with later stages of CKD.

Once again, these trials have demonstrated a clear and consistent reduction in the risk of progressive albuminuria in people with established nephropathy (generally microalbuminuria) at entry. Although none of the studies had adequate power to assess the impact of tight glycemic control on the development of ESRD (or reduced GFR), encouraging results from a number of trials suggest this would be an important priority area for future research.

Does Intensive Glycemic Therapy Provide Benefits in Patients with Later Stages of CKD?

Current studies addressing the effect of intensive glycemic control have failed to include meaningful numbers of patients with more advanced CKD (i.e. reduced GFR) with much of the evidence coming from studies involving diabetic patients in the general population or patients with early stage kidney disease. There is currently no definitive randomized controlled trial evidence to suggest that intensive glycemic control reduces the risk of ESRD development, although the benefits described above suggest this is possible. As such, the effect of intensive glycemic control in advanced CKD patients, and particularly those with ESKD, remains unclear. In addition, due to decreased clearance of insulin and other drugs, as well as loss of renal gluconeogenesis, patients with ESKD may be susceptible to increased risk of hypoglycemia [19]. The potentially devastating impact of severe hypoglycemia has been highlighted by the ACCORD and ADVANCE studies [14, 20].

Observational studies assessing the effects of intensive glycemic control in dialysis patients have produced conflicting results [21, 22]. As with other risk factors in the ESRD population, including blood pressure and obesity, HbA1c was observed to be inversely related to mortality in a recent large observational study of 23,618 patients. However, after adjusting for relevant risk factors, mortality was shown to be highest in groups with the highest HbA1c levels [23]. This paradoxical relationship may be another example of the numerous U-shaped relationships between various risk factors and adverse outcomes observed in the dialysis population due to confounding of the association between these risk factors and survival, and/or reverse causality. There are currently no large randomized controlled trials completed or underway assessing the effects of glycemic control on survival outcomes in the dialysis population.

Conclusion

There is strong evidence that intensive glycemic control in diabetic patients with relatively preserved kidney function reduces the incidence and progression of microvascular outcomes, including new onset microalbuminuria. The results are consistent across both type 1 and 2 diabetic patients. However, there is currently no evidence to suggest that intensive therapy improves outcomes among patients with later stages of CKD, due at least in part to the long period of time required for the development of ESRD. With the critical lack of randomized controlled trials in people with ESRD, and demonstrated risks associated with severe hypoglycemic episodes, individualized therapy to avoid serious hypoglycemia is the optimal approach. Renal outcomes are clearly linked to cardiovascular outcomes, and the issue as to whether intensive glycemic control reduces the risk of long-term cardiovascular disease remains to be determined. Additional information may be gained through long-term follow-up of trials such as UKPDS and ADVANCE, but trials focusing on people with established DN that have sufficient power to detect a difference in hard renal outcomes would be extremely valuable.

References

1   Klein R, Klein BE, Moss SE: Relation of glycemic control to diabetic microvascular complications in diabetes mellitus. Ann Int Med 1996;124:90–96.

2   Newman DJ, Mattock MB, Dawnay AB, Kerry S, McGuire A, Yaqoob M, Hitman GA, Hawke C: Systematic review on urine albumin testing for early detection of diabetic complications. Health Technol Assess 2005;9:xiii-163.

3   O'Seaghdha CM, Perkovic V, Lam TH, McGinn S, Barzi F, Gu DF, Cass A, Suh I, Muntner P, Giles GG, Ueshima H, Woodward M, Huxley R: Blood pressure is a major risk factor for renal death. An analysis of 560352 participants from the Asia-Pacific region. Hypertension 2009;54:509–515.

4   Stratton IM, Adler AI, Neil A, Matthews DR, Manley SE, Cull CA, Hadden D, Turner RC, Holman RR: Association of glycaemia with macrovascular and microvascular complications of type 2 diabetes (UKPDS 35): prospective observational study. BMJ 2000;321:405–412.

5   DCCT Group: The effect of intensive treatment of diabetes on the development and progression of long-term complications in insulin-dependent diabetes mellitus. The Diabetes Control and Complications Trial Research Group. N Engl J Med 1993;329:977–986.

6   ADVANCE Study Group: Intensive blood glucose control and vascular outcomes in patients with type 2 diabetes. N Engl J Med 2008;358:2560–2572.

7   UKPDS Study Group: Intensive blood-glucose control with sulphonylureas or insulin compared with conventional treatment and risk of complications in patients with type 2 diabetes (UKPDS 33). UK Prospective Diabetes Study (UKPDS) Group. Lancet 1998;352:837–853.

8   KDOQI: Clinical practice guidelines and clinical practice recommendations for diabetes and chronic kidney disease. Am J Kidney Dis 2007;49(suppl 2):S1–S180.

9   Reichard P, Britz A, Cars I, Nilsson BY, Sobocinsky-Olsson B, Rosenqvist U: The Stockholm Diabetes Intervention Study (SDIS): 18 months' results. Acta Med Scand 1988;224:115–122.

10 Reichard P, Nilsson BY, Rosenqvist U: The effect of long-term intensified insulin treatment on the development of microvascular complications of diabetes mellitus. N Engl J Med 1993;329:304–309.

11 EDIC Study Group: Sustained effect of intensive treatment of type 1 diabetes mellitus on development and progression of diabetic nephropathy: the Epidemiology of Diabetes Interventions and Complications (EDIC) study. JAMA 2003;290:2159–2167.

12 Holman RR, Paul SK, Bethel MA, Matthews DR, Neil A: 10-year follow-up of intensive glucose control in type 2 diabetes. N Engl J Med 2008;359:1577–1589.

13 Levin SR, Coburn JW, Abraira C, Henderson WG, Colwell JA, Emanuele NV, Nuttall FQ, Sawin CT, Comstock JP, Silbert CK: Effect of intensive glycemic control on microalbuminuria in type 2 diabetes. Diabetes Care 2000;23:1478–1485.

14 ACCORD Study Group: Effects of intensive glucose lowering in type 2 diabetes. N Engl J Med 2008;358:2545–2559.

15 Ismail-Beigi F, Craven T, Banerji MA, Calles J, Cohen RM, Cuddihy R, Cushman WC, Genuth S, Grimm RH Jr, Hamilton BP, Hoogwerf B, Karl D, Katz L, Krikorian A, O'Connor P, Pop-Busui R, Schubart U, Simmons D, Taylor H, Thomas A, Weiss D, Hramiak I: Effect of intensive treatment of hyperglycaemia on microvascular outcomes in type 2 diabetes: an analysis of the ACCORD randomised trial. Lancet 2010;376:419–430.

16 Ohkubo Y, Kishikawa H, Araki E, Miyata T, Isami S, Motoyoshi S, Kojima Y, Furuyoshi N, Shichiri M: Intensive insulin therapy prevents the progression of diabetic microvascular complications in Japanese patients with non-insulin-dependent diabetes mellitus: a randomized prospective 6-year study. Diabetes Res Clin Pract 1995;28:103–117.

17 VADT Investigators: Glucose control and vascular complications in veterans with type 2 diabetes. N Engl J Med 2009;360:129–139.

18 Shichiri M, Kishikawa H, Ohkubo Y, Wake N: Long-term results of the Kumamoto Study on optimal diabetes control in type 2 diabetic patients. Diabetes Care 2000;23:B21–B9.

19 Muhlhauser I, Toth G, Sawicki PT, Berger M: Severe hypoglycemia in type I diabetic patients with impaired kidney function. Diabetes Care 1991;14:344–346.

20 Zoungas S, Patel A, Chalmers J, de Galan BE, Li Q, Billot L, Woodward M, Ninomiya T, Neal B, MacMahon S, Grobbee DE, Kengne AP, Marre M, Heller S, Group AC, Zoungas S, Patel A, Chalmers J, de Galan BE, Li Q, Billot L, Woodward M, Ninomiya T, Neal B, MacMahon S, Grobbee DE, Kengne AP, Marre M, Heller S: Severe hypoglycemia and risks of vascular events and death. N Engl J Med 2010;363:1410–1418.

21 Oomichi T, Emoto M, Tabata T, Morioka T, Tsujimoto Y, Tahara H, Shoji T, Nishizawa Y, Oomichi T, Emoto M, Tabata T, Morioka T, Tsujimoto Y, Tahara H, Shoji T, Nishizawa Y: Impact of glycemic control on survival of diabetic patients on chronic regular hemodialysis: a 7-year observational study. Diabetes Care 2006;29:1496–1500.

22 Williams ME, Lacson E Jr, Teng M, Ofsthun N, Lazarus JM: Hemodialyzed type I and type II diabetic patients in the US: characteristics, glycemic control, and survival. Kidney Int 2006;70:1503–1509.

23 Kalantar-Zadeh K, Kopple JD, Regidor DL, Jing J, Shinaberger CS, Aronovitz J, McAllister CJ, Whellan D, Sharma K: Hemoglobin A1c and survival in maintenance hemodialysis patients. Diabetes Care 2007;30:1049–1055.

Alan Cass, MD, PhD, FRACP
Renal & Metabolic Division, The George Institute for Global Health
Sydney Medical School, University of Sydney
Camperdown, NSW 2050 (Australia)
Tel. +61 2 9993 4553, Fax +61 2 9993 4502, E-Mail acass@georgeinstitute.org.au

Lai KN, Tang SCW (eds): Diabetes and the Kidney.
Contrib Nephrol. Basel, Karger, 2011, vol 170, pp 209–216

Nuclear Hormone Receptors as Therapeutic Targets

Moshe Levi[a] · Xiaoxin Wang[a] · Devasmita Choudhury[b]

[a]Division of Renal Diseases and Hypertension, Department of Medicine, University of Colorado Denver, Aurora, Colo., and [b]Division of Nephrology, Department of Medicine, University of Texas Southwestern Medical Center Dallas and VA Medical Center, Dallas, Tex., USA

Abstract

In spite of excellent glucose and blood pressure control, including administration of angiotensin-converting enzyme inhibitors and/or angiotensin II receptor blockers, diabetic nephropathy (DN) still develops and progresses. The development of additional protective therapeutic interventions is, therefore, a major priority. Nuclear hormone receptors regulate carbohydrate metabolism, lipid metabolism, the immune response, inflammation and development of fibrosis. The increasing prevalence of DN has led to intense investigation of the role that nuclear hormone receptors may have in slowing or preventing the progression of renal disease. Several nuclear receptor-activating ligands (agonists) have been shown to have a renal protective effect in the context of DN. This review will discuss the evidence regarding the beneficial effects of the activation of the vitamin D receptor (VDR) and the farnesoid X receptor (FXR) in preventing the progression of DN, and will describe how the discovery and development of compounds that modulate the activity of VDR and FXR may provide potential additional therapeutic approaches in the management of DN.

The pathogenesis of diabetic nephropathy (DN) is multifactorial. Hypertension, abnormal carbohydrate metabolism, abnormal lipid metabolism, accumulation of lipids, upregulation of profibrotic growth factors (including, renin, angiotensin II, transforming growth factor-β and vascular endothelial growth factor), upregulation of proinflammatory cytokines (including nuclear factor-κB – NF-κB, monocyte/macrophage chemoattractant protein 1, tumor necrosis factor and interleukin-1β), increased oxidative stress, and increased production of advanced glycation end products all have an important role in the pathogenesis and progression of DN [1–5].

In spite of all the beneficial interventions implemented in patients with diabetes, including tight glucose control, tight blood pressure control, and angiotensin II receptor blockade, renal injury progresses in most of these patients. Additional treatment modalities that modulate the pathogenetic pathways involved in DN are therefore needed to slow the progression of renal failure in patients with diabetes.

Studies in humans with obesity, type 1 or type 2 diabetes mellitus and in animal models of these diseases have reported an accumulation of lipids (triglycerides and cholesterol) in the kidney, which sometimes occurs even in the absence of abnormalities in serum lipid levels. The accumulation of lipids is associated with the development of glomerulosclerosis, tubulointerstitial fibrosis, and the progression of diabetic renal disease [6–11]. The accumulation of triglycerides and cholesterol in the kidney is mediated by increased expression and activity of the transcriptional factors the sterol regulatory element binding proteins 1 and 2 (SREBP-1 and SREBP-2), which are master regulators of fatty acid and cholesterol synthesis. We have therefore become interested in identifying nuclear receptors that are potential negative regulators of SREBPs and also inflammation, oxidative stress, and fibrosis and therefore may slow down the progression of diabetic kidney disease.

This review discusses two nuclear hormone receptors that play a role in the pathogenesis of DN, including the vitamin D receptor (VDR) and the farnesoid X receptor (FXR).

Vitamin D Receptor

Vitamin D_3 is acquired either from dietary sources or is generated via solar ultraviolet irradiation of 7-dehydrocholesterol in the skin. Vitamin D_3 is then processed into the active hormone, 1,25 dihydroxyvitamin D_3 [1,25(OH)$_2$D$_3$], via two consecutive hydroxylation reactions [12]. The first such reaction takes place in the liver and is catalyzed by 25-hydroxylase and the second is catalyzed by 1α-hydroxylase, which is expressed predominantly in the kidney. Extrarenal 1,25(OH)$_2$D$_3$ can also be produced locally in a number of cell types that express VDR, notably cells of the skin, immune system, colon, pancreas, and vasculature. Locally produced 1,25(OH)$_2$D$_3$ does not contribute significantly to circulating 1,25(OH)$_2$D$_3$ levels, but it retains the capacity to be active in the cells and tissues where they are produced.

The activity of 1,25(OH)$_2$D$_3$ is mediated by VDR. 1,25(OH)$_2$D$_3$-VDR has multiple physiological and pathological roles that extend beyond the regulation of mineral metabolism, including the regulation of renal and cardiovascular functions. Naturally occurring, low-affinity VDR ligands that have been identified over the past 2 years (for example lithocholic acid, curcumin, and polyunsaturated fatty acids) may trigger 1,25(OH)$_2$D$_3$-independent activation of VDR in specific tissues [13]. 1,25(OH)$_2$D$_3$ can also induce rapid nongenomic effects on target cells in the vitamin D endocrine system regulated by vitamin D.

Although VDR may not be involved in this pathway as rapid actions of 1,25-(OH)$_2$D$_3$ can be elicited in the absence of VDR, the nongenomic pathway regulated by 1,25(OH)$_2$D$_3$ can interact with genomic pathway to modulate VDR-dependent gene expression [14]. 1,25(OH)$_2$D$_3$ rapid activation of cytosolic kinases via nongenomic pathway has been shown to lead to the phosphorylation of critical coactivators of VDR, resulting in modulation of VDR-dependent gene transcription [14]. VDR may also exert direct effects on the nucleus and form complexes with the p65 subunit of NF-κB and thus cause repression of p65-mediated gene transcription. This expression inhibition results in an anti-inflammatory effect [15].

VDR agonists have been shown to be protective in mouse models of diabetes mellitus. Studies in mice with type 1 diabetes mellitus and type 2 diabetes mellitus have shown that the renal protective effects of VDR agonists are mediated by inhibition or antagonism of the renin-angiotensin system [16–18] and NF-κB [15, 19], which are major mediators of increased intraglomerular pressure, podocyte damage, oxidative stress, inflammation, and fibrosis. In addition, VDR agonists activate hepatocyte growth factor and prevent renal interstitial myofibroblast activation [20]. In contrast, VDR knockout mice have been shown to be more susceptible to streptozotocin (STZ)-induced diabetic kidney disease in part via activation of the renin-angiotensin system [21]. In this regard, VDR agonists in combination therapy with an angiotensin-converting enzyme inhibitor (ACEI) or an angiotensin II type 1 receptor blocker have synergistic effects in prevention of kidney disease in rats with subtotal nephrectomy [22] and in mice with diabetes [23].

Recently, we investigated if treatment of mice with diet-induced obesity (DIO) and insulin resistance with the VDR agonist doxercalciferol (1α-hydroxyvitamin D$_2$) prevents renal disease. DIO and insulin resistance in mice is associated with proteinuria, renal mesangial expansion, accumulation of extracellular matrix proteins, and activation of oxidative stress, proinflammatory cytokines, profibrotic growth factors, and the sterol regulatory element-binding proteins, SREBP-1 and SREBP-2 that mediate increases in fatty acid and cholesterol synthesis [10]. Renal VDR expression was decreased in the DIO mice and treatment with doxercalciferol increased VDR expression in the kidney. Our results indicate that treatment of DIO mice with the VDR agonist decreases proteinuria, podocyte injury, mesangial expansion, and extracellular matrix protein accumulation. The VDR agonist also decreases macrophage infiltration, oxidative stress, proinflammatory cytokines and profibrotic growth factors. Furthermore, the VDR agonist also prevents the activation of the renin angiotensin aldosterone system including the angiotensin II type 1 receptor and the mineralocorticoid receptor. An additional novel finding of our study is that activation of VDR results in decreased accumulation of neutral lipids (triglycerides and cholesterol) and expression of adipophilin in the kidney by decreasing SREBP-1 and SREBP-2 expression and target enzymes that mediate fatty acid and cholesterol synthesis [24].

Treatment of DIO mice with doxercalciferol also increased the expression of the bile acid-activated nuclear hormone receptor FXR and bile acid-activated G protein-coupled receptor TGR5. Previous studies have indicated that bile acids activate VDR [25]. Recent studies also indicate that VDR agonists can increase portal bile acid concentration and trigger FXR-like effects in the liver and the intestine in rats. In further support of our findings, VDR knockout mice have decreased FXR expression in the ileum, and treatment of mice with calcitriol increased FGF15 expression in the ileum, which is a known FXR target [26]. These results warrant further investigation into how these two nuclear receptors cross-talk.

VDR agonists have also been shown to decrease proteinuria in human subjects with DN. A recent large-scale randomized controlled study (VITAL) in patients with type 2 diabetes mellitus who had already been treated with ACEI or angiotensin II type 1 receptor blocker showed that treatment with paricalcitol resulted in a significant reduction in urinary albumin [27].

These studies therefore indicate that in addition to animal models of obesity, insulin resistance and diabetes, VDR agonists are also highly beneficial in human subjects with diabetic or nondiabetic kidney disease.

Farnesoid X Receptor

FXR is an adopted orphan nuclear receptor named after farnesol, an intermediate in the mevalonate biosynthetic pathway, which was found to weakly activate FXR at supraphysiological concentrations. Subsequently, endogenous bile acids were found to bind to and activate FXR, and FXR was found to have a critical role in the regulation of bile acid metabolism [28–30]. The hydrophobic bile acid chenodeoxycholic acid and its conjugated forms are the most potent endogenous agonists of FXR, whereas the hydrophilic bile acids ursodeoxycholic acid and muricholic acid do not activate this receptor. Ligand-activated FXR can bind to and activate or repress the transcription of a large number of genes either as an FXR-retinoid X receptor heterodimer or as a monomer.

FXR is highly expressed in the liver, intestine, adrenal gland, and with particularly highest expression in the kidney [31]. Activation of FXR has an important role in maintaining glucose, lipid, and bile acid homeostasis in the enterohepatic system [32–33]. Activation of FXR also prevents atherosclerosis in LDL receptor knockout mice and ApoE knockout mice and vascular calcification in ApoE knockout mice with chronic kidney disease [34].

In mouse kidney, FXR has been detected in both isolated glomeruli and proximal tubules. The expression level in proximal tubule cells is much higher than in glomeruli [35]. In addition, FXR is expressed in both cultured mouse mesangial cells and podocytes [35].

FXR (Nr1h4) knockout mice with STZ-induced type 1 diabetes develop increased renal injury compared with wild-type C57BL/6 mice with STZ-induced type 1 diabetes [36]. These mice demonstrate increases in proteinuria, glomerulosclerosis, tubulointerstitial fibrosis, mesangial expansion, podocyte damage and glomerular basement membrane thickening. There is also increased macrophage infiltration, as well as increased expression of SREBP-1 and SREBP-2 that result in increased renal triglyceride and cholesterol content. In contrast, treatment of *db/db* mice with type 2 diabetes [35], DBA/2J mice with DIO and insulin resistance [37], and DBA/2J mice with STZ-induced type 1 diabetes mellitus [36] with an FXR agonist INT-747 (6α-ethyl-chenodeoxycholic acid) has marked renal protective effects. These experimental models of DN showed improvements in proteinuria, glomerulosclerosis, tubulointerstitial fibrosis, and macrophage infiltration following treatment with FXR-activating agonists. These renal protective effects are mediated by effects on lipid metabolism, oxidative stress, and on the production of proinflammatory cytokines and profibrotic growth factors [35–37]. FXR agonists inhibit expression of SREBP-1 and carbohydrate response element-binding protein in the kidney resulting in decreased fatty acid synthesis and triglyceride accumulation [35–37]. FXR agonists also inhibit SREBP-2 resulting in decreased cholesterol synthesis and accumulation in the kidney.

These studies suggest that FXR agonists can prevent the progression of kidney disease in mouse models of type 1 diabetes mellitus, DIO and insulin resistance, and type 2 diabetes mellitus. Since FXR agonists are now in clinical trials to examine the physiological effects of FXR activation in humans with nonalcoholic fatty liver disease, it will be possible in the near future to test their efficacy in humans with renal disease as well.

Conclusion

The provided evidence suggests that the VDR and the FXR through their multiple actions in regulating lipid and energy metabolism, inflammation, oxidative stress and fibrosis exhibit a great potential for preventing the progression of DN (fig. 1).

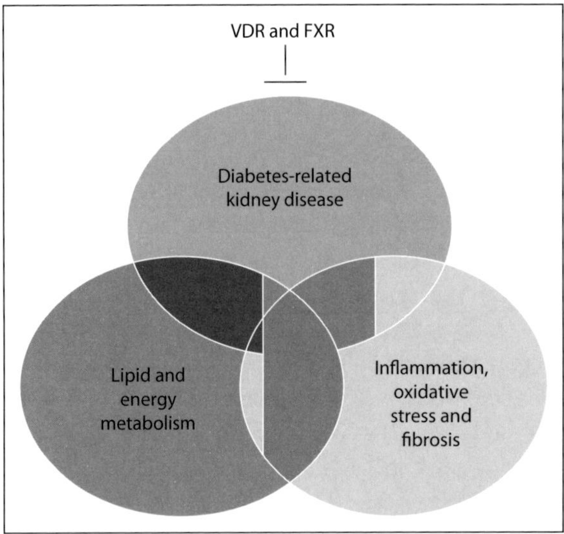

Fig. 1. Metabolic actions of VDR and FXR in preventing the progression of DN.

Acknowledgements

The authors thank Jolene Richardson and Amber Leider for expert assistance in preparation of this review. The original studies from the authors' lab cited in this review have been supported by grants from NIH (U01 DK076134 and R01 AG026529) and VA Merit Review.

References

1 Qian Y, Feldman E, Pennathur S, Kretzler M, Brosius FC 3rd: From fibrosis to sclerosis: mechanisms of glomerulosclerosis in diabetic nephropathy. Diabetes 2008;57:1439–1445.

2 Forbes JM, Coughlan MT, Cooper ME: Oxidative stress as a major culprit in kidney disease in diabetes. Diabetes 2008; 57:1446–1454.

3 Zhu Y, Usui HK, Sharma K: Regulation of transforming growth factor beta in diabetic nephropathy: implications for treatment. Semin Nephrol 2007;27:153–160.

4 Gurley SB, Coffman TM: The renin-angiotensin system and diabetic nephropathy. Semin Nephrol 2007;27:144–152.

5 Weinberg JM: Lipotoxicity. Kidney Int 2006;70:1560–1566.

6 Kimmelstiel P, Wilson C: Intercapillary lesions in the glomeruli of the kidney. Am J Pathol 1936;12:83–98.

7 Wilens SL, Elster SK: The role of lipid deposition in renal arteriolar sclerosis. Am J Med Sci 1950;219:183–196.

8 Sun L, Halaihel N, Zhang W, Rogers T, Levi M: Role of sterol regulatory element-binding protein 1 in regulation of renal lipid metabolism and glomerulosclerosis in diabetes mellitus. J Biol Chem 2002;277:18919–18927.

9 Proctor G, Jiang T, Iwahashi M, Wang Z, Li J, Levi M: Regulation of renal fatty acid and cholesterol metabolism, inflammation, and fibrosis in Akita and OVE26 mice with type 1 diabetes. Diabetes 2006;55:2502–2509.

10 Jiang T, Wang Z, Proctor G, Moskowitz S, Liebman SE, Rogers T, et al: Diet-induced obesity in C57BL/6J mice causes increased renal lipid accumulation and glomerulosclerosis via a sterol regulatory element-binding protein-1c-dependent pathway. J Biol Chem 2005;280:32317–32325.

11 Wang Z, Jiang T, Li J, Proctor G, McManaman JL, Lucia S, et al: Regulation of renal lipid metabolism, lipid accumulation, and glomerulosclerosis in FVBdb/db mice with type 2 diabetes. Diabetes 2005;54: 2328–2335.

12 Plum LA, DeLuca HF: Vitamin D, disease and therapeutic opportunities. Nat Rev Drug Discov 2010;9:941–955.

13 Haussler MR, Haussler CA, Bartik L, Whitfield GK, Hsieh JC, Slater S, et al: Vitamin D receptor: molecular signaling and actions of nutritional ligands in disease prevention. Nutr Rev 2008;66(suppl 2):S98–S112.

14 Ordonez-Moran P, Munoz A: Nuclear receptors: genomic and non-genomic effects converge. Cell Cycle 2009;8:1675–1680.

15 Tan X, Wen X, Liu Y: Paricalcitol inhibits renal inflammation by promoting vitamin D receptor-mediated sequestration of NF-kappaB signaling. J Am Soc Nephrol 2008;19:1741–1752.

16 Zhang Y, Kong J, Deb DK, Chang A, Li YC: Vitamin D receptor attenuates renal fibrosis by suppressing the renin-angiotensin system. J Am Soc Nephrol 2010;21:966–973.

17 Freundlich M, Quiroz Y, Zhang Z, Zhang Y, Bravo Y, Weisinger JR, et al: Suppression of renin-angiotensin gene expression in the kidney by paricalcitol. Kidney Int 2008; 74:1394–1402.

18 Kong J, Qiao G, Zhang Z, Liu SQ, Li YC: Targeted vitamin D receptor expression in juxtaglomerular cells suppresses renin expression independent of parathyroid hormone and calcium. Kidney Int 2008;74:1577–1581.

19 Deb DK, Chen Y, Zhang Z, Zhang Y, Szeto FL, Wong KE, et al: 1,25-Dihydroxyvitamin D3 suppresses high glucose-induced angiotensinogen expression in kidney cells by blocking the NF-kappaB pathway. Am J Physiol Renal Physiol 2009;296:F1212–F1218.

20 Tan X, Li Y, Liu Y: Therapeutic role and potential mechanisms of active vitamin D in renal interstitial fibrosis. J Steroid Biochem Mol Biol 2007;103:491–496.

21 Zhang Z, Sun L, Wang Y, Ning G, Minto AW, Kong J, et al: Renoprotective role of the vitamin D receptor in diabetic nephropathy. Kidney Int 2008;73:163–171.

22 Mizobuchi M, Morrissey J, Finch JL, Martin DR, Liapis H, Akizawa T, et al: Combination therapy with an angiotensin-converting enzyme inhibitor and a vitamin D analog suppresses the progression of renal insufficiency in uremic rats. J Am Soc Nephrol 2007;18:1796–1806.

23 Deb DK, Sun T, Wong KE, Zhang Z, Ning G, Zhang Y, et al: Combined vitamin D analog and AT1 receptor antagonist synergistically block the development of kidney disease in a model of type 2 diabetes. Kidney Int 2010;77:1000–1009.

24 Wang XX, Jiang T, Shen Y, Santamaria H, Solis N, Arbeeny CM, et al: The vitamin D receptor agonist doxercalciferol modulates dietary fat induced renal disease and renal lipid metabolism. Am J Physiol Renal Physiol 2011, Epub ahead of print.

25 Makishima M, Lu TT, Xie W, Whitfield GK, Domoto H, Evans RM, et al: Vitamin D receptor as an intestinal bile acid sensor. Science 2002;296:1313–1316.

26 Schmidt DR, Holmstrom SR, Fon Tacer K, Bookout AL, Kliewer SA, Mangelsdorf DJ: Regulation of bile acid synthesis by fat-soluble vitamins A and D. J Biol Chem 2010;285:14486–14494.

27 de Zeeuw D, Agarwal R, Amdahl M, Audhya P, Coyne D, Garimella T, et al: Selective vitamin D receptor activation with paricalcitol for reduction of albuminuria in patients with type 2 diabetes (VITAL study): a randomised controlled trial. Lancet 2010;376:1543–1551.

28 Makishima M, Okamoto AY, Repa JJ, Tu H, Learned RM, Luk A, et al: Identification of a nuclear receptor for bile acids. Science 1999;284:1362–1365.

29 Parks DJ, Blanchard SG, Bledsoe RK, Chandra G, Consler TG, Kliewer SA, et al: Bile acids: natural ligands for an orphan nuclear receptor. Science 1999;284: 1365–1368.

30 Wang H, Chen J, Hollister K, Sowers LC, Forman BM: Endogenous bile acids are ligands for the nuclear receptor FXR/BAR. Mol Cell 1999;3:543–553.

31 Bookout AL, Jeong Y, Downes M, Yu RT, Evans RM, Mangelsdorf DJ: Anatomical profiling of nuclear receptor expression reveals a hierarchical transcriptional network. Cell 2006;126:789–799.

32 Thomas C, Pellicciari R, Pruzanski M, Auwerx J, Schoonjans K: Targeting bile-acid signalling for metabolic diseases. Nat Rev Drug Discov 2008;7:678–693.

33 Lefebvre P, Cariou B, Lien F, Kuipers F, Staels B: Role of bile acids and bile acid receptors in metabolic regulation. Physiol Rev 2009;89:147–191.

34 Miyazaki-Anzai S, Levi M, Kratzer A, Ting TC, Lewis LB, Miyazaki M: Farnesoid X receptor activation prevents the development of vascular calcification in ApoE–/– mice with chronic kidney disease. Circ Res 2010;106:1807–1817.

35 Jiang T, Wang XX, Scherzer P, Wilson P, Tallman J, Takahashi H, et al: Farnesoid X receptor modulates renal lipid metabolism, fibrosis, and diabetic nephropathy. Diabetes 2007;56:2485–2493.

36 Wang XX, Jiang T, Shen Y, Caldas Y, Miyazaki-Anzai S, Santamaria H, et al: Diabetic nephropathy is accelerated by farnesoid X receptor deficiency and inhibited by farnesoid X receptor activation in a type 1 diabetes model. Diabetes 2010;59:2916–2927.

37 Wang XX, Jiang T, Shen Y, Adorini L, Pruzanski M, Gonzalez FJ, et al: The farnesoid X receptor modulates renal lipid metabolism and diet-induced renal inflammation, fibrosis, and proteinuria. Am J Physiol Renal Physiol 2009;297:F1587–F1596.

Moshe Levi, MD
Division of Renal Diseases and Hypertension
Department of Medicine, University of Colorado Denver
12700 East 19th Avenue
Aurora, CO 80045 (USA)
Tel. +1 303 724 4825, E-Mail Moshe.Levi@ucdenver.edu

                                                                 Levi · Wang · Choudhury

Lai KN, Tang SCW (eds): Diabetes and the Kidney.
Contrib Nephrol. Basel, Karger, 2011, vol 170, pp 217–227

Antifibrotic Treatment and Other New Strategies for Improving Renal Outcomes

Anna Mathew[a,b] · Robyn Cunard[a,b] · Kumar Sharma[a,b]

[a]Center for Renal Translational Medicine, Division of Nephrology-Hypertension, Department of Medicine, University of California San Diego and [b]Veterans Affairs San Diego Healthcare System, Veterans Medical Research Foundation, San Diego, Calif., USA

Abstract

Diabetic nephropathy (DN) is clinically characterized by proteinuria and hypertension. Investigations suggest that matrix accumulation and inflammatory processes contribute to the pathological features of this progressive disease. This chapter reviews novel targeted approaches to the treatment of DN, with the goal of slowing the progression and improving renal function. Many studies support the use of agents that block the renin-angiotensin-aldosterone system in DN. Novel, oral agents that are promising in early clinical studies are agents such as pirfenidone and bardoxolone as they are associated with early improvement in renal function in patients with advanced diabetic kidney disease. Additionally, strategies that inhibit inflammatory cytokines, chemokines, adhesion molecules and mediators of the innate immune response may provide novel targets for the treatment of DN. Larger clinical studies are eagerly awaited to determine if new agents that specifically block kidney fibrosis and inflammation will delay, arrest and possibly reverse progressive renal failure.

Clinically, progressive diabetic nephropathy (DN) is characterized by heavy proteinuria, high blood pressure and decline in the glomerular filtration rate (GFR). Pathologically, the degree of mesangial matrix accumulation, tubulointerstitial fibrosis and inflammation portend the decline in renal function. Evaluating treatment approaches from the viewpoint of blocking matrix accumulation and inflammation within the kidney may offer more potent and targeted approaches to prevent decline and possibly improve renal function in advanced DN.

Renin-Angiotensin-Aldosterone System

The renin-angiotensin-aldosterone system (RAAS) pathway has been the central component of therapeutic measures to prevent progression of diabetic kidney disease. In addition to angiotensin-converting enzyme inhibitors (ACEI) and angiotensin II receptor blockers (ARBs), direct renin inhibition using agents like aliskiren are showing increasing potential to prevent the progression of DN. Renin may promote kidney disease via the activation of the mitogen-activated protein kinases, ERK1 and ERK2, transforming growth factor-β (TGF-β) secretion and hypertrophic and proliferative effects independently of angiotensin II [1, 2]. In the diabetic transgenic (mRen-2)27 rat, which has greatly enhanced tissue RAS and plasma prorenin levels, aliskiren and perindopril were equally effective in reducing blood pressure and albuminuria, but aliskiren reduced tubulointerstitial fibrosis to a greater extent than perindopril [3]. In a separate study, using the same model system, aliskiren also reduced renal expression of TGF-β and collagen I, and attenuated the expression of the renin receptor [4].

In two placebo-controlled studies, aliskiren was as effective as the ARB irbesartan, and their combination was even more effective than monotherapy in reducing baseline albuminuria. The Aliskiren in the Evaluation of Proteinuria in Diabetes study is a multinational, randomized, double-blind, placebo-controlled study that included 599 patients with hypertension, type 2 diabetes and nephropathy. The double-blinded phase compared the combination of aliskiren and losartan with losartan plus placebo. The results showed that the combination of aliskiren and losartan was 20% more effective (p < 0.001) than losartan alone in reducing the mean urinary albumin-creatinine ratio [5]. Also, there is an ongoing international phase 3 trial looking at aliskiren 300 mg OD versus placebo in hypertensive diabetics looking at renal outcomes of time to dialysis, transplantation and creatinine >6 mg/dl or sustained doubling of creatinine above upper limit of the central laboratory (NCT00549757).

There is accumulating interest in a local aldosterone axis within the kidney and specifically within podocytes [6]. Several studies have suggested that aldosterone may be produced within the kidney independent of the adrenal gland [7]. Additionally, podocyte expression of the mineralocorticoid receptor (MR) appears to be activated even without exposure to aldosterone [8]. As aldosterone blockade may be more effective than an ARB to reduce proteinuria in patients on ACEI therapy [9], new approaches that block the MR or block local aldosterone production may be beneficial. However, systemic aldosterone or MR blockade can cause hyperkalemia necessitating close follow-up of electrolytes.

Profibrotic Growth Factors

Of the many growth factors that have been implicated as the direct effectors of the RAAS, TGF-β appears to be the central player mediating the fibrotic pathway in the diabetic kidney of type 1 and type 2 diabetes. In the streptozotocin-induced diabetes model, anti-TGF-β antibody was able to reduce hypertrophy and mRNA production of matrix proteins [10]. The protective effects seem to be independent of albuminuria reduction as anti-TGF-β therapy in the *db/db* mice prevented expansion of mesangial matrix, but not albuminuria. This suggests a renoprotective effect of anti-TGF-β antibodies independent of albuminuria [11]. The finding that an antifibrotic approach may confer renoprotection at a histologic level but may not reduce albuminuria, has been identified with several antifibrotic and anti-inflammatory approaches [reviewed in 12].

Studies have demonstrated synergistic effects of the RAAS blockade and anti-TGF-β antibody on the progression of proteinuria and renal injury in experimental DN and other models. Phase I/II trials of human anti-TGF-β antibody (CAT-192) in 45 patients with scleroderma showed no evidence of antifibrotic effects; however, clinical studies with new neutralizing anti-TGF-β antibodies are in early phase [13]. Several large pharmaceutical companies are in the process of pursuing clinical trials with anti-TGF-β antibodies for progressive kidney disease, and these results are eagerly awaited (NCT00125385, NCT00464321). Strategies to manipulate TGF-β activity include upregulating Betaglycan, a soluble TGF-β coreceptor and decorin, a leucine-rich proteoglycan that can bind active TGF-β. Betaglycan has been shown to reduce structural and functional renal injury [14], and the decorin knockout mice exhibit enhanced diabetic kidney disease [15]. Another strategy would be to inhibit thrombospondin activity, as this molecule is stimulated in diabetic kidney disease and can enhance activation of latent TGF-β [16].

Pirfenidone [5-methyl-1-phenyl-2(1H)-pyridone] is a small synthetic molecule with potent antifibrotic and hydroxyl ion scavenging properties. In renal cells, pirfenidone blocks the TGF-β promoter, TGF-β protein secretion, and phosphorylation of the downstream TGF-β target SMAD 2 [17]. Pirfenidone reduces tubulointerstitial and glomerular lesions, collagen and TGF-β production in animal models of diabetes, subtotal nephrectomy and unilateral obstruction. Additionally, pirfenidone treatment significantly reduced mesangial matrix expansion and expression of renal matrix genes in *db/db* mice with DN without affecting albuminuria. Using a combined proteomics and protein-protein interaction networks, pirfenidone was found to inhibit the eukaryotic initiation factor 4E protein. Pirfenidone in clinical studies has shown promising effects in patients with idiopathic and advanced focal segmental glomerulosclerosis. In a recently completed placebo controlled randomized clinical trial with 77 patients with diabetic kidney disease and proteinuria, pirfenidone showed improvement in estimated GFR after one-year follow-up [18]. The major reported side effects were gastrointestinal symptoms and photosensitivity. These findings provide support for a larger clinical trial to determine the renoprotective benefit of this convenient oral drug.

Another novel pyridine agent, fluorofenodine, has recently been studied in a model of renal cells stimulated by angiotensin II. The data showed that fluorofenodine reduced the production of reactive oxygen species and inhibited TGF-β expression more potently than pirfenidone [19]. Finally, anthraquinone compounds isolated from rhubarb have shown a potential to improve diabetic kidney injury: in vitro, they reduced hypertrophy and extracellular matrix expansion in tubular epithelial cells; in vivo, they attenuated renal alterations in *db/db* mice [20, 21].

Bone morphogenic protein 7 (BMP-7) is a member of the TGF-β superfamily, and has been shown to play an important role in diabetic kidney disease [22]. Overexpression of BMP-7 and inhibition of gremlin (a BMP-7 antagonist) ameliorate features of diabetic kidney disease [23]. Another potential target is SMP-534, a molecule that inhibits p38 signaling downstream of TGF-β. In the *db/db* mice model, SMP-534 decreased extracellular matrix and ameliorated progression of glomerular fibrosis greater than losartan alone and had an additive effect when combined with losartan [24, 25].

Tranilast [N-(3,4-dimethoxycinnamoyl) anthranilic acid] suppresses collagen synthesis by interfering with the actions of TGF-β. In one study, tranilast reduced the slope of the reciprocal of serum creatinine and decreased urinary proteinuria and urinary type IV collagen [26]. Similar observations were made in a study on proteinuric (<1 g) diabetics with normal creatinine; there was no change in creatinine after one year of treatment with tranilast; however, the urinary protein and urinary type IV collagen levels decreased from baseline levels in the treatment group [27]. In diabetic rat models, tranilast decreased albuminuria, TGF-β excretion, and mesangial expansion [28].

Nrf2 is a transcription factor controlling antioxidant genes that help maintain redox homeostasis. Nrf2 also inhibits TGF-β mRNA production. Nrf2-deficient mice have increased reactive oxygen species production and renal injury. Bardoxolone is an anti-inflammatory agent that acts via the Nrf2 pathway. A recent study presented at the American Society of Nephrology meeting in 2010, revealed encouraging benefits at week 24 of the ongoing 52-week trial with 200 diabetic proteinuric patients, mostly chronic kidney disease stage 3b. Bardoxolone treatment was associated with a fairly rapid increase in GFR that was seen as early as 4 weeks, with a continuous increase up to 12 weeks. The increase was sustained through week 24, with the average increase in GFR of 10.1 ml/min/1.73 m^2, compared with no change in the placebo group (NCT00811889).

Connective tissue growth factor (CTGF) is a downstream partner of TGF-β and may also play an independent role in the progression of renal fibrosis. Plasma CTGF level is an early marker of renal disease and may be a good predictor of end-stage renal disease in patients with DN [29]. Its inhibition by anti-sense oligonucleotides, siRNA or neutralizing antibodies can prevent the extracellular matrix expansion associated with several experimental models of kidney disease. CTGF is highly expressed in podocytes during the development of diabetic kidney disease and, in contrast to anti-TGF-β antibodies, there appears to be a reduction in albuminuria with anti-CTGF treatment in early phase studies. Diabetic mice models with CTGF attenuation have decreased fibrosis, proteinuria and progression of nephropathy [30–32]. In a recent phase I study, anti-CTGF antibodies were administered intravenously every 14 days for four doses and then followed up at day 62 and 365. Urinary albumin/creatinine ratio (ACR) decreased significantly from mean pretreatment ACR of 48 mg/g to mean posttreatment (day 56) ACR of 20 mg/g (p = 0.027) without evidence for a dose-response relationship and minimal side effects [33]. There seems to be minimal infusion adverse events and no significant drug-attributable adverse effects over 1 year of follow-up. Changes in albuminuria were promising but require validation in a prospective, randomized, blinded study.

Hepatocyte growth factor (HGF) is considered an endogenous inhibitor of TGF-β with multiple renoprotective properties that may find therapeutic application in the prevention of chronic kidney disease. Multiple animal studies have demonstrated that the administration of recombinant HGF protein or a corresponding gene therapy ameliorates several forms of chronic renal dysfunction [34]. A combination of exogenous HGF and an ARB (losartan) exhibited a synergistic efficacy in slowing the progression of renal fibrosis in a model of murine obstructive nephropathy [35]. Increased vascular endothelial growth factor (VEGF) production causes glomerular hypertrophy and is associated with proteinuria [36]. Semaxanib (SU5416) is a VEGF inhibitor that reduces albuminuria in mouse models of DN [37].

Alternate modes of drug delivery such as ultrasound microbubbles may be useful to target the kidney with specific neutralizing antibodies against key mediators involved in the progression of DN such as TGF-β. Although this technology was recently used in a model of liver fibrosis in rats, more studies are needed in DN [38].

Advanced Glycation End Product Pathway

The receptor for advanced glycation end products (RAGE) activates multiple intracellular pathways and amplifies the inflammatory effects of AGEs. Recently, the soluble form of the RAGE was shown to decrease AGE-induced vascular injury [39]. Anti-RAGE antibody administered in a type 1 diabetic animal model decreased the urine albumin excretion, kidney weight and creatinine clearance [40]. Pyridoxamine is a form of vitamin B_6 that blocks formation of preglycated amodori compounds, advanced lipoxidation protein end-products and scavenges reactive carbonyl species [41]. Pyridoxamine decreased proteinuria,

TGF-β and laminin 1 mRNA expression in animal models of type 2 diabetes [42]. In phase 2 clinical studies, pyridoxamine showed benefit in decreasing serum creatinine and proteinuria, and a phase 2b trial is currently underway [43].

Other Novel Pathways

Studies suggest that vitamin D analogs downregulate the RAAS pathway. In murine models of diabetes, paricalcitol, an active analog of vitamin D, acts synergistically with losartan in decreasing albuminuria, glomerular inflammation and sclerosis both in type 1 and type 2 models [44]. Similar results were seen with doxercalciferol, another vitamin D analog, which impedes progression of diabetic kidney disease and downregulates the intrarenal RAAS [45]. A recent clinical study, VITAL, showed reduction in albuminuria in diabetic patients taking paricalcitol (2 μg/day), when compared to placebo [46].

The sodium glucose cotransporter (SGLT2) is a high-capacity, low-affinity proximal tubular cell transporter that reabsorbs filtered glucose. Mutations in this transporter lead to renal glycosuria. In diabetic animal modes, there appears to be upregulation of the SGLT2 receptors and GLUT2 mRNA. SGLT2 antagonism provides an interesting therapeutic option to decrease glucose loads in diabetes. Nonspecific SGLT2 inhibitors like phlorizin have disturbing gastrointestinal side effects. More specific inhibitors like dapagliflozin and sergliflozin are being tested in small clinical trials. The effect on the progression of diabetic kidney disease remains to be studied [47].

Glucagon-like peptide is a gut hormone that improves insulin secretion in pancreatic cells, and its agonist exendin-4 (Eventide) ameliorates albuminuria, glomerular hyperfiltration, hypertrophy and essential matrix expansion in diabetic rats without changing blood pressure or body weight. Exendin-4 also prevented macrophage infiltration, and decreased protein levels of intercellular adhesion molecule-1 (ICAM-1) and type IV collagen, as well as decreasing oxidative stress and nuclear factor-κB activation in kidney tissue [48]. Long-term treatment of *db/db* mice with exendin-4 decreased albuminuria. Glomerular hypertrophy, mesangial matrix expansion, TGF-β_1 expression, and type IV collagen accumulation and associated glomerular lipid accumulation were significantly decreased [49]. Taken together, exendin-4 treatment seems to ameliorate DN as well as the metabolic abnormalities.

Inflammatory Pathway

The progression and manifestations of DN are likely associated with activation of the innate and adaptive immune systems [50]. Therefore, investigators have employed multiple strategies to interfere with inflammatory signaling networks. The innate immune response is activated by Toll-like receptors (TLRs) that bind to pattern recognition peptides, promoting a rapid response to microbial pathogens. TLRs are present on many intrinsic renal cells and infiltrating leukocytes, and TLR expression is augmented in type 1 and type 2 diabetic patients [51] and in the kidneys of diabetic rodents. Given the link between diabetes and inflammation, modulation of TLR expression may be an attractive target for treatment of DN.

Overexpression of cytokines including tumor necrosis factor-α (TNF-α), interleukin (IL)-1β, IL-6 and IL-18 has been implicated in the progression of DN [51]. In rodent models of diabetes, reduction in these cytokines has improved renal outcomes. Multiple cytokines signal through STAT-3; reduction in STAT-3 activity dampens inflammation (IL-6, MCP-1, activated NF-κB, TGF-β, ICAM-1) and abnormal matrix synthesis in early DN. Suppressors of cytokine signaling negatively regulate activation of cytokine-stimulated JAK/STAT pathways in DN and may prove to be another efficient therapeutic target [52]. The immunosuppressant mycophenolate mofetil and pentoxifylline (decreases TNF-α) reduce cytokine elaboration and ameliorate manifestations of diabetic kidney disease.

Chemokines are molecules that drive the tissue migration of inflammatory cells. One of the most actively studied chemokines is monocyte chemotactic protein-1 (MCP-1)/CCL2. MCP-1 attracts and activates monocytes and macrophages, and plays a central role in promoting renal injury. Blockade of MCP-1 or its receptor CCR2 improves DN, and pharmacologic blockers of MCP-1 activity are currently in preclinical human studies. Inhibition of the RAAS (ACEI, ARBs, spironolactone) and nuclear receptor ligands (thiazolidinediones, vitamin D) can also block MCP-1 activity and improve renal abnormalities in diabetic kidney disease. Other chemokines that may play a role in the progression of chronic inflammatory diseases include fractaline/CX3CL1, CXCL12, IL-8, macrophage inflammatory protein-1α, interferon-γ inducible protein-10 and RANTES [53].

Migration and infiltration of inflammatory cells into tissues requires leukocyte adhesion molecules, including ICAM-1 and vascular adhesion molecule. Inhibitors of ICAM-1 are currently being developed for treatment in chronic inflammatory disease. Furthermore, uric acid upregulates ICAM-1 expression. Thus, modulators of gout including xanthine oxidase inhibitors and colchicine are being employed in human and rodent studies to investigate progression of diabetic kidney disease.

The endoplasmic reticulum (ER) stress pathway is activated during cellular stress and is designed to repress protein synthesis and increase ER chaperone content to facilitate efficient trafficking of proteins through the ER. This pathway is activated in the diabetic kidney, and currently efforts are under way to identify novel regulators of this pathway with the goal of improving the progression of diabetic kidney disease [54].

Conclusion

There are many new promising targets for therapy that may retard the progression and improve renal structure and function in DN. Two oral agents, pirfenidone and bardoxolone, appear especially promising as renal function improvement was found in early phase trials. These studies suggest that renal functional improvement may occur in patients even with advanced disease but will need validation in larger trials. The results of ongoing clinical studies are eagerly awaited to determine if new agents that specifically block kidney fibrosis and inflammation will delay or arrest progressive renal failure.

References

1   Nguyen G, Delarue F, Burckle C, Bouzhir L, Giller T, Sraer JD: Pivotal role of the renin/prorenin receptor in angiotensin II production and cellular responses to renin. J Clin Invest 2002;109:1417–1427.

2   Huang J: The important role of angiotensin receptor blocker on treatment of chronic heart failure (in Chinese). Zhonghua Xin Xue Guan Bing Za Zhi 2006;34:773–774.

3   Kelly DJ, Zhang Y, Moe G, Naik G, Gilbert GE: Aliskiren, a novel renin inhibitor, is renoprotective in a model of advanced diabetic nephropathy in rats. Diabetologia 2007; 50:2398–2404.

4   Feldman DL, Jin L, Xuan H, Contrepas A, Zhou Y, Webb RL, Mueller DN, Feldt S, Cumin F, Maniara W, et al: Effects of aliskiren on blood pressure, albuminuria, and (pro)renin receptor expression in diabetic TG(mRen-2)27 rats. Hypertension 2008;52:130–136.

5   Parving HH, Persson F, Lewis JB, Lewis EJ, Hollenberg NK: Aliskiren combined with losartan in type 2 diabetes and nephropathy. N Engl J Med 2008;358:2433–2446.

6   Calhoun DA, Sharma K: The role of aldosteronism in causing obesity-related cardiovascular risk. Cardiol Clin 2010;28:517–527.

7   Siragy HM, Xue C: Local renal aldosterone production induces inflammation and matrix formation in kidneys of diabetic rats. Exp Physiol 2008;93:817–824.

8   Shibata S, Nagase M, Yoshida S, Kawarazaki W, Kurihara H, Tanaka H, Miyoshi J, Takai Y, Fujita T: Modification of mineralocorticoid receptor function by Rac1 GTPase: implication in proteinuric kidney disease. Nat Med 2008;14:1370–1376.

9   Mehdi UF, Adams-Huet B, Raskin P, Vega GL, Toto, RD: Addition of angiotensin receptor blockade or mineralocorticoid antagonism to maximal angiotensin-converting enzyme inhibition in diabetic nephropathy. J Am Soc Nephrol 2009;20:2641–2650.

10   Sharma K, Jin Y, Guo Y, Ziyadeh FN: Neutralization of TGF-beta by anti-TGF-beta antibody attenuates kidney hypertrophy and the enhanced extracellular matrix gene expression in STZ-induced diabetic mice. Diabetes 1996;45:522–530.

11   Ziyadeh FN, Hoffman BB, Han DC, Iglesias-De La Cruz MC, Hong SW, Isono M, Chen S, McGowan TA, Sharma K: Long-term prevention of renal insufficiency, excess matrix gene expression, and glomerular mesangial matrix expansion by treatment with monoclonal antitransforming growth factor-beta antibody in db/db diabetic mice. Proc Natl Acad Sci U S A 2000;97:8015–8020.

12   Declèves AE, Sharma K: New pharmacological treatment for improving renal outcomes in diabetes. Nat Rev Nephrol 2010;6: 371–380.

13   Denton CP, Merkel PA, Furst DE, Khanna D, Emery P, Hsu VM, Silliman N, Streisand J, Powell J, Akesson A, et al: Recombinant human anti-transforming growth factor beta1 antibody therapy in systemic sclerosis: a multicenter, randomized, placebo-controlled phase I/II trial of CAT-192. Arthritis Rheum 2007;56:323–333.

14   Juarez P, Vilchis-Landeros MM, Ponce-Coria J, Mendoza V, Hernandez-Pando R, Bobadilla NA, Lopez-Casillas F: Soluble betaglycan reduces renal damage progression in db/db mice. Am J Physiol Renal Physiol 2007;292:F321–F329.

15   Merline R, Lazaroski S, Babelova A, Tsalastra-Greul W, Pfeilschifter J, Schluter KD, Gunther A, Iozzo RV, Schaefer RM, Schaefer L: Decorin deficiency in diabetic mice: aggravation of nephropathy due to overexpression of profibrotic factors, enhanced apoptosis and mononuclear cell infiltration. J Physiol Pharmacol 2009;60(suppl 4):5–13.

16   Daniel C, Schaub K, Amann K, Lawler J, Hugo C: Thrombospondin-1 is an endogenous activator of TGF-beta in experimental diabetic nephropathy in vivo. Diabetes 2007;56:2982–2989.

17   RamachandraRao SP, Zhu Y, Ravasi T, McGowan TA, Toh I, Dunn SR, Okada S, Shaw MA, Sharma K: Pirfenidone is renoprotective in diabetic kidney disease. J Am Soc Nephrol 2009;20:1765–1775.

18   Sharma K, Mathew AV, Fervenza FC, Cho M, Pflueger A, Dunn SR, Francos B, Sharma S, Falkner B, McGowan TA, Donohue M, Ramachandrarao SP, Xu R, Kopp JB: Pirfenidone for diabetic nephropathy: a randomized placebo controlled clinical trial. J Am Soc Nephrol 2011, in press.

19   Peng ZZ, Hu GY, Shen H, Wang L, Ning WB, Xie YY, Wang NS, Li BX, Tang YT, Tao LJ: Fluorofenidone attenuates collagen I and transforming growth factor-beta1 expression through a nicotinamide adenine dinucleotide phosphate oxidase-dependent way in NRK-52E cells. Nephrology (Carlton) 2009;14:565–572.

20   Zheng JM, Zhu JM, Li LS, Liu ZH: Rhein reverses the diabetic phenotype of mesangial cells over-expressing the glucose transporter (GLUT1) by inhibiting the hexosamine pathway. Br J Pharmacol 2008;153:1456–1464.

21   Gao Q, Qin WS, Jia ZH, Zheng JM, Zeng CH, Li LS, Liu ZH: Rhein improves renal lesion and ameliorates dyslipidemia in db/db mice with diabetic nephropathy. Planta Med 2010;76:27–33.

22   Sugimoto H, Grahovac G, Zeisberg M, Kalluri R: Renal fibrosis and glomerulosclerosis in a new mouse model of diabetic nephropathy and its regression by bone morphogenic protein-7 and advanced glycation end product inhibitors. Diabetes 2007;56:1825–1833.

23   Mitu GM, Wang S, Hirschberg R: BMP7 is a podocyte survival factor and rescues podocytes from diabetic injury. Am J Physiol Renal Physiol 2007;293:F1641–F1648.

24   Sugaru E, Nakagawa T, Ono-Kishino M, Nagamine J, Tokunaga T, Kitoh M, Hume WE, Nagata R, Taiji M: Enhanced effect of combined treatment with SMP-534 (antifibrotic agent) and losartan in diabetic nephropathy. Am J Nephrol 2006;26:50–58.

25   Sugaru E, Nakagawa T, Ono-Kishino M, Nagamine J, Tokunaga T, Kitoh M, Hume WE, Nagata R, Taiji M: Amelioration of established diabetic nephropathy by combined treatment with SMP-534 (antifibrotic agent) and losartan in db/db mice. Nephron Exp Nephrol 2007;105:e45–e52.

26   Soma J, Sato K, Saito H, Tsuchiya Y: Effect of tranilast in early-stage diabetic nephropathy. Nephrol Dial Transplant 2006;21:2795–2799.

27   Soma J, Sugawara T, Huang YD, Nakajima J, Kawamura M: Tranilast slows the progression of advanced diabetic nephropathy. Nephron 2002;92:693–698.

28   Akahori H, Ota T, Torita M, Ando H, Kaneko S, Takamura T: Tranilast prevents the progression of experimental diabetic nephropathy through suppression of enhanced extracellular matrix gene expression. J Pharmacol Exp Ther 2005;314:514–521.

29   Nguyen TQ, Tarnow L, Jorsal A, Oliver N, Roestenberg P, Ito Y, Parving HH, Rossing P, van Nieuwenhoven FA, Goldschmeding R: Plasma connective tissue growth factor is an independent predictor of end-stage renal disease and mortality in type 1 diabetic nephropathy. Diabetes Care 2008;31:1177–1182.

30 Yokoi H, Mukoyama M, Mori K, Kasahara M, Suganami T, Sawai K, Yoshioka T, Saito Y, Ogawa Y, Kuwabara T, et al: Overexpression of connective tissue growth factor in podocytes worsens diabetic nephropathy in mice. Kidney Int 2008;73:446–455.

31 Yokoi H, Mukoyama M, Nagae T, Mori K, Suganami T, Sawai K, Yoshioka T, Koshikawa M, Nishida T, Takigawa M, et al: Reduction in connective tissue growth factor by antisense treatment ameliorates renal tubulointerstitial fibrosis. J Am Soc Nephrol 2004;15: 1430–1440.

32 Guha M, Xu ZG, Tung D, Lanting L, Natarajan R: Specific down-regulation of connective tissue growth factor attenuates progression of nephropathy in mouse models of type 1 and type 2 diabetes. FASEB J 2007; 21:3355–3368.

33 Adler SG, Schwartz S, Williams ME, Arauz-Pacheco C, Bolton WK, Lee T, Li D, Neff TB, Urquilla PR, Sewell KL: Phase 1 study of anti-CTGF monoclonal antibody in patients with diabetes and microalbuminuria. Clin J Am Soc Nephrol 2010;5:1420–1428.

34 Mizuno S, Nakamura T: Suppressions of chronic glomerular injuries and TGF-beta 1 production by HGF in attenuation of murine diabetic nephropathy. Am J Physiol Renal Physiol 2004;286:F134–F143.

35 Yang J, Dai C, Liu Y: Hepatocyte growth factor gene therapy and angiotensin II blockade synergistically attenuate renal interstitial fibrosis in mice. J Am Soc Nephrol 2002;13: 2464–2477.

36 Liu E, Morimoto M, Kitajima S, Koike T, Yu Y, Shiiki H, Nagata M, Watanabe T, Fan J: Increased expression of vascular endothelial growth factor in kidney leads to progressive impairment of glomerular functions. J Am Soc Nephrol 2007;18:2094–2104.

37 Mendel DB, Laird AD, Smolich BD, Blake RA, Liang C, Hannah AL, Shaheen RM, Ellis LM, Weitman S, Shawver LK, et al: Development of SU5416, a selective small molecule inhibitor of VEGF receptor tyrosine kinase activity, as an anti-angiogenesis agent. Anticancer Drug Des 2000;15:29–41.

38 Deelman LE, Decleves AE, Rychak JJ, Sharma K: Targeted renal therapies through microbubbles and ultrasound. Adv Drug Deliv Rev 2010;62:1369–1377.

39 Yamamoto Y, Doi T, Kato I, Shinohara H, Sakurai S, Yonekura H, Watanabe T, Myint KM, Harashima A, Takeuchi M, et al: Receptor for advanced glycation end products is a promising target of diabetic nephropathy. Ann N Y Acad Sci 2005;1043: 562–566.

40 Jensen LJ, Denner L, Schrijvers BF, Tilton RG, Rasch R, Flyvbjerg A: Renal effects of a neutralising RAGE-antibody in long-term streptozotocin-diabetic mice. J Endocrinol 2006;188:493–501.

41 Onorato JM, Jenkins AJ, Thorpe SR. Baynes JW: Pyridoxamine, an inhibitor of advanced glycation reactions, also inhibits advanced lipoxidation reactions. Mechanism of action of pyridoxamine. J Biol Chem 2000;275: 21177–21184.

42 Tanimoto M, Gohda, T, Kaneko S, Hagiwara S, Murakoshi M, Aoki T, Yamada K, Ito T, Matsumoto M, Horikoshi S, et al: Effect of pyridoxamine (K-163), an inhibitor of advanced glycation end products, on type 2 diabetic nephropathy in KK-A(y)/Ta mice. Metabolism 2007;56:160–167.

43 Williams ME, Bolton WK, Khalifah RG, Degenhardt TP, Schotzinger RJ, McGill JB: Effects of pyridoxamine in combined phase 2 studies of patients with type 1 and type 2 diabetes and overt nephropathy. Am J Nephrol 2007;27:605–614.

44 Deb DK, Sun T, Wong KE, Zhang Z, Ning G, Zhang Y, Kong J, Shi H, Chang A, Li YC: Combined vitamin D analog and AT1 receptor antagonist synergistically block the development of kidney disease in a model of type 2 diabetes. Kidney Int 2010;77:1000–1009.

45 Zhang Y, Deb DK, Kong J, Ning G, Wang Y, Li G, Chen Y, Zhang Z, Strugnell S, Sabbagh Y, et al: Long-term therapeutic effect of vitamin D analog doxercalciferol on diabetic nephropathy: strong synergism with AT1 receptor antagonist. Am J Physiol Renal Physiol 2009;297:F791–F801.

46 de Zeeuw D, Agarwal R, Amdahl M, Audhya P, Coyne D, Garimella T, Parving HH, Pritchett T, Remuzzi G, Ritz E, Andress D: Selective vitamin D receptor activation with paricalcitol for reduction of albuminuria in patients with type 2 diabetes (VITAL study): a randomised controlled trial. Lancet 2010; 376:1543–1551.

Mathew · Cunard · Sharma

47 Vallon V, Sharma K: Sodium-glucose transport: role in diabetes mellitus and potential clinical implications. Curr Opin Nephrol Hypertens 2010;19:425–431.

48 Kodera R, Shikata K, Kataoka HU, Takatsuka T, Miyamoto S, Sasaki M, Kajitani N, Nishishita S, Sarai K, Hirota D, et al: Glucagon-like peptide-1 receptor agonist ameliorates renal injury through its anti-inflammatory action without lowering blood glucose level in a rat model of type 1 diabetes. Diabetologia 2011, Epub ahead of print.

49 Park CW, Kim HW, Ko SH, Lim JH, Ryu GR, Chung HW, Han SW, Shin SJ, Bang BK., Breyer MD, et al: Long-term treatment of glucagon-like peptide-1 analog exendin-4 ameliorates diabetic nephropathy through improving metabolic anomalies in db/db mice. J Am Soc Nephrol 2007;18:1227–1238.

50 Pickup JC, Crook MA: Is type II diabetes mellitus a disease of the innate immune system? Diabetologia 1998;41:1241–1248.

51 Rivero A, Mora C, Muros M, Garcia J, Herrera H, Navarro-Gonzalez JF: Pathogenic perspectives for the role of inflammation in diabetic nephropathy. Clin Sci (Lond) 2009; 116:479–492.

52 Ortiz-Munoz G, Lopez-Parra V, Lopez-Franco O, Fernandez-Vizarra P, Mallavia B, Flores C, Sanz A, Blanco J, Mezzano S, Ortiz A, Egido J, Gomez-Guerrero C: Suppressors of cytokine signaling abrogate diabetic nephropathy. J Am Soc Nephrol 2010;21:763–772.

53 Ruster C, Wolf G: The role of chemokines and chemokine receptors in diabetic nephropathy. Front Biosci 2008;13:944–955.

54 Sharma K, Cunard R: The endoplasmic reticulum stress response and diabetic kidney disease. American Journal of Physiology Renal 2011, in press.

Kumar Sharma, MD
Center for Renal Translational Medicine, Division of Nephrology-Hypertension
Department of Medicine, University of California San Diego
San Diego, CA 92093 (USA)
Tel. +1 858 822 0860, Fax +1 858 822 7483, E-Mail Kusharma@ucsd.edu

Lai KN, Tang SCW (eds): Diabetes and the Kidney.
Contrib Nephrol. Basel, Karger, 2011, vol 170, pp 228–236

Kidney Regeneration: Any Prospects?

Paola Romagnani

Excellence Centre for Research, Transfer and High Education for the Development of de novo
Therapies, and Meyer Children's Hospital, Nephrology Unit, Department of Clinical Pathophysiology,
University of Florence, Florence, Italy

Abstract

Chronic kidney disease is a leading cause of mortality and morbidity in western countries
which affects about 11% of the adult population. With the increasing rate of chronic kidney disease and limited alternatives for its treatment, potential regenerative approaches
for kidney damage are urgently needed, but are limited by the complexity of this organ.
Bone marrow-derived stem cells as well as mesenchymal stem cells were envisioned for
the development of this type of treatment. However, most studies suggested that these
cells cannot differentiate into renal epithelial cells, and concluded that their beneficial
effects are probably related to secretion of growth factors. In addition, a long-term partial
maldifferentiation of injected mesenchymal stem cells into adipocytes accompanied by
glomerular sclerosis was reported. The incapacity of bone marrow-derived stem cells to
differentiate into renal cells suggested that turnover of resident renal epithelial cells may
be related to the existence of potential stem/progenitor cells within the adult human kidney. Consistently, renal progenitors with the potential to differentiate into podocytes as
well as tubular cells were recently identified at the urinary pole of the Bowman's capsule
in adult kidneys. The discovery of renal progenitors that encourage regeneration and
promote functional repair of glomerular injury demonstrates that prevention and treatment of glomerulosclerosis may be possible. In addition, converging evidence suggests
that the outcome of glomerular disorders depends on a balance between injury and
regeneration provided by renal progenitors. In summary, understanding how self-renewal
and fate decision of renal progenitors may be perturbed or modulated will be of crucial
importance to obtain novel pharmacological tools for prevention and treatment of diabetic nephropathy, as well as other causes of glomerulosclerosis.

Chronic kidney disease (CKD) is a leading cause of mortality and morbidity
in Western countries which affects about 11% of the adult population [1]. It
can progress to end-stage renal disease (ESRD), which has no cure and requires

renal replacement therapy, i.e. dialysis or renal transplantation. Although tubulointerstitial injury can be responsible for CKD, following tubular injury the kidney undergoes a regenerative response which involves the proliferation of other tubular cells and that leads in most cases to recovery of renal function. By contrast, glomerulosclerosis is the major pathologic process responsible for progression to ESRD in human. Underlying conditions that predispose to glomerulosclerosis include diabetes, hypertension, IgA nephropathy, focal segmental glomerulosclerosis (FSGS), and other immune-mediated forms of glomerular injury. In aggregate, these conditions constitute 90% of ESRD at a cost of approximately USD 20 billion per year in the US [1]. However, worldwide diabetic nephropathy (DN) is the primary diagnosis in 25–50% of people starting renal replacement therapy for ESRD, and 44% of all new patients starting dialysis in 2009 in the US had DN [2].

The glomerular tuft comprises three resident cell types, mesangial cells, endothelial cells, and podocytes [3, 4]. Primary injury to each of these cell types is associated with glomerular disease [3, 4]. However, injury to endothelial and mesangial cells can be repaired by proliferation of adjacent cells [3, 4]. By contrast, podocytes are highly differentiated neuron-like cells that cannot divide, which explains why podocyte injury is key in driving glomerulosclerosis [3, 4]. A large body of evidence from experimental models suggests that 20–40% podocyte loss results in mild proteinuria and a scarring response, until at greater than 60% podocyte loss, proteinuria becomes severe and glomeruli become globally sclerotic and nonfiltering. Several reports also deal with podocyte depletion in type 2 DN [5–7]. The averaged values from these two studies for percentage of podocyte loss (relative to normal glomeruli) in individuals with type 2 diabetes and relatively normal renal function were as follows: normoalbuminuria was associated with 16% podocyte loss, microalbuminuria was associated with 24% podocyte loss, and macroalbuminuria was associated with 36% podocyte loss [5, 7]. Type 1 diabetic glomerulopathy and IgA nephropathy have both been reported to show similar degrees of podocyte depletion from glomeruli in proportion to injury [5–7, 8]. All these studies therefore support the concept of a direct relationship between the progression of DN and podocyte loss. Although podocytes cannot divide, severe podocyte loss does not always lead to glomerulosclerosis. Remission of the disease and regression of renal lesions have been observed in patients with chronic glomerular disorders and severe proteinuria. Regression of diabetic renal disease with glomerular architecture remodeling has been observed in a few patients with type 1 diabetes after 10 years of normoglycemia induced by a pancreatic transplant [9], suggesting that if the cause of injury is removed, glomeruli can regenerate. In addition, strong evidence suggests that reduction of proteinuria favors glomerular regeneration. Indeed, improvement of glomerular filtration rate was reported in a subset of patients treated with angiotensin-converting enzyme inhibitors in the follow-up of the REIN study [10]. In animal models, regression of glomerular sclerosis associates with increased podocyte number, suggesting that novel podocytes can be generated, which triggers the development of a novel capillary network [10]. Taken together, these results suggest that renal injury can potentially be repaired, even when it is related to chronic glomerular disorders driving progressive glomerulosclerosis.

In most adult tissues, regenerative processes are enabled by the presence of stem/progenitor cells [11]. Stem cells can be defined by three main criteria: (a) ability to self-renew, which is a prerequisite for sustaining the stem cell pool; (b) ability to generate at the single cell level differentiated progeny cells, in general of multiple lineages, and (c) the ability to functionally reconstitute a given tissue in vivo [11]. In this chapter, we highlight the burst of knowledge leading to preclinical studies describing the possible use of bone marrow-derived stem cells (BMSC) for treatment of chronic renal injury in general and of DN in particular. In addition, we describe the recent identification of a stem cell compartment also in adult kidneys, and how this changes the perspective of treatment of CKD.

Bone Marrow-Derived Stem Cells

Several studies have examined the possibility that BMSC may represent a source for cell replacement in adult kidney, with heterogeneous results [12–24]. Initial studies reported that unfractionated BMSC can differentiate into new tubular epithelial cells in functionally important numbers after kidney injury [12, 13], while glomerular regeneration was not analyzed. Subsequent work by several groups has shown that tubular cell replacement by BMSC is rare, calling into question the possible role of BMSC as a source for cell replacement in adult kidney [12, 14, 15, 17–19]. In addition, some studies suggested that in a model of autosomal recessive Alport syndrome, improved renal function and a marked reduction in renal histological damage can be achieved through BM transplantation from wild-type donors [16, 20]. However, the extent of podocyte cell engraftment by BM cells was extremely low [16, 20], suggesting that the benefit of administered BMSC was derived from the paracrine action of the injected population. In further studies, some authors have addressed the possible utility of selected populations of BMSC, such as mesenchymal stem cells (MSCs). The role of MSCs in renal protection has been first analyzed by isolating MSCs from mouse BM and transplanting them into the mice with cisplatin-induced acute chemical injury of the kidneys [12, 17, 18]. Improvement in renal structure and a significant increase in tubular cell proliferation which were accompanied by the acceleration of renal function recovery indicated by lower levels of blood urea nitrogen were detected at 4 and 11 days after MSC treatment, suggesting stimulation of tubular cell growth by factors released by MSCs, such as insulin-like growth factor-1 [12, 17]. Thus, MSCs protect the kidney from toxic injury by secreting factors that limit apoptosis and enhance proliferation of the endogenous tubular cells [12, 17–19].

More recently, MSCs have also been used for the treatment of DN in NOD/SCID and C57Bl/6 mice which succumb to diabetes mellitus after application of multiple low doses of streptozotocin (STZ) [21–24]. About 30–60 days after STZ injection, kidneys of treated mice showed the presence of abnormal glomeruli characterized by increased deposits of extracellular matrix protein in the mesangium, hyalinosis and increased number of macrophages in the glomeruli [21–24]. Data obtained from studies using NOD/SCID mice transplanted with human MSCs and C57Bl/6 mice that received murine MSCs indicate that injected MSCs engraft in damaged kidneys and regulate the immune response resulting in an efficient treatment of DN, with improvement of kidney function and regeneration of glomerular structure [21–24]. However, one month after MSC treatment, only a few MSCs were detected in kidneys, suggesting that they were unable to proliferate [21, 24], so an alternative scenario for improvement of kidney function could be the ability of MSCs to scavenge cytotoxic molecules or to promote neovascularization [21–24]. In addition, some investigators also proposed that successful MSC treatment of DN may also be explained by their differentiation into insulin-producing β-cells followed by decrease in glycemia and glycosuria, factors important for damaging renal cells [22]. Taken together, these data indicate that MSC transplantation can improve kidney function in diabetic animals. However, recent studies suggested that beneficial effects linked to MSC injection can be offset by a long-term partial maldifferentiation of intraglomerular MSCs into adipocytes accompanied by glomerular sclerosis [19], thus calling into question the potential benefit of MSC treatment for chronic glomerular disorders.

Renal Stem Cells

The inability of the podocyte to proliferate and replace injured cells, as well as the incapacity of BMSC to differentiate into novel renal cells suggested the existence of potential stem/progenitor cells within the adult human kidney. Recently, we provided the first evidence that the adult human kidney contains a population of stem/progenitor cells. Interestingly, CD24+CD133+ progenitors are physically located at the urinary pole of the Bowman's capsule, the only place in the kidney which appears to be contiguous with both tubular cells and glomerular podocytes [25–29]. In addition, we demonstrated that CD133+CD24+ renal progenitors are a heterogeneous and hierarchical population of undifferentiated and more differentiated cells that are arranged in a precise sequence within the Bowman's capsule of adult human kidneys [25–29]. A subset of more undifferentiated cells expressing renal progenitor markers in the absence of podocyte markers is localized at the urinary pole of Bowman's capsule and acts as bipotent

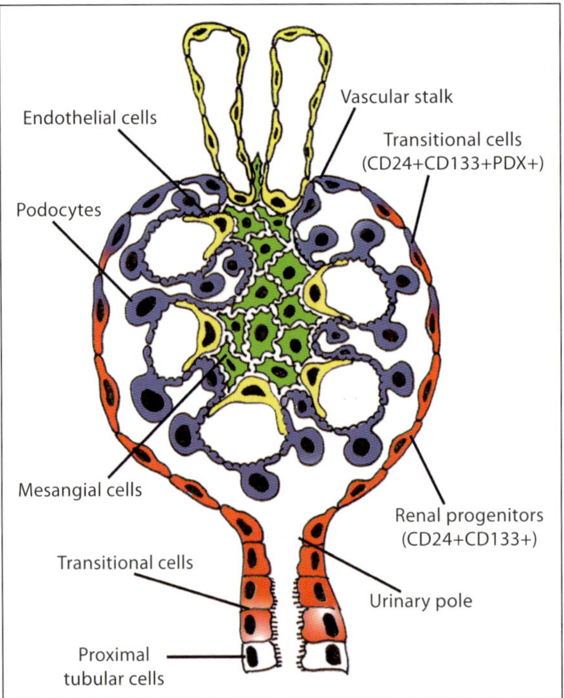

Fig. 1. A stem cell compartment in adult human kidney. CD24+CD133+ renal progenitors are localized at the urinary pole and are in close contiguity with podocytes at one extremity (the vascular stalk) and with tubular renal cells at the other extremity. A transitional cell population (CD24+CD133+PDX+) displays features of either renal progenitors or podocytes and localizes between the urinary pole and the vascular pole. At the vascular stalk of the glomerulus, the transitional cells are localized in close continuity with cells that lack progenitor markers, but exhibit the podocyte markers and the phenotypic features of differentiated podocytes. From Romagnani [26], with permission.

progenitor for both tubular cells and podocytes and exhibits self-renewal potential [25–29] (fig. 1). A transition population expressing both renal progenitors and podocyte markers is localized between the urinary and the vascular pole of Bowman's capsule, exhibits a committed differentiative potential only toward the podocyte lineage and lacks self-renewal properties [25–29] (fig. 1). Finally, differentiated cells that do not express renal progenitor markers but display high levels of podocyte-specific markers localize at the vascular pole of Bowman's capsule [25–29]. These findings in humans were also confirmed in a parallel study performed in rodents by Appel et al. [28] demonstrating that transitional cells with morphologic and immunohistochemical features of both parietal epithelial cells and podocytes could be detected in proximity of the glomerular vascular stalk. More importantly, using genetic tagging of parietal epithelial cells in rodents, the same authors unequivocally demonstrated that podocytes are recruited from parietal epithelial cells, which proliferate and differentiate from the urinary to the vascular stalk generating novel podocytes [28] (fig. 1). The discovery that renal progenitors of the Bowman's capsule represent a potential source for podocyte generation provides the basis for a novel concept that injured podocytes can be replaced. Consistently, CD24+CD133+ renal progenitors isolated from adult human kidneys generated novel podocytes and reduced proteinuria once injected in into SCID mice models of FSGS [27]. However, recent evidence also suggests that in patients with FSGS the regenerative process can sometimes be inadequate because of an inefficient proliferative response of renal stem cells [28]. In addition, an excessive proliferative response of renal progenitors to podocyte injury can even generate disorders [28]. Accordingly, aberrant renal progenitor proliferation initiates extracapillary lesions in crescentic glomerulonephritis or collapsing glomerulopathy and is involved in the generation of the tip lesion, a histopathological lesion characterized by podocyte prominence, capsular adhesion, and/or intracapillary foam cells at or adjacent to the urinary pole that is frequently observed in patients affected by DN [28, 30].

The discovery of CD24+CD133+ progenitors and their localization at the urinary pole of the Bowman's capsule suggests that these cells might participate also in tubular regeneration, progressively migrating and differentiating in proximal tubular cells at the tubuloglomerular junction (fig. 1). Accordingly, CD24+CD133+ renal progenitors isolated from adult human kidneys generated novel tubular structures and reduced the morphologic and functional kidney damage once injected in mice affected by acute renal failure, suggesting that these cells may participate in tubular regeneration in adult human kidneys [25–27, 29] especially when proliferation of neighboring unwounded tubular cells is exhausted, as it may occur during chronic tubular injury. Interestingly, injury to the glomerulotubular junction is a common feature of chronic tubulointerstitial injury, as already reported for obstructive nephropathy, chronic allograft nephropathy and even for DN [29, 30].

Taken altogether, these findings suggest that the kidney contains a stem cell compartment, deputed to regeneration of podocytes and tubular cells. Recent evidence suggests that a modulation of the regenerative properties of endogenous renal progenitors may help to achieve remission of renal disorders and avoid their progression toward ESRD.

Conclusion

Until now, regenerative approaches for kidney damage were limited by the complexity of the organ. Bone-marrow derived stem cells, as well as MSCs were envisioned for the development of stem cell treatment of chronic glomerular disorders [12–24]. However, most studies suggested that these cells cannot differentiate into renal epithelial cells, and concluded that their beneficial effects are probably related to secretion of growth factors [12, 17–19]. By contrast, recent evidence suggests that adult kidney contains a stem cell compartment localized at the urinary pole of the Bowman's capsule [25–29]. These renal progenitors can differentiate into glomerular as well as tubular epithelial cells. Converging evidence suggests that the outcome of glomerular disorders depends on the balance between injury and regeneration provided by renal progenitors [29]. Thus, modulation of the regenerative potential of renal progenitors through pharmacological agents represents a novel promising tool for prevention and treatment of glomerulosclerosis [29].

Acknowledgements

The research leading to these results has received funding from the European Community under the European Community's Seventh Framework Programme (FP7/2007–2013), grant number 223007, and from the European Research Council Starting Grant under the European Community's Seventh Framework Programme (FP7/2007–2013), ERC grant number 205027. This study was also supported by the Tuscany Ministry of Health and the Ministero della Salute.

References

1 Meguid El Nahas A, Bello AK: Chronic kidney disease: The global challenge. Lancet 2005;365:331–340.

2 Tang SC: Diabetic nephropathy: a global and growing threat. Hong Kong Med J 2010;16: 244–245.

3 Wiggins RC: The spectrum of podocytopathies: a unifying view of glomerular diseases. Kidney Int 2007;71:1205–1214.

4 Kriz W, LeHir M: Pathways to nephron loss starting from glomerular diseases-insights from animal models. Kidney Int 2005;67: 404–419.

5 Pagtalunan ME, Miller PL, Jumping-Eagle S, Nelson RG, Myers BD, Rennke HG, Coplon NS, Sun L, Meyer TW: Podocyte loss and progressive glomerular injury in type II diabetes. J Clin Invest 1997;99: 342–348.

6 White KE, Bilous RW, Marshall SM, El Nahas M, Remuzzi G, Piras G, De Cosmo S, Viberti G: The European Study for the Prevention of Renal Disease in Type I diabetes (ESPRIT): podocyte number in normotensive type I diabetic patients with albuminuria. Diabetes 2002;51:3083–3089.

7 Dalla Vestra M, Masiero A, Roiter AM, Saller A, Crepaldi G, Fioretto P: Is podocyte injury relevant in diabetic nephropathy? Studies in patients with type 2 diabetes. Diabetes 2003; 52:1031–1035.

8 Lemley KV, Lafayette RA, Safai M, Derby G, Blouch K, Squarer A, Myers BD: Podocytopenia and disease severity in IgA nephropathy. Kidney Int 2002;61: 1475–1485.

9 Fioretto P, Steffes MW, Sutherland DE, Goetz FC, Mauer M: Reversal of lesions of diabetic nephropathy after pancreas transplantation. N Engl J Med 1998;339:69–75.

10 Remuzzi G, Benigni A, Remuzzi A: Mechanisms of progression and regression of renal lesions of chronic nephropathies and diabetes. J Clin Invest 2006;116:288–296.

11 Blanpain C, Horsley V, Fuchs E: Epithelial stem cells: turning over new leaves. Cell 2007;128:445–458.

12 Sagrinati C, Ronconi E, Lazzeri E, Lasagni L, Romagnani P: Stem-cell approaches for kidney repair: choosing the right cells. Trends Mol Med 2008;14:277–285.

13 Kale S, Karihaloo A, Clark PR, Kashgarian M, Krause DS, Cantley LG: Bone marrow stem cells contribute to repair of the ischemically injured renal tubule. J Clin Invest 2003;112:42–49.

14 Lin F, Moran A, Igarashi P: Intrarenal cells, not bone marrow-derived cells, are the major source for regeneration in post ischemic kidney. J Clin Invest 2005;115:1756–1764.

15 Duffield JS, Park KM, Hsiao LL, Kelley VR, Scadden DT, Ichimura T, Bonventre JV: Restoration of tubular epithelial cells during repair of the postischemic kidney occurs independently of bone marrow-derived stem cells. J Clin Invest 2005;115:1743–1755.

16 Sugimoto H, Mundel TM, Sund M, Xie L, Cosgrove D, Kalluri R: Bone-marrow-derived stem cells repair basement membrane collagen defects and reverse genetic kidney disease. Proc Natl Acad Sci USA 2006;103:7321–7326.

17 Imberti B, Morigi M, Tomasoni S, Rota C, Corna D, Longaretti L, Rottoli D, Valsecchi F, Benigni A, Wang J, Abbate M, Zoja C, Remuzzi G: Insulin-like growth factor-1 sustains stem cell mediated renal repair. J Am Soc Nephrol 2007;18:2921–2928.

18 Tögel F, Weiss K, Yang Y, Hu Z, Zhang P, Westenfelder C: Vasculotropic, paracrine actions of infused mesenchymal stem cells are important to the recovery from acute kidney injury. Am J Physiol Renal Physiol 2007;292:F1626–F1635.

19 Kunter U, Rong S, Boor P, Eitner F, Müller-Newen G, Djuric Z, van Roeyen CR, Konieczny A, Ostendorf T, Villa L, Milovanceva-Popovska M, Kerjaschki D, Floege J: Mesenchymal stem cells prevent progressive experimental renal failure but maldifferentiate into glomerular adipocytes. J Am Soc Nephrol 2007;18:1754–1764.

20 Prodromidi EI, Poulsom R, Jeffery R, Roufosse CA, Pollard PJ, Pusey CD, Cook HT: Bone marrow-derived cells contribute to podocyte regeneration and amelioration of renal disease in a mouse model of Alport syndrome. Stem Cells 2006;24:2448–2455.

21 Ninichuk V, Gross O, Segerer S, Hoffmann R, Radomska E, Buchstaller A, Huss R, Akis N, Schlöndorff D, Anders HJ: Systemic administration of multipotent mesenchymal stromal cells reverts hyperglycemia and prevents nephropathy in type 1 diabetic mice. Biol Blood Marrow Transplant 2008;14: 631–640.

22 Lee RH, Seo MJ, Reger RL, Spees JL, Pulin AA, Olson SD, Prockop DJ: Multipotent stromal cells from human marrow home to and promote repair of pancreatic islets and renal glomeruli in diabetic NOD/scid mice. Proc Natl Acad Sci USA 2006;103: 17438–17443.

23 Amin AH, Abd Elmageed ZY, Nair D, Partyka MI, Kadowitz PJ, Belmadani S, Matrougui K: Modified multipotent stromal cells with epidermal growth factor restore vasculogenesis and blood flow in ischemic hind-limb of type II diabetic mice. Lab Invest 2010;90:985–996.

24 Volarevic V, Arsenijevic N, Lukic ML, Stojkovic M: Mesenchymal stem cell treatment of complications of diabetes mellitus. Stem Cells 2010, Epub ahead of print.

25 Sagrinati C, Netti GS, Mazzinghi B, Lazzeri E, Liotta F, Frosali F, Ronconi E, Meini C, Gacci M, Squecco R, Carini M, Gesualdo L, Francini F, Maggi E, Annunziato F, Lasagni L, Serio M, Romagnani S, Romagnani P: Isolation and characterization of multipotent progenitor cells from the Bowman's capsule of adult human kidneys. J Am Soc Nephrol 2006;17:2443–2456.

26 Romagnani P: Toward the identification of a 'renopoietic system'? Stem Cells 2009;27: 2247–2253.

27 Ronconi E, Sagrinati C, Angelotti ML, Lazzeri E, Mazzinghi B, Ballerini L, Parente E, Becherucci F, Gacci M, Carini M, Maggi E, Serio M, Vannelli GB, Lasagni L, Romagnani S, Romagnani P: Regeneration of glomerular podocytes by human renal progenitors. J Am Soc Nephrol 2009;20:322–332.

28 Appel D, Kershaw D, Smeets B, Yuan G,
    Fuss A, Frye B, Elger M, Kriz W, Floege J:
    Moeller MJ: Recruitment of podocytes from
    glomerular parietal epithelial cells. J Am Soc
    Nephrol 2009;20:333–343.

29 Lasagni L, Romagnani P: Glomerular epithe-
    lial stem cells: the good, the bad and the ugly.
    J Am Soc Nephrol 2010;21:1612–1619.
30 Najafian B, Kim Y, Crosson JT, Mauer M:
    Atubular glomeruli and glomerulotubular
    junction abnormalities in diabetic nephro-
    pathy. J Am Soc Nephrol 2003;14:908–917.

Paola Romagnani, MD, PhD
Excellence Centre for Research, Transfer and High Education for the development
of de novo Therapies, and Meyer Children's Hospital, Nephrology Unit,
Department of Clinical Pathophysiology, University of Florence
Viale Pieraccini 6
IT–50139 Florence (Italy)
Tel. +39 554271356, Fax +39 554271357, E-Mail p.romagnani@dfc.unifi.it

Lai KN, Tang SCW (eds): Diabetes and the Kidney.
Contrib Nephrol. Basel, Karger, 2011, vol 170, pp 237–246

Effects of Pancreas Transplantation on the Prevention and Reversal of Diabetic Nephropathy

Paola Fioretto[a] · Michael Mauer[b]

[a]Department of Medical and Surgical Sciences, University of Padova Medical School, Padova, Italy; [b]Departments of Pediatrics and Medicine, University of Minnesota Medical School, Minneapolis, Minn., USA

Abstract

Pancreas transplantation (PTx) is the only available treatment which is able to restore normoglycemia without exposing patients to the risks of severe hypoglycemia, thus allowing testing the effects of very long-term euglycemia in preventing, halting and reversing diabetic nephropathy (DN). Pancreas and islet transplantation in animal models have been shown to prevent, ameliorate or reverse the development of DN lesions. PTx, performed simultaneously or shortly after kidney transplantation in patients with type 1 diabetes prevents the recurrence of diabetic glomerulopathy lesions in the renal allograft. To test whether DN lesions are reversible in humans, we studied renal structure before and 5 and 10 years after PTx in nonuremic patients with long-term type 1 diabetes, with mild to advanced DN lesions at baseline. Diabetic glomerular lesions were not significantly changed at 5 years after PTx, but were markedly improved after 10 years. Indeed, in most patients glomerular structure had returned to normal at 10-year follow-up. These pancreas transplant studies also showed that remodeling of the tubulointerstitium and decrease in interstitial collagen was possible. Thus, the lesions of DN are reversible by long-term normoglycemia, and that it is possible in humans associated with substantial architectural remodeling and healing of glomerular, tubular, and interstitial structures.

Diabetic nephropathy (DN) is the single most important disorder leading to end-stage renal disease (ESRD) in adults in the US, Japan and Europe [1, 2]. The dramatic increase in the proportion of ESRD patients affected by diabetes in the last decades has led to increased research focus on the pathogenesis of renal lesions in diabetes and on strategies aimed at preventing, slowing and possibly halting DN. Unfortunately, by the time clinical or laboratory abnormalities of DN become manifest, renal lesions in type 1 diabetes (T1DM) are far advanced [3], and current treatments may slow but usually cannot arrest progression towards end-stage renal disease [4–7]. DN is secondary to the long-term metabolic abnormalities of diabetes. The most important structural renal changes underlying functional abnormalities in T1DM occur in the glomeruli [2, 3]. The two major early glomerular lesions, mesangial expansion and increased thickness of the glomerular basement membrane (GBM), are not present at the onset of T1DM, but can be demonstrated after a few years [8]. Glycemic control has a strong impact on the development of renal lesions and dysfunction [9, 10], and on the rate of clinical progression of DN [11]. Thus, the kidneys of nondiabetic members of identical twins discordant for T1DM were structurally normal, and in each twin pair GBM and mesangial measures were greater in the diabetic twin [12]. Moreover, normal kidneys from nondiabetic donors that are transplanted into T1DM patients develop all of the lesions of DN [13, 14]. Interventions aimed at improving metabolic control clearly influence the development of DN: T1DM randomized to receive maximized glycemic control in the first 5 years following kidney transplantation do not develop mesangial matrix expansion, while this occurs in those randomized to standard control [15]. The Diabetes Control and Complications Trial documented that T1DM patients randomized to strict control had lower incidences of microalbuminuria and proteinuria after 7–8 years of follow-up [10]. Similar trends were seen for the prevention of microalbuminuria and proteinuria in the large United Kingdom Prospective Diabetes Study in type 2 diabetic patients [16]. Finally, the crucial role of glycemic control in DN was proven by the dramatic reversal of established DN lesions in the native kidneys of T1DM patients with long-term normoglycemia following successful pancreas transplantation (PTx) [17].

Pancreas Transplantation and Diabetic Renal Disease

PTx offers the unique opportunity to evaluate the impact of prolonged normoglycemia on the different stages of DN, without the risk of severe hypoglycemia allowing test of the capability of long-term normoglycemia to prevent, halt and reverse DN. However, PTx is not wholly adequate for the study of the potential beneficial effects on abnormalities of renal function since current immunosuppressive agents, especially cyclosporine, decrease glomerular filtration rate and albumin excretion rate and increase blood pressure; these are the cardinal clinical manifestations of DN. For these reasons, studies of renal structure are more informative in evaluating the effects of PTx in preventing, slowing or reversing DN, and these will be the focus of this chapter.

Before considering the effects of PTx, it is necessary to keep in mind that DN is a disease largely resulting from the accumulation of extracellular matrix

(ECM) in the mesangium, GBM and tubular basement membrane (TBM), and interstitium [2, 3, 18–20]. Increased quantities of the normal site-specific ECM components are present in the diabetic kidney, including the α_3- and α_4-type IV collagen chains and laminins in the GBM and α_1- and α_2-chains of types IV and VI collagen, laminins, and fibronectin in mesangial matrix [21–23]. In contrast, interstitial (types I and III) collagens are absent from the glomerulus throughout most of the natural history of DN, becoming evident only in advanced Kimmelstiel-Wilson nodules and in the final stages of global glomerular sclerosis in the context of Kimmelstiel-Wilson nodules [21].

Thus, in diabetes there is a progressive ECM accumulation in all renal compartments; whether this is due to increased production, decreased removal, or both is currently unknown. This contrasts with the observation in normal subjects that mesangial fractional volume [Vv(Mes/glom)] and GBM thickness do not change between ages 16 and 60 years, which reflect a remarkable constancy in the balance of glomerular ECM production and removal in healthy adult life [24].

Regardless of the pathway(s), whereby the balance between ECM production and removal is altered in DN, if reversal of lesions is possible, it needs to operate by switching renal cellular function towards a net decrease in ECM.

Simultaneous Pancreas and Kidney and Pancreas after Kidney Transplantation: Prevention of Development and Progression of Early Renal Lesions

PTx, in most instances, is performed in uremic T1DM patients at the time of renal transplant (SPK) or, less frequently, shortly after kidney transplant (PAK). Thus, it is possible to test whether PTx prevents the development of diabetic glomerular lesions, since the renal graft has never been exposed to hyperglycemia in the normal kidney. In 1985, Bohman et al. [25] demonstrated prevention of early lesions of diabetic glomerulopathy in 2 SPK patients. These authors subsequently studied kidney biopsies performed in 9 patients 1–4 years after SPK and compared them with biopsies from 29 diabetic and 4 nondiabetic kidney alone transplant recipients (KA) [26]. The width of GBM was in the normal range in all SPK patients, while it was increased in most diabetic KA patients. They reported no mesangial expansion in SPK patients; however, baseline biopsies were not performed and, given the variability in glomerular structure among normal controls, it cannot be definitely concluded that the development of early mesangial lesions was completely prevented. In addition, Vv(Mes/glom) was estimated by light microscopy, and this method does not provide sufficient precision to detect early mesangial expansion. Later, this same group [27] described a larger cohort of 20 SPK and 34 KA recipients followed for 1–6.5 years confirming the previous observations, with the same drawbacks. In contrast, in a small number of patients (11 SPK recipients followed for 2–4 years and 6 diabetic KA recipients followed for 2–6 years), Nyberg et al. [28] suggested that, despite SPK, early diabetic glomerular changes were reoccurring. There was in fact a similar increase in GBM width in both groups, while estimates of mesangium, by electron microscopy, were still in the normal range. However, kidney biopsies were not performed at the time of SPK, and it is therefore impossible to estimate the rate of development of very early diabetic glomerulopathy from these studies. In addition, of 11 SKP patients, only 6 had normal HbA1c values and only 2 had normal HbA1c values and glucose tolerance tests. Thus, one logical explanation for these odd results is that patients still had mild diabetes [28]. Overall, despite some design limitations, studies in SPK recipients suggest that when PTx is entirely successful, recurrence of diabetic glomerulopathy lesions is prevented.

In addition to the early work of Bohman et al. [25], only one study has been performed in PAK patients. PTx was performed 1–7 years after KA [29]. Renal biopsies were performed before successful PTx and 4 years later, and were compared to those obtained at similar times from T1DM KA recipients. Mesangial fractional volume [Vv(Mes/glom)] was lower in the PAK than in the KA recipients; in contrast, the width of GBM was not significantly different in the 2 groups. PAK patients had smaller glomeruli, suggesting that the glomerular enlargement of diabetes is reversible. Thus, PTx performed within a few years after KA can halt progression of diabetic glomerulopathy lesions.

Pancreas Transplantation Alone: Evidence for Reversibility

Islet transplantation in diabetic rats results in rapid reversal of established mesangial matrix and cell expansion but not of GBM thickening [30]. A more recent study suggested that glomerular lesions are reversible after shorter but not longer diabetes duration in rats [31], but previous studies had shown reversal in rats with long-standing diabetes and well-established diabetic glomerulopathy [32–34]. However, lesions in rats do not entirely parallel those in man. Thus, in rats, diabetic mesangial expansion is due equally to expansion of its matrix and cell components [30, 33], while in man this is predominantly due to matrix accumulation [20]. Cure of diabetes in rats results in rapid reversal of mesangial expansion, which is very different from the pace of events in humans [32] (see below). Finally, as noted above, unlike in humans, amelioration of GBM thickening did not parallel that of mesangial expansion in rats. Thus, animal models of development and reversal of DN remain, at best, approximations of the human condition.

Reversal of DN in T1DM patients is best addressed by studying recipients of PTx alone (PTA). In a report on 13 PTA recipients, we documented that, despite 5 years of normoglycemia, there was no amelioration of established DN lesions [35]. In fact, the width of GBM, abnormal before PTA, was

Fioretto · Mauer

unchanged after 5 years. The mesangial fractional volume actually increased further over these 5 years due to a decrease in glomerular volume (GV), while the total mesangial volume per glomerulus (TM) remained unchanged. Both GV and Vv(Mes/glom), and thus TM, increased over 5 years in a group of persistently diabetic patients. Thus, it appeared that the mesangium stopped expanding in the PTA recipients but continued in the untreated patients [35]. Nevertheless, this study showed that DN lesions were not ameliorated by 5 years of normoglycemia.

Eight of the original cohort of 13 PTA recipients were restudied after 10 years of normoglycemia [17]; 2 of 13 patients underwent kidney transplant 6 and 8 years after PTA, 2 lost graft function and required insulin therapy, and one refused the 10-year studies. Renal function tests and kidney biopsies were performed before and at 5 and 10 years after transplantation in the 8 available patients. At PTA, they were 33 ± 3 years old with 22 ± 5 years T1DM and HbA1c of 8.7 ± 1.5%. Renal function was difficult to interpret, given the effects of cyclosporine on glomerular filtration rate (GFR) and albuminuria. Indeed, there was a significant correlation between decline in GFR and cyclosporine dose and blood levels in the first posttransplant year [36]. As discussed above, there was no benefit on glomerular structure at 5 years after PTA [35]. In contrast, we observed dramatic reversal of diabetic glomerulopathy lesions in all 8 patients [17] at 10 years. The widths of GBM and TBM, unchanged at 5 years, were decreased at 10 years, the values falling into the normal range in several patients while approaching normal in the others. The mesangial fractional and mesangial matrix fractional volumes, increased from baseline to 5 years, were lower at 10 years than at baseline or 5 years. GV, decreased from baseline to 5 years, was stable thereafter. Total mesangial and total mesangial matrix volumes per glomerulus, unchanged at 5 years, were markedly decreased at 10 years (fig. 1). Light microscopy revealed remarkable remodeling of glomerular architecture including the total disappearance of Kimmelstiel-Wilson nodules and reopening of glomerular capillaries previously compressed by mesangial expansion (fig. 2). Thus, this study proved that diabetic glomerular and tubular lesions are reversible in humans.

The reasons for the long delay before reversal of DN lesions became evident are unknown. Perhaps renal cells have 'metabolic memory' for the diabetic state [37], or perhaps ECM molecules are heavily glycosylated and thus more resistant to proteolysis until replaced by less glycosylated molecules [38]. The long time necessary for these diabetic lesions to disappear is consistent with their slow development [2, 3, 7, 20]. In fact, diabetic renal lesions develop over at least one decade before they can cause any functional abnormality [39]. Regardless of the mechanism(s), at some point after PTA, rates of ECM removal begin to exceed ECM production rates. This active healing process is clearly different from the normal state since, as noted above, renal ECM production and removal remain in near perfect balance during adult life [24].

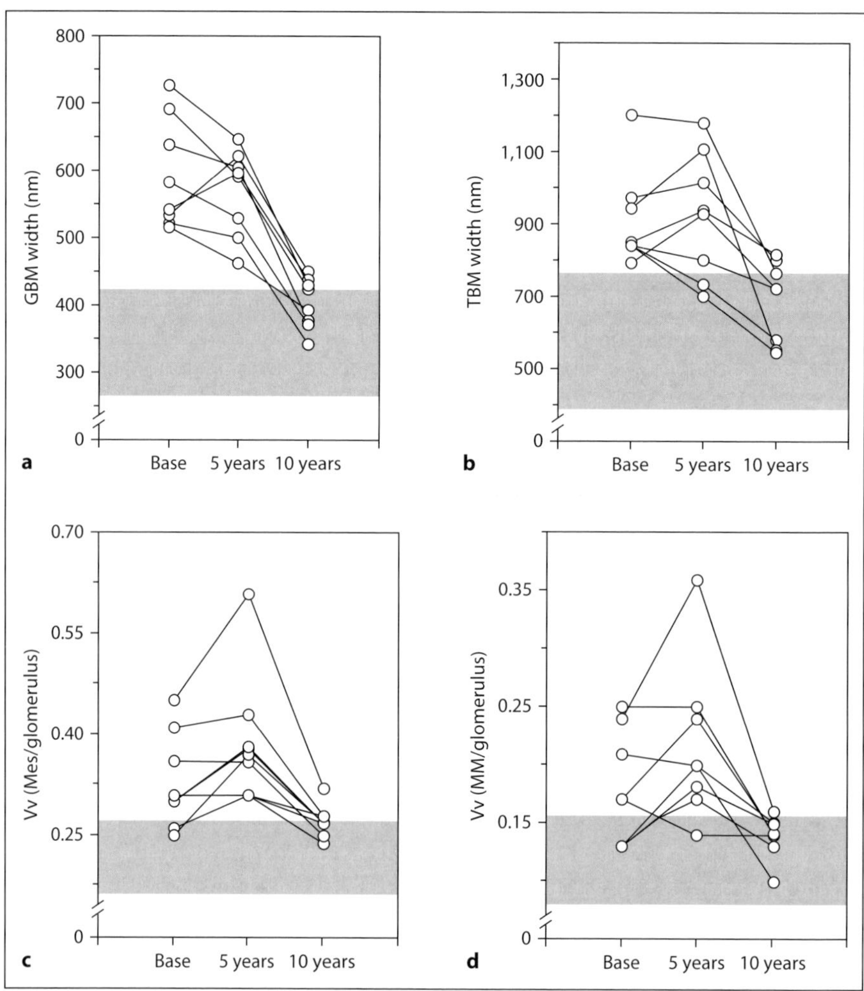

Fig. 1. Thickness of the GBM (**a**) and TBM (**b**), mesangial fractional volume (**c**) and mesangial matrix fractional volume (**d**) at baseline and 5 and 10 years after PTA. The shaded area represents the normal ranges obtained in 66 age- and sex-matched normal controls (mean ± 2 SD). Data for individual patients are connected by lines. From Fioretto et al. [17], with permission.

All PTA recipients were receiving cyclosporine, known to be nephrotoxic [40, 41]. We reported an increase in interstitial expansion and in the proportion of atrophic tubules at 5 years after PTA in these patients related to cyclosporine treatment [36]. More recently, analyzing the biopsies 10 years after PTA revealed remodeling of interstitial and tubular lesions [42]. Indeed, the worsening of interstitial fibrosis and tubular atrophy observed at 5 years after PTA, likely consequent to cyclosporine therapy [36], had significantly improved by 10 years

Fioretto · Mauer

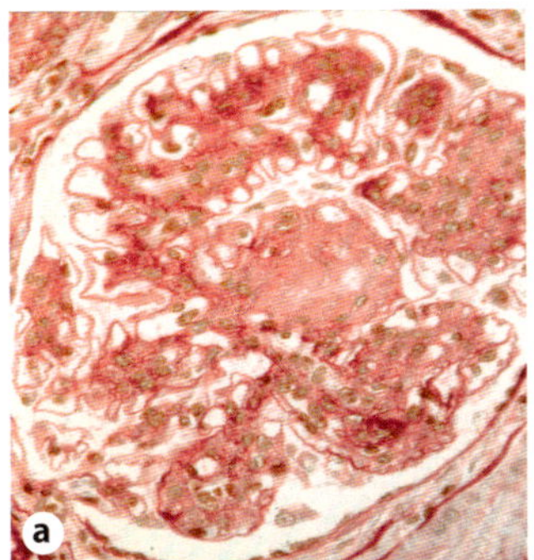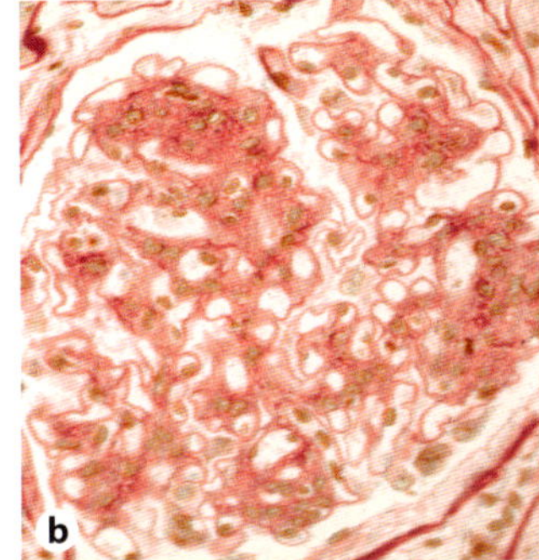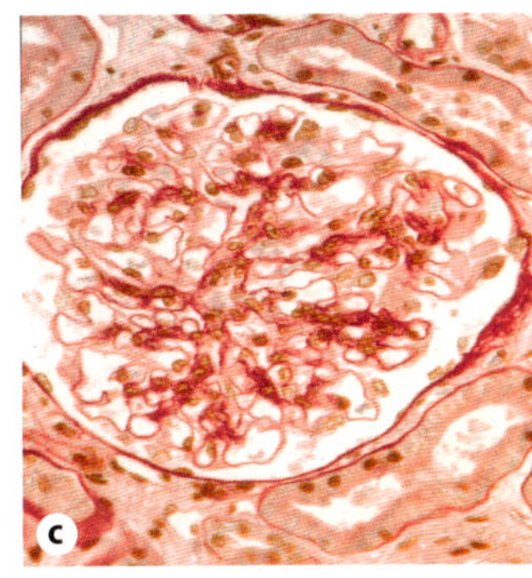

Fig. 2. Glomeruli before and after PTx. **a** Typical glomerulus from the baseline biopsy, with diffuse diabetic glomerulopathy and Kimmelstiel-Wilson nodules. **b** Typical glomerulus 5 years after PTA shows persistence of diabetic glomerulopathy. **c** Typical glomerulus 10 years after PTA, with marked resolution of diabetic glomerulopathy lesions and substantially normal glomerular structure. From Fioretto et al. [17], with permission.

after PTA [42]. Thus, cortical tubules and interstitium can undergo substantial remodeling, with a decrease in interstitial fractional volume and the fractional volume of atrophic tubules, and a reduction in renal interstitial fibrillar collagen [42]. While the decrease in interstitial collagen has some parallels with the glomerular remodeling, these studies suggested that atrophic tubules underwent reabsorption rather than healing. Whether these tubulointerstitial improvements resulted from prolonged normoglycemia, decreased cyclosporine dose or both could not be dissected. Regardless of the mechanisms, at some point after PTA, glomerular, tubular and interstitial cells changed their behavior towards ECM removal and architectural remodeling in remarkable demonstrations of the recooperative capacity of the kidney.

Conclusion

Studies in PTx recipients have been crucial in our understanding of the pathophysiology of DN. We have learned that DN lesions not only can be prevented and halted, but, once established, can also be reversed. It is currently unknown whether other therapeutic interventions can lead to the same outcome. These studies have proven that the kidney has intrinsic mechanisms for reorganizing cellular functions to reverse these alterations upon removal of the injury or deleterious stimulus in processes that in some ways recapitulate renal development, i.e. the renal cells 'know' what normal renal structure is and will head there given the correct physiologic environment. It is likely that there are points of no return, i.e. levels of structural damage beyond which healing is no longer possible, but we were astonished by the severity of diabetic glomerulopathy lesions which were amenable to repair. Further studies aimed at understanding the cellular and molecular mechanisms involved in these healing processes could lead to new directions in the prevention and treatment of DN.

Acknowledgements

This work was supported by grants from the National Institutes of Health (DK13083), the National Center for Research Resources (MO1-KK00400), and by an endowment from the Kroc Research Foundation. During the initial stages of this work, Dr. Fioretto was supported by Research Fellowship and Career Development Awards from the Juvenile Diabetes Foundation International. We are indebted to the patients who so generously participated in these studies.

References

1   US Renal Data System: US Renal Data System Annual Report. Bethesda, National Institutes of Health, National Institute of Diabetes and Digestive and Kidney Diseases, 2006.

2   Mauer M, Fioretto P, Woredekal Y, et al: Diabetic nephropathy; in Schrier RW (ed): Disease of the Kidney and Urinary Tract. Philadelphia, Lippincott Williams & Wilkins, 2001, pp 2083–2127.

3   Mauer SM, Steffes MW, Ellis EN, et al: Structural-functional relationships in diabetic nephropathy. J Clin Invest 1984;74:1143–1155.

4   Lewis EJ, Hunsicker LG, Bain RP, et al: The effect of angiotensin-converting enzyme inhibition on diabetic nephropathy. N Engl J Med 1993;329:1456–1462.

5   Lewis EJ, Hunsicker LG, Clarke WR, et al, for the Collaborative Study Group: Renoprotective effect of the angiotensin-receptor antagonist irbesartan in patients with nephropathy due to type 2 diabetes. N Engl J Med 2001;345:851–860.

6   Brenner BM, Cooper ME, De Zeeuw D, et al, for the Renal Study Investigators: Effects of losartan on renal and cardiovascular outcomes in patients with type 2 diabetes and nephropathy. N Engl J Med 2001;345: 861–869.

7   Mauer M, Zinman B, Gardiner R, Suissa S, Sinaiko A, Strand T, Drummond K, Donnelly S, Goodyer P, Gubler MC, Klein R: Renal and retinal effects of Enalapril and Losartan in type 1 diabetes. N Engl J Med 2009;361: 40–51.

8   Østerby R: Early phases in the development of diabetic glomerulopathy. Acta Med Scand 1975;475:1–84.

9   Pirart J: Diabetes mellitus and its degenerative complications: a prospective study of 4,400 patients observed between 1947 and 1973. Diabetes Metab 1977;3:173–182.

10   Diabetes Complications and Control Trial Research Group: The effect of intensive treatment of diabetes on the development and progression of long-term complications of insulin dependent diabetes mellitus. N Engl J Med 1993;329:977–986.

11   Parving H-H, Rossing P, Hommel E, et al: Angiotensin-converting enzyme inhibition in diabetic nephropathy: ten years' experience. Am J Kidney Dis 1995;26:99–107.

12   Steffes MW, Sutherland DER, Goetz FC, et al: Studies of kidney and muscle biopsy specimens from identical twins discordant for type I diabetes mellitus. N Engl J Med 1985; 312:1282–1287.

13   Mauer SM, Goetz FC, McHugh LE, et al: Long-term study of normal kidneys transplanted into patients with type I diabetes. Diabetes 1989;38:516–523.

14 Mauer SM, Steffes MW, Connett J, et al: The development of lesions in the glomerular basement membrane and mesangium after transplantation of normal kidneys into diabetic patients. Diabetes 1983;32:948–952.

15 Barbosa J, Steffes MW, Sutherland DER, et al: The effect of glycemic control on early diabetic renal lesions. A 5-year randomized controlled clinical trial of insulin-dependent diabetic kidney transplant recipients. JAMA 1994;272:600–606.

16 UKPDS Group: Intensive blood-glucose control with sulphonylureas or insulin compared with conventional treatment and risk of complications in patients with type 2 diabetes. Lancet 1998;352:837–853.

17 Fioretto P, Steffes MW, Sutherland DER, et al: Reversal of lesions of diabetic nephropathy after pancreas transplantation. N Engl J Med 1998;339:69–75.

18 Fioretto P, Steffes MW, Mauer SM: Glomerular structure in non-proteinuric insulin-dependent diabetic patients with various levels of albuminuria. Diabetes 1994;43:1358–1364.

19 Brito PL, Fioretto P, Drummond K, et al: Proximal tubular basement membrane width in insulin-dependent diabetes mellitus. Kidney Int 1998;53:754–761.

20 Steffes MW, Bilous RW, Sutherland DER, et al: Cell and matrix components of the glomerular mesangium in type I diabetes. Diabetes 1992;41:679–684.

21 Falk RJ, Scheinman JI, Mauer SM, et al: Polyantigenic expansion of basement membrane constituents in diabetic nephropathy. Diabetes 1983;32:34–39.

22 Zhu D, Kim Y, Steffes MW, et al: Glomerular distribution of type IV collagen in diabetes by high resolution quantitative immunochemistry. Kidney Int 1994;45:425–433.

23 Moriya T, Groppoli TJ, Kim Y, et al: Quantitative immunoelectron microscopy of type VI collagen in glomeruli in type I diabetic patients. Kidney Int 2001;59:317–323.

24 Steffes MW, Barbosa J, Basgen JM, Sutherland DER, Najarian JS, Mauer SM: Quantitative glomerular morphology of the normal human kidney. Lab Invest 1983;49:82–86.

25 Bohman S-O, Tyden G, Wilczek H, Lundgren G, Jaremko G, Gunnarsson R, Ostman J, Groth G: Prevention of kidney graft diabetic nephropathy by pancreas transplantation in man. Diabetes 1985;34:306–308.

26 Bohman S-O, Wilczek H, Tyden G, Jaremko G, Lundgren G, Groth G: Recurrent diabetic nephropathy in renal allografts placed in diabetic patients and protective effect of simultaneous pancreas transplantation. Transplant Proc 1987;1:2290–2293.

27 Wilczek HE, Jaremko G, Tyden G, Groth CG: Pancreatic graft protects a simultaneously transplanted kidney from developing diabetic nephropathy: a 1 to 6 year follow-up study. Transplant Proc 1993;1:1314–1315.

28 Nyberg G, Holdaas H, Brekke IB, et al: Glomerular ultrastructure in kidneys transplanted simultaneously with a segmental pancreas to patients with type 1 diabetes. Nephrol Dial Transplant 1996;11:1029–1033.

29 Bilous RW, Mauer SM, Sutherland DER, et al: The effects of pancreas transplantation on the glomerular structure of renal allografts in patients with insulin-dependent diabetes. N Engl J Med 1989;32:80–85.

30 Mauer SM, Sutherland DER, Steffes MW, et al: Pancreatic islet transplantation. Effects on the glomerular lesions of experimental diabetes in the rat. Diabetes 1974;23:748–753.

31 Pugliese G, Pricci F, Pesce C, et al: Early, but not advanced, glomerulopathy is reversed by pancreatic islet transplants in experimental diabetic rats: correlation with glomerular extracellular matrix mRNA levels. Diabetes 1997;46:1198–2006.

32 Mauer SM, Steffes MW, Sutherland DER, et al: Studies of the rate of regression of the glomerular lesions in diabetic rats treated with pancreatic islet transplantation. Diabetes 1975;24:280–285.

33 Steffes MW, Brown DM, Basgen JM, et al: Amelioration of mesangial volume and surface alterations following islet transplantation in diabetic rats. Diabetes 1980;29:509–515.

34 Steffes MW, Brown DM, Basgen JM, et al: Glomerular basement membrane thickness following islet transplantation in the diabetic rat. Lab Invest 1979;41:116–118.

35   Fioretto P, Mauer SM, Bilous RW, et al:
     Effects of pancreas transplantation on glom-
     erular structure in insulin-dependent dia-
     betic patients with their own kidneys. Lancet
     1993;342:1193–1196.
36   Fioretto P, Steffes MW, Mihach MJ, et al:
     Cyclosporine associated lesions in native kid-
     neys of diabetic pancreas transplant recipi-
     ents. Kidney Int 1995;48:489–495.
37   Roys S, Sala R, Cagliero E, et al:
     Overexpression of fibronectin induced by
     diabetes or high glucose: phenomenon with
     a memory. Proc Nat Acad Sci USA 1990;87:
     404–408.
38   Brownlee M, Cerami A, Vlassara H:
     Advanced glycosylation end products in
     tissue and the biochemical basis of diabetic
     complications. N Engl J Med 1988;318:
     1315–1322.
39   Krolewski AS, Warram JH, Christleb AR, et
     al: The changing natural history of nephrop-
     athy in type I diabetes. Am J Med 1985;78:
     785–794.
40   Myers BD, Sibley R, Newton L, et al:
     The long-term course of cyclosporine-
     associated chronic nephropathy. Kidney Int
     1988;33:590–600.
41   Feutren G, Mihatsch MJ: Risk factors for
     cyclosporine induced nephropathy in
     patients with autoimmune diseases. N Engl
     J Med 1992;326:1654–1660.
42   Fioretto P, Sutherland DER, Najafian B,
     Mauer M: Remodeling of renal interstitial
     and tubular lesions in pancreas transplant
     recipients. Kidney Int 2006;69:907–912.

Paola Fioretto, MD
Department of Medical and Surgical Sciences, University of Padova Medical School
via Giustiniani 2
IT–35128 Padova (Italy)
Tel. +39 049 8211879, Fax +39 049 8212151, E-Mail paola.fioretto@unipd.it

Lai KN, Tang SCW (eds): Diabetes and the Kidney.
Contrib Nephrol. Basel, Karger, 2011, vol 170, pp 247–255

New-Onset Diabetes and Nephropathy after Renal Transplantation

Joelle Guitard[a] · Lionel Rostaing[a,b] · Nassim Kamar[a,b]

[a]Department of Nephrology, Dialysis, and Organ Transplantation, CHU Rangueil, and
[b]INSERM U563, IFR-BMT, CHU Purpan, Toulouse, France

Abstract

New-onset diabetes mellitus after transplantation (NODAT) is an important complication after kidney transplantation, responsible for increased mortality mainly related to cardiovascular events and infections. It has been also identified as an independent risk factor associated with graft loss. The varying definitions of diabetes used in the literature, and the prevalence and incidence of NODAT, are difficult to establish: it varies from 2 to 50% in the kidney transplant population. However, using the World Health Organization and the American Diabetes Association criteria, it occurs in ~15% of transplant recipients at 1 year in the USA, and ~5–10% in Europe. The main recipient-related risk factors include old age, high body weight and high body mass index (BMI) before transplantation, Afro-American or Hispanic ethnicity, hepatitis C virus (HCV) infection, and impaired fasting glucose level before transplantation. The other significant risk factors include male donor, acute rejection episodes, cytomegalovirus (CMV) infection and the immunosuppressive regimen, especially the use of tacrolimus (FK506), steroids, or mammalian target for rapamycin inhibitors. While some risk factors cannot be modified, others, such as recipient body weight, BMI, HCV or CMV infection, and immunosuppressive regimen can be controlled with an appropriate diet or treatment. Screening for impaired fasting glucose should be done before transplantation in dialysis or end-stage renal disease patients to identify patients at risk for NODAT. In patients with increased BMI, weight loss before transplantation reduces the risk of NODAT. HCV infection should be treated before transplantation. Prolonged anti-CMV prophylaxis may be used. Cyclosporine A is preferred to tacrolimus in patients at high risk of NODAT. A steroid-free regimen or early steroid withdrawal is encouraged to reduce the risk of NODAT.

The introduction of novel immunosuppressive drugs has allowed a significant decrease in the incidence of allograft rejection after renal transplantation.

Patient and graft survival rates, however, have not significantly improved [1], partly because of the increased burden of posttransplant complications, including new-onset diabetes mellitus after transplantation (NODAT).

Harmful Effect of New-Onset Diabetes Mellitus after Renal Transplantation

Increased mortality has been reported in patients with NODAT compared to patients without NODAT [2], mainly due to cardiovascular complications, but also because of an increased incidence of infections. Compared to kidney transplant patients without NODAT, diabetic patients have a higher incidence and greater severity of infections, which results in increased posttransplant mortality [2].

Cosio et al. [3] have shown that the 5-year cumulative incidence of cardiovascular events was correlated with fasting glucose level at 1 year after transplantation [3]. The incidence of cardiovascular events was significantly higher in patients with diabetes or a moderately elevated fasting glucose level compared to those with a normal fasting glucose level at 1 year after transplantation. In addition, a multivariate analysis identified fasting glucose levels at 1, 4, and 12 months after transplantation, as independent predictive factors for a cardiovascular event occurring after kidney transplantation [3]. In a prospective study, Ducloux et al. [4] had shown that NODAT is an independent risk factor for the appearance of atheromatous disease in kidney transplant patients (RR = 1.34, 95% CI 1.04–2.18). Pretransplantation diabetes is also a predictive factor for all-cause and cardiovascular mortality after kidney transplantation [5].

NODAT also accounts for lower graft survival [2] and is an independent risk factor associated with graft loss [6]. At 5 years, graft function, as defined by serum creatinine level, is significantly lower in NODAT patients compared to patients without NODAT [6]. Authentic biopsy-proven diabetic nephropathy has been reported in patients who develop NODAT.

Moreover, classic complications of diabetes mellitus, such as ketoacidosis, hyperosmotic diabetic coma, or peripheral neuropathy, have been described in renal-allograft patients with NODAT [6]. NODAT also has an impact on the quality of life of transplant patients and represents an important economic burden [7].

Prevalence and Incidence of New-Onset Diabetes Mellitus after Renal Transplantation

The varying definitions of diabetes used in the literature and the prevalence and incidence of NODAT are difficult to establish.

According to the World Health Organization (WHO) and the American Diabetes Association (ADA), diabetes is currently defined by the association of diabetes symptoms (polyuria, polydipsia, weight loss) and a fasting glucose level ≥7 mM (1.26 g/dl), or a nonfasting glucose level of ≥11.1 mM (2 g/dl), or the use of insulin and/or oral antidiabetic drugs. Impaired glucose tolerance is defined by a fasting glucose level between 6.1 and 7 mM. Abnormal glucose levels must be assessed twice to confirm diagnosis.

The prevalence and incidence of diabetes is higher in transplant patients than in the general population, and is estimated to be between 2 and 50% according to the literature [8]. NODAT usually occurs in the first 3 post-transplantation months [8]. Using the USRDS registry, Kasiske et al. [2] estimated the cumulative incidence of NODAT in patients who had received a first renal allograft between 1996 and 2000. Patients were considered diabetic if reported to the Medicare system. The cumulative incidences were 9.1, 16, and 24% at 3, 12, and 36 months, respectively [2]. Using the WHO/ADA criteria for diabetes, Cosio et al. [3] reported a prevalence of impaired glucose tolerance and NODAT in 33 and 13% of patients, respectively, at 1 year after transplantation [3].

A French multicenter observational study that included 527 patients identified using WHO criteria reported a NODAT prevalence of 7%, with a median time until NODAT onset of 1.6 months [9]. Impaired fasting glucose was found in 7.9% of patients [9]. Diabetes is probably underestimated in the majority of studies. For example, in 122 renal transplant patients in whom fasting glucose levels were between 5.6 and 6.1 mM, and who had undergone an oral glucose tolerance test (OGT), the occurrence of NODAT, impaired fasting glucose, or high fasting glucose levels were diagnosed in 10, 14, and 18% of patients, respectively [10].

Similarly, with the help of an OGT, Porrini et al. [11] reported a 33% incidence of impaired fasting glucose and a 20% incidence of NODAT in 154 nondiabetic renal transplant patients receiving a tacrolimus-based immunosuppressive regimen in the first year after kidney transplantation [11]. Conversely, a normal fasting glucose level and a normal OGT at day 5 after transplantation are independent protective factors against NODAT [12].

Risk Factors for New-Onset Diabetes Mellitus after Transplantation

Numerous studies have identified risk factors associated with NODAT (table 1) [13]. The main recipient-related risk factors include old age, high body weight and high body mass index (BMI) before transplantation, Afro-American or Hispanic ethnicity, hepatitis C virus (HCV) infection, and an impaired fasting glucose level before transplantation. Weight gain during the first posttransplantation year does not seem to be associated with NODAT. The polymorphism

Table 1. Main risk factors for new-onset diabetes mellitus after kidney transplantation

Recipients' factors	– older age
	– male gender (controversial)
	– family history of diabetes
	– weight at transplantation
	– high BMI before transplantation
	– maximum BMI before transplantation
	– Afro-American or Hispanic ethnicity
	– low educational status
	– HCV infection
	– fasting glucose level before transplantation
	– impaired fasting glucose level before transplantation
	– metabolic syndrome:
	high triglyceride level
	low high-density lipoprotein level
Transplantation factors	– male donor
	– cadaveric donor (controversial)
	– number of HLA mismatches (controversial)
	– occurrence of at least one episode of acute rejection
	– CMV infection
	– immunosuppressive regimen:
	use of tacrolimus
	tacrolimus trough levels
	steroid pulses
	high steroid daily dosage
	high cumulative dose of steroids
	use of mTOR inhibitors

HLA = Human leukocyte antigen; mTOR = mammalian target for rapamycin.

of the promoter region of interleukin-6, in the position 174(G→C), has been associated not only with type 2 diabetes and insulin resistance, but also with NODAT [14]. The rs7903146 polymorphism of transcription factor-7-like 2 was also significantly associated with NODAT in a population of 1,229 kidney transplant patients (OR 1.55, p = 0.02, for the CT genotype, and OR 1.79, p = 0.04, for the TT genotype) [15].

The other significant risk factors associated with transplantation are gender (male donors), acute rejection episodes, cytomegalovirus (CMV) infection, and the immunosuppressive regimen, especially the use of tacrolimus (FK506), steroids, and mammalian target for rapamycin inhibitors. An induction therapy with basiliximab has been recently identified as a risk factor for NODAT [16]. Basiliximab, a monoclonal antibody directed against the receptor of interleukin-2 (CD25), inhibits CD4+ CD25+ lymphocytes that include

Guitard · Rostaing · Kamar

regulatory CD25+ CD4+ FOXP3+ T cells. Apparently, these subpopulations play a role in the development of diabetes, in particular type 1 diabetes [16]. Although there is no clear pathophysiological explanation, it has been shown that albuminuria on day 5 after transplantation [12], and albuminuria at months 3 and 6 after transplantation [17], are associated with NODAT. Adiponectin, an anti-inflammatory protein synthesized by adipocytes, reduces insulin resistance and atheromatosis. Elevated adiponectin levels appear to prevent NODAT [18].

Although some risk factors cannot be modified, others, such as recipient body weight and BMI, HCV or CMV infection, and the immunosuppressive regimen, could be controlled with an appropriate diet or modified treatment.

Measures to Reduce New-Onset Diabetes Mellitus after Transplantation Risk Factors?

Screening for Impaired Fasting Glucose before Transplantation
Screening for impaired fasting glucose should be done before transplantation in dialysis or end-stage renal disease patients to identify patients at risk for NODAT, in order to ensure meticulous posttransplantation surveillance of glycemic levels and to adapt the immunosuppressive regimen.

Reducing Excess Weight before Transplantation
In patients with increased BMI, weight loss before transplantation reduces the risk of NODAT, cardiovascular events, and postoperative complications.

Treatment of HCV Infection before Transplantation
A meta-analysis has shown that the relative risk of NODAT is 3.97-fold greater (CI 95%: 1.83–8.61) in HCV-positive renal transplant patients compared to HCV-negative renal transplant patients [19]. In HCV-positive patients, NODAT may occur more frequently in patients treated with tacrolimus compared to those receiving cyclosporine A (CsA; 57.8 vs. 7.7%, p < 0.0001) [20]. However, the mechanism by which HCV favors NODAT is not well established. HCV may affect glucose metabolism by inducing hepatocyte necrosis. Moreover, a link between HCV infection and insulin resistance has been established in vitro and in vivo. Finally, HCV could induce dysregulation in the hepatocyte production of cytokines and insulin receptors.

As treatment of HCV infection is contraindicated after kidney transplantation because of the high risk of interferon therapy-induced acute rejection, it is recommended to treat all prospective HCV-positive, RNA-positive, renal transplantation candidates in order to achieve viral eradication before transplantation [21].

Reducing the Risk of Cytomegalovirus Infection
The risk of CMV primary infection or reactivation can be reduced by extended prophylaxis of CMV infection in patients at risk for NODAT.

Selection of Immunosuppressive Regimen
Calcineurin Inhibitors
Tacrolimus seems more prone to induce NODAT than CsA. In a meta-analysis that evaluated the effect of different calcineurin inhibitors (CNIs) on the incidence of NODAT, Heisel et al. [22] found a higher incidence of NODAT in patients receiving FK506 than patients receiving CsA (9.8 vs. 2.7%, RR = 3.7, CI 95%: 2.6–15.36, p < 0.0001). The DIRECT Study is an international multicenter randomized study that compared the incidence of NODAT in de novo renal transplant recipients who received either CsA or FK506 [23]. Amongst the nondiabetic patients before transplantation, the incidence of NODAT at 6 months after transplantation was 26% in the CsA arm and 33.6% in the FK506 arm (p = 0.0046). At 6 months, 8.9% of patients receiving CsA and 16.8% of patients receiving FK506 had received treatment for diabetes (p = 0.005) [23].

The economic impact of immunosuppressive regimen-induced diabetes was evaluated to be USD 2,025 and USD 3,308 per patient receiving tacrolimus at 1 and 2 years after transplantation, respectively, while similar figures for patients receiving CsA were only USD 1,137 and USD 1,611, respectively [7].

Both CsA and FK506 induce diabetes by reducing insulin secretion and also, to a lesser extent, by inducing insulin resistance [24]. Insulin resistance is increased by concomitant administration of steroids [24]. Asberg et al. [25] have shown that, at 1 year after transplantation, insulin sensitivity was significantly improved in patients who had never received CNIs compared to patients treated with CNIs. The different tissue localizations of the molecular targets of CsA and FK506 could account for the greater impact of FK506 on NODAT. Indeed, FKBP12, the target of FK506, is highly expressed in β-cells in the islets of Langerhans, whereas CsAs, which target cyclophilin, is mostly expressed in the heart, liver, and kidney.

Conversion from tacrolimus to cyclosporine in patients with NODAT significantly improves fasting glucose levels and glycated hemoglobin compared to patients maintained on tacrolimus [26]. At 1-year after conversion, remission of NODAT, as defined by fasting glucose levels <1.26 g/l without any antidiabetic treatment, was observed in 42% of converted patients versus 0% of patients maintained on tacrolimus (p = 0.001) [26]. The concomitant interruption of steroids also increases the advantage of conversion. Conversely, Luan et al. [27] recently showed that a switch from CsA to FK506 at 11.4 months (range: 0.3–83.8) after transplantation in 171 patients did not increase the incidence of NODAT or impaired glucose tolerance when compared to 533 patients maintained on CsA, during a mean follow-up time of 31.5 ± 21.6 months.

Steroid Avoidance or Steroid Reduction

Cumulative dose and daily steroid dosages are associated with an increased incidence of NODAT [13], essentially through enhanced insulin resistance. Immunosuppressive regimens based on steroid minimization have shown a lower incidence of NODAT. At 6 months after transplantation, the incidence of NODAT was 0.4% in patients receiving a steroid-free immunosuppressive regimen compared to 5.4% in patients receiving steroids [28]. Steroid withdrawal within the first week after transplantation significantly reduced the incidence of NODAT at 1 year compared to patients who were continued on steroids [29] or compared to patients in whom steroids were withdrawn between 3 and 6 months after transplantation [1]. Conversely, comparison study of early steroid withdrawal by day 7 versus continuation of low-dose steroids in de novo renal transplant patients did not reveal any significant difference in the incidence of NODAT at 5 years after transplantation [30]. However, the number of patients developing NODAT that required insulin therapy was significantly lower in the group of patients with early steroid withdrawal compared to the 'low dose' group [30]. Finally, a recent meta-analysis has shown that steroid withdrawal between 3 and 6 months after transplantation does not reduce the incidence of NODAT [1]. Thus, only a steroid-free regimen or early steroid withdrawal reduces the incidence of diabetes.

Conclusion

The incidence of de novo diabetes after renal transplantation is higher than that in the general population, and accounts for a decrease in patient and graft survival rate. Among the identified risk factors, some can be controlled in order to reduce the incidence of NODAT.

References

1   Pascual J, Galeano C, Royuela A, Zamora J: A systematic review on steroid withdrawal between 3 and 6 months after kidney transplantation. Transplantation 2010;90:343–349.

2   Kasiske BL, Snyder JJ, Gilbertson D, Matas AJ: Diabetes mellitus after kidney transplantation in the United States. Am J Transplant 2003;3:178–185.

3   Cosio FG, Kudva Y, van der Velde M, Larson TS, Textor SC, Griffin MD, Stegall MD: New onset hyperglycemia and diabetes are associated with increased cardiovascular risk after kidney transplantation. Kidney Int 2005; 67:2415–2421.

4   Ducloux D, Kazory A, Chalopin JM: Posttransplant diabetes mellitus and atherosclerotic events in renal transplant recipients: a prospective study. Transplantation 2005;79: 438–443.

5   Kuo HT, Sampaio MS, Vincenti F, Bunnapradist S: Associations of pretransplant diabetes mellitus, new-onset diabetes after transplant, and acute rejection with transplant outcomes: an analysis of the Organ Procurement and Transplant Network/United Network for Organ Sharing (OPTN/UNOS) database. Am J Kidney Dis 2010;56:1127–1139.

6   Miles AM, Sumrani N, Horowitz R, Homel P, Maursky V, Markell MS, Distant DA, Hong JH, Sommer BG, Friedman EA: Diabetes mellitus after renal transplantation: as deleterious as non-transplant-associated diabetes? Transplantation 1998;65:380–384.

7   Woodward RS, Schnitzler MA, Baty J, Lowell JA, Lopez-Rocafort L, Haider S, Woodworth TG, Brennan DC: Incidence and cost of new onset diabetes mellitus among U.S. waitlisted and transplanted renal allograft recipients. Am J Transplant 2003;3:590–598.

8   Montori VM, Basu A, Erwin PJ, Velosa JA, Gabriel SE, Kudva YC: Posttransplantation diabetes: a systematic review of the literature. Diabetes Care 2002;25:583–592.

9   Kamar N, Mariat C, Delahousse M, Dantal J, Al Najjar A, Cassuto E, Lefrancois N, Cointault O, Touchard G, Villemain F, Di Giambattista F, Benhamou PY: Diabetes mellitus after kidney transplantation: a French multicentre observational study. Nephrol Dial Transplant 2007;22:1986–1993.

10 Sharif A, Moore RH, Baboolal K: The use of oral glucose tolerance tests to risk stratify for new-onset diabetes after transplantation: an underdiagnosed phenomenon. Transplantation 2006;82:1667–1672.

11 Porrini E, Moreno JM, Osuna A, Benitez R, Lampreabe I, Diaz JM, Silva I, Dominguez R, Gonzalez-Cotorruelo J, Bayes B, Lauzurica R, Ibernon M, Moreso F, Delgado P, Torres A: Prediabetes in patients receiving tacrolimus in the first year after kidney transplantation: a prospective and multicenter study. Transplantation 2008;85:1133–1138.

12 Kuypers DR, Claes K, Bammens B, Evenepoel P, Vanrenterghem Y: Early clinical assessment of glucose metabolism in renal allograft recipients: diagnosis and prediction of post-transplant diabetes mellitus (PTDM). Nephrol Dial Transplant 2008;23:2033–2042.

13 Rodrigo E, Fernandez-Fresnedo G, Valero R, Ruiz JC, Pinera C, Palomar R, Gonzalez-Cotorruelo J, Gomez-Alamillo C, Arias M: New-onset diabetes after kidney transplantation: risk factors. J Am Soc Nephrol 2006;17:S291–S295.

14 Bamoulid J, Courivaud C, Deschamps M, Mercier P, Ferrand C, Penfornis A, Tiberghien P, Chalopin JM, Saas P, Ducloux D: IL-6 promoter polymorphism –174 is associated with new-onset diabetes after transplantation. J Am Soc Nephrol 2006;17:2333–2340.

15 Ghisdal L, Baron C, Le Meur Y, Lionet A, Halimi JM, Rerolle JP, Glowacki F, Lebranchu Y, Drouet M, Noel C, El Housni H, Cochaux P, Wissing KM, Abramowicz D, Abramowicz M: TCF7L2 polymorphism associates with new-onset diabetes after transplantation. J Am Soc Nephrol 2009;20:2459–2467.

16 Aasebo W, Midtvedt K, Gretland Valderhaug T, Leivestad T, Hartmann A, Varberg Reisaeter A, Jenssen T, Holdaas H: Impaired glucose homeostasis in renal transplant recipients receiving basiliximab. Nephrol Dial Transplant 2010:25:1289–1293.

17 Roland M, Gatault P, Al-Najjar A, Doute C, Barbet C, Chatelet V, Laouad I, Marliere JF, Nivet H, Buchler M, Lebranchu Y, Halimi JM: Early pulse pressure and low-grade proteinuria as independent long-term risk factors for new-onset diabetes mellitus after kidney transplantation. Am J Transplant 2008;8:1719–1728.

18 Bayes B, Granada ML, Pastor MC, Lauzurica R, Salinas I, Sanmarti A, Espinal A, Serra A, Navarro M, Bonal J, Romero R: Obesity, adiponectin and inflammation as predictors of new-onset diabetes mellitus after kidney transplantation. Am J Transplant 2007;7:416–422.

19 Fabrizi F, Martin P, Dixit V, Bunnapradist S, Kanwal F, Dulai G: Post-transplant diabetes mellitus and HCV seropositive status after renal transplantation: meta-analysis of clinical studies. Am J Transplant 2005;5:2433–2440.

20 Bloom RD, Rao V, Weng F, Grossman RA, Cohen D, Mange KC: Association of hepatitis C with posttransplant diabetes in renal transplant patients on tacrolimus. J Am Soc Nephrol 2002;13:1374–1380.

21 Kidney Disease: Improving Global Outcomes (KDIGO): Prevention, diagnosis, evaluation and treatment of hepatitis C in chronic kidney disease. Kidney Int 2008;72(suppl 109):S1–S99.

22 Heisel O, Heisel R, Balshaw R, Keown P: New onset diabetes mellitus in patients receiving calcineurin inhibitors: a systematic review and meta-analysis. Am J Transplant 2004;4:583–595.

23 Vincenti F, Friman S, Scheuermann E, Rostaing L, Jenssen T, Campistol JM, Uchida K, Pescovitz MD, Marchetti P, Tuncer M, Citterio F, Wiecek A, Chadban S, El-Shahawy M, Budde K, Goto N: Results of an international, randomized trial comparing glucose metabolism disorders and outcome with cyclosporine versus tacrolimus. Am J Transplant 2007;7:1506–1514.

24 van Hooff JP, Christiaans MH, van Duijnhoven EM: Tacrolimus and posttransplant diabetes mellitus in renal transplantation. Transplantation 2005;79:1465–1469.

25 Asberg A, Midtvedt K, Voytovich MH, Line PD, Narverud J, Reisaeter AV, Morkrid L, Jenssen T, Hartmann A: Calcineurin inhibitor effects on glucose metabolism and endothelial function following renal transplantation. Clin Transplant 2009;23:511–518.

26 Ghisdal L, Bouchta NB, Broeders N, Crenier L, Hoang AD, Abramowicz D, Wissing KM: Conversion from tacrolimus to cyclosporine A for new-onset diabetes after transplantation: a single-centre experience in renal transplanted patients and review of the literature. Transpl Int 2008;21:146–151.

27 Luan FL, Zhang H, Schaubel DE, Miles CD, Cibrik D, Norman S, Ojo AO: Comparative risk of impaired glucose metabolism associated with cyclosporine versus tacrolimus in the late posttransplant period. Am J Transplant 2008;8:1871–1877.

28 Rostaing L, Cantarovich D, Mourad G, Budde K, Rigotti P, Mariat C, Margreiter R, Capdevilla L, Lang P, Vialtel P, Ortuno-Mirete J, Charpentier B, Legendre C, Sanchez-Plumed J, Oppenheimer F, Kessler M: Corticosteroid-free immunosuppression with tacrolimus, mycophenolate mofetil, and daclizumab induction in renal transplantation. Transplantation 2005;79:807–814.

29 Walczak D, Calvert D, Oberholzer J, Testa G, Sankary H, Benedetti E: Effect of race on the occurrence of posttransplant diabetes in kidney transplant recipients treated with early steroid discontinuation. Transplant Proc 2005;37:819–821.

30 Woodle ES, First MR, Pirsch J, Shihab F, Gaber AO, Van Veldhuisen P: A prospective, randomized, double-blind, placebo-controlled multicenter trial comparing early (7 day) corticosteroid cessation versus long-term, low-dose corticosteroid therapy. Ann Surg 2008;248:564–577.

Nassim Kamar, MD, PhD
Department of Nephrology, Dialysis, and Organ Transplantation
CHU Rangueil, TSA 50032
FR–31059 Toulouse Cedex 9 (France)
Tel. +33 5 61 32 23 35, Fax +33 5 61 32 39 89, E-Mail kamar.n@chu-toulouse.fr

Lai KN, Tang SCW (eds): Diabetes and the Kidney.
Contrib Nephrol. Basel, Karger, 2011, vol 170, pp 256–263

Into the Future: Prevention of Diabetes

Yolanda E. Bogaert · Robert W. Schrier

Department of Medicine, University of Colorado School of Medicine, Denver,
Aurora, Colo., USA

Abstract

With the increasing prevalence of obesity and an aging population, type 2 diabetes mellitus (T2DM) is rapidly becoming the most common chronic disease in the US. In the US alone, over 1.5 million new cases by the Center for Disease Control are diagnosed per year. Concomitant with the rising epidemic in diabetes are the resulting severe complications such as renal and cardiac dysfunction, amputations, impairment in vision and increased mortality. Since T2DM has such an enormous cost in terms of quality of life and mortality as well as economic cost, prevention of T2DM is an important and appropriate target. Interventions for T2DM range from lifestyle modifications such as weight loss and diet to pharmacological agents and surgical intervention such as bariatric surgery. Of the various lifestyle modifications studied, weight loss has consistently been shown to be an effective mechanism of T2DM prevention. Regarding dietary interventions, only the increased consumption of green, leafy vegetables reduces the risk of T2DM in contrast to the popular belief that a diet high in fruits and vegetables reduces this risk. Pharmacological agents investigated in preventing diabetes primarily target the same pathological pathways found in T2DM. Metformin has consistently been shown to reduce the incidence of T2DM and improve impaired glucose tolerance. α_2-Glucosidase inhibitors are also effective in prevention of T2DM, but are limited by their gastrointestinal side effects. Interestingly, although ramipril showed no effect on development of diabetes, valsartan did reduce the incidence of diabetes in patients with impaired glucose tolerance and cardiovascular disease. Although single intervention studies are important in delineating whether an intervention or agent is effective, a multi-faceted approach has been shown to be more effective. Importantly, such an approach is pragmatic and urgently needed to stem the tide of T2DM and its significant human and financial cost.

Rationale for Prevention Strategies Based on Pathophysiology

Type 2 diabetes mellitus (T2DM) is characterized by both insulin resistance (decreased insulin action on cells) as well as β-cell dysfunction (characterized by impaired insulin secretion). The particular manifestations of T2DM within individuals can vary widely as the balance between these two processes also varies. Although insulin resistance manifests earlier in the pathology towards T2DM, it is the presence of β-cell dysfunction which usually distinguishes those in whom glucose intolerance worsens from those in whom it remains stable [1]. Interventions for preventing T2DM, such as drugs, weight loss and physical activity, primarily target these two pathways.

Studies of Lifestyle Modification

It is clear from several randomized control trials that lifestyle modification, weight loss and/or increased activity or exercise has a dramatic impact on T2DM. It is difficult to compare these studies as their designs are so disparate and many combined diet and exercise, but the clear overriding message is that these strategies do prevent T2DM, are cost-effective and have minimal side effects but the primary limitation is continued adherence.

Diet and/or Exercise
The Finnish Diabetes Prevention Study (DPS) is a randomized study of 522 overweight, middle-aged adults with impaired glucose tolerance (IGT). The lifestyle intervention included dietary and exercise components. The intervention group lost an average of 4.2 kg in the first year of the study and had a 58% reduction in diabetes incidence during the 4 study years. Although the study was not designed to analyze specific components, patients who did vigorous leisure-time physical activities had a significantly reduced risk of diabetes [2].

The largest and most comprehensive lifestyle modification study was the US Diabetes Prevention Program (DPP) which randomized 3,234 overweight adults with IGT and fasting glucose values of 5.3 mM or higher to intensive lifestyle intervention, metformin or placebo with standard lifestyle recommendations [3]. The intensive lifestyle intervention was intended to achieve a 7% body weight over 24 weeks. In order to achieve this goal, they performed 140 min of moderate–intense physical activity per week, maintained a low-fat, reduced-calorie diet and received intensive one-to-one counseling. The DPP lifestyle intervention was associated with a 58% reduction in the incidence of diabetes, compared with placebo plus standard lifestyle recommendations. Participants who achieved exercise but not weight loss goals still experienced reduction in diabetes risk (44% compared with placebo). Interestingly, the effectiveness of the DPP lifestyle intervention was greatest in older participants (age 60–85 years).

Studies in other countries have confirmed the effectiveness of lifestyle modification programs. Even with minimal weight loss (0–2 kg), risk reductions of 67% in a study from Japan [4] and 28% in a study from India [5] were achieved with the use of those programs, supporting the effectiveness of lifestyle modification alone across nationalities and ethnicities.

Regarding the effectiveness of diet alone, a recent meta-analysis found with summary estimates that a greater intake of green leafy vegetables was associated with a 14% (CI 0.77–0.97) reduction in risk of T2DM (p = 0.01). However, if patients increased the consumption of all vegetables (not just leafy and green), fruit or fruit and vegetables combined there was no significant benefit in reducing the risk of T2DM [6].

Weight Loss

As already noted, the DPP study was designed to achieve a 7% body weight loss over 24 weeks in the intensive lifestyle intervention group. The lifestyle intervention group did have a mean weight loss of 7% within the first year and overall mean weight loss of 5.6% during follow-up with a 58% reduction in the incidence of diabetes, compared to placebo. Of note, weight loss was the predominant predictor of reduced diabetes incidence, with a 16% reduction in risk per kg of weight loss [3]. It is difficult to separate out diet and exercise from weight loss, especially with regard to importance for diabetes prevention, but a multivariate analysis of DPS determined that only weight reduction remained a significant variable in diabetes prevention.

Surgical Modification

Bariatric Surgery

The Swedish Obese Subjects study was a prospective trial of more than 2,000 individuals receiving multiple bariatric procedures (most common, vertical banded gastroplasty) compared to standard of care [7]. The odds ratio for diabetes in the surgically treated group was 0.14 at 2 years' and 0.25 at 10 years' follow-up. The incidence of diabetes was related to the amount of weight lost. Patients with greater than 12% loss of their initial body weight did *not* develop diabetes compared to a 9% incidence in the nonsurgical arm who gained weight. A more recent study utilizing laparoscopic gastric banding in patients with established diabetes resulted in remission of diabetes in 64% and major improvements in glycemic control in another 26% [8]. Biliopancreatic diversion has also been reported to normalize glucose tolerance in obese patients with either IGT or established diabetes [9].

Surgical Removal of Fat

Liposuction has been shown to be ineffective in preventing diabetes. This was felt to be the case as only peripheral/subcutaneous fat was removed. Therefore,

Fabbrini et al. [10] evaluated whether surgical removal of omental (visceral) fat would improve insulin sensitivity and cardiovascular risk factors in obese adults. They performed omentectomy in 22 obese subjects following post-Roux-en-y gastric bypass surgery and in 7 obese subjects with T2DM to see if insulin sensitivity improved. Roux-en-y gastric bypass improved insulin sensitivity significantly but the addition of omentectomy had no effect. Likewise, omentectomy alone did not improve insulin sensitivity or minimize use of diabetic medications in the obese subjects with T2DM.

Studies of Pharmacological Interventions

Metformin

The DPP study is the largest randomized control trial to date evaluating the effectiveness of metformin in prevention of diabetes. They enrolled high-risk individuals with IGT and randomized them to metformin (850 mg b.i.d.; n = 1,073) or placebo (n = 1,082), with metformin reducing the risk of diabetes by 31% compared to placebo. Metformin appeared most effective in patients with a BMI >35 (50% reduction), but did not have beneficial effects in participants aged 60–85 years [11].

α_2-Glucosidase Inhibitors

Acarbose is also effective in preventing diabetes. STOP-NIDDM (Study to Prevent Non-Insulin-Dependent Diabetes Mellitus) [12] showed that acarbose treatment was associated with a 25% reduction in diabetes incidence in high-risk individuals with IGT (n = 1,429). However, there was an approximate 25% dropout rate in the acarbose arm secondary to gastrointestinal side effects, which obviously limits this medication in practice.

The efficacy of voglibose, another α_2-glucosidase inhibitor, in reducing incidence of T2DM was evaluated in a multicenter, double-blind parallel group trial in high-risk Japanese patients with IGT on a standard diet and taking regular exercise. Voglibose 0.2 mg three times a day (n = 897) versus placebo (n = 883) had a 50% lower risk of progression to T2DM (p = 0.0014) [13]. Voglibose has been studied extensively in Europe and Japan, but is currently not FDA approved in the US for treatment of T2DM.

Weight Loss Agents

Various weight loss agents have been tried in small studies with mixed results. Orlistat, an intestinal lipase inhibitor, decreased the incidence of diabetes significantly (40–52%) in obese patients with IGT. However, the mechanism of this drug was confounded by its effect on weight loss, which was greater than with lifestyle change alone. In addition, the dropout rate approached 50% from side effects.

Thiazolidinediones
Thiazolidinediones have shown to be effective in reducing the incidence of diabetes in individuals with IGT, impaired fasting glucose or both [14]. The CANOE study randomized 207 patients with IGT combination rosiglitazone (2 mg) and metformin (500 mg) twice daily or matching placebo for a median of 3.9 years. Incident diabetes occurred in significantly fewer individuals in the active treatment group (n = 14, 14%) than in the placebo group (n = 41, 39%; p < 0.0001) [15]. However, given the concern over higher cardiovascular incidence and mortality, these drugs should not be recommended for prevention of diabetes, especially as the population of patients in which these drugs would be used generally has a higher risk of cardiovascular disease (CVD).

Inhibitors of the Renin-Angiotensin-Aldosterone System
Rationale for use of renin-angiotensin-aldosterone system (RAAS) inhibition for prevention of diabetes comes from studies demonstrating increased RAAS activity in obese individuals and in individuals with prediabetes. The effect of ramipril (up to 15 mg per day) versus placebo in development of diabetes was evaluated in a large, double-blinded, placebo-controlled, randomized trial in patients with impaired fasting glucose levels or IGT. Ramipril showed no effect on development of diabetes or median fasting plasma glucose level, but did cause reduction in plasma glucose levels after an oral glucose load [16].

In another study, patients with IGT and either CVD or risk factors for CVD were randomized to either the angiotensin receptor blocker valsartan or placebo, together with lifestyle changes for 5 years, and valsartan reduced the incidence of diabetes by 14% [17].

Liraglutide
Liraglutide is a glucagon-like peptide-1 receptor agonist which is FDA approved for the treatment of T2DM. Liraglutide was shown in a double-blind, placebo-controlled 20-week trial, with open-label orlistat comparator in 564 obese individuals to significantly reduce weight and improve the incidence of prediabetes (84–96% reduction in the incidence of type 2 diabetes with use of 1.8–3.0 mg/day). Of note, participants on liraglutide lost significantly more weight than those on orlistat (p = 0.003 for liraglutide 2.4 mg and p < 0.0001 for liraglutide 3.0 mg) [18]. Whether liraglutide will be effective in actually preventing diabetes needs further study.

Does Pharmacological Treatment Truly Reduce or Merely Mask the Onset of Diabetes?

One caveat to whether pharmacological intervention truly prevents diabetes is that these positive results may reflect masking true incidence of diabetes as new diabetes in many studies was defined by fasting plasma glucose. This issue was clearly recognized in the DPP study in which a formal 1- to 2-week washout period was employed following completion of the trial. When an oral glucose tolerance test was performed after washout, diabetes was more frequently diagnosed in the metformin group than in the placebo recipients, although this was not significant (HR 1.49; 95% CI 0.93–2.38). Even with inclusion of these patients, metformin reduced the incidence of diabetes by 25%.

Multifaceted Approach

A prevention study using a multifaceted and aggressive treatment approach for the metabolic syndrome and/or prediabetes would make it difficult to determine which intervention(s) are most important; however, such a multifaceted approach has been shown to be more effective in reducing mortality in patients with diabetes than any single intervention [19].

Conclusion

The rising prevalence and incidence of diabetes [20] and its high risk of complications mandates that strategies for prevention of T2DM are aggressively pursued. Major issues in prevention strategies include long-term patient compliance, economic burden of implementation, risk/benefit ratio of pharmacological intervention as well as efficacy of prevention of T2DM. The clear message is that intensive lifestyle intervention is a cost-effective, low-risk and effective method of reducing the incidence of diabetes in high-risk individuals, but the primary limitations are long-term adherence and feasibility of implementation in the general population.

References

1 Kahn SE: The relative contributions of insulin resistance and beta-cell dysfunction to the pathophysiology of type 2 diabetes. Diabetologia 2003;46:3–19.

2 Tuomilehto J, Lindström J, Eriksson JG, Valle TT, Hämäläinen H, Ilanne-Parikka P, Keinänen-Kiukaanniemi S, Laakso M, Louheranta A, Rastas M, Salminen V, Uusitupa M, Finnish Diabetes Prevention Study Group: Prevention of type 2 diabetes mellitus by changes in lifestyle among subjects with impaired glucose tolerance. N Engl J Med 2001;344:1343–1350.

3   Hamman RF, Wing RR, Edelstein SL,
    Lachin JM, Bray GA, Delahanty L, Hoskin
    M, Kriska AM, Mayer-Davis EJ, Pi-Sunyer
    X, Regensteiner J, Venditti B, Wylie-Rosett
    J: Effect of weight loss with lifestyle inter-
    vention on risk of diabetes. Diabetes Care
    2006; 29:2102–2107.
4   Kosaka K, Noda M, Kuzuya T: Prevention
    of type 2 diabetes by lifestyle intervention:
    a Japanese trial in IGT males. Diabetes Res
    Clin Pract 2005;67:152–162.
5   Ramachandran A, Snehalatha C, Mary S,
    Mukesh B, Bhaskar AD, Vijay V, Indian
    Diabetes Prevention Programme (IDPP):
    The Indian Diabetes Prevention Programme
    shows that lifestyle modification and met-
    formin prevent type 2 diabetes in Asian
    Indian subjects with impaired glucose
    tolerance (IDPP-1). Diabetologia 2006;49:
    289–297.
6   Carter P, Gray LJ, Troughton J, Khunti K,
    Davies MJ: Fruit and vegetable intake and
    incidence of type 2 diabetes mellitus: sys-
    tematic review and meta-analysis. BMJ 2010;
    341:c4229.
7   Sjöström L, Lindroos AK, Peltonen M,
    Torgerson J, Bouchard C, Carlsson B,
    Dahlgren S, Larsson B, Narbro K, Sjöström
    CD, Sullivan M, Wedel H, Swedish Obese
    Subjects Study Scientific Group: Lifestyle,
    diabetes, and cardiovascular risk factors 10
    years after bariatric surgery. N Engl J Med
    2004;351:2683–2693.
8   Dixon JB, Dixon AF, O'Brien PE:
    Improvements in insulin sensitivity and
    beta-cell function (HOMA) with weight
    loss in the severely obese. Homeostatic
    model assessment. Diabet Med
    2003;20:127–134.
9   Mari A, Manco M, Guidone C, Nanni G,
    Castagneto M, Mingrone G, Ferrannini E:
    Restoration of normal glucose tolerance in
    severely obese patients after bilio-pancreatic
    diversion: role of insulin sensitivity and
    beta cell function. Diabetologia 2006;49:
    2136–2143.

10 Fabbrini E, Tamboli RA, Magkos F,
    Marks-Shulman PA, Eckhauser AW,
    Richards WO, Klein S, Abumrad NN:
    Surgical removal of omental fat does
    not improve insulin sensitivity and car-
    diovascular risk factors in obese adults.
    Gastroenterology 2010;139: 448–455.
11 Knowler WC, Barrett-Connor E, Fowler
    SE, Hamman RF, Lachin JM, Walker EA,
    Nathan DM: Diabetes Prevention Program
    Research Group. Reduction in the incidence
    of type 2 diabetes with lifestyle interven-
    tion or metformin. N Engl J Med 2002;346:
    393–403.
12 Chiasson JL, Josse RG, Gomis R, Hanefeld
    M, Karasik A, Laakso M, STOP-NIDDM
    Trial Research Group: Acarbose for preven-
    tion of type 2 diabetes mellitus: the STOP-
    NIDDM randomised trial. Lancet 2002;359:
    2072–2077.
13 Kawamori R, Tajima N, Iwamoto Y,
    Kashiwagi A, Shimamoto K, Kaku K,
    Voglibose Ph-3 Study Group: Voglibose for
    prevention of type 2 diabetes mellitus: a
    randomized, double-blind trial in Japanese
    individuals with impaired glucose tolerance.
    Lancet 2009;373:1607–1614.
14 DREAM (Diabetes Reduction Assessment
    with ramipril and rosiglitazone Medication)
    Trial Investigators, Gerstein HC, Yusuf
    S, Bosch J, Pogue J, Sheridan P, Dinccag
    N, Hanefeld M, Hoogwerf B, Laakso M,
    Mohan V, Shaw J, Zinman B, Holman RR:
    Effect of rosiglitazone on the frequency of
    diabetes in patients with impaired glucose
    tolerance or impaired fasting glucose: a ran-
    domized controlled trial. Lancet 2006;368:
    1096–1105.
15 Zinman B, Harris SB, Neuman J, Gerstein
    HC, Retnakaran RR, Raboud J, Qi Y, Hanley
    AJ: Low-dose combination therapy with
    rosiglitazone and metformin to prevent type
    2 diabetes mellitus (CANOE trial): a double-
    blind randomized controlled study. Lancet
    2010;376:103–111.
16 DREAM Trial Investigators, Bosch J, Yusuf S,
    Gerstein HC, Pogue J, Sheridan P, Dagenais
    G, Diaz R, Avezum A, Lanas F, Probstfield
    J, Fodor G, Holman RR: Effect of ramipril
    on the incidence of diabetes. N Engl J Med
    2006;355:1551–1562.

Bogaert · Schrier

17 NAVIGATOR Study Group, McMurray JJ, Holman RR, Haffner SM, Bethel MA, Holzhauer B, Hua TA, Belenkov Y, Boolell M, Buse JB, Buckley BM, Chacra AR, Chiang FT, Charbonnel B, Chow CC, Davies MJ, Deedwania P, Diem P, Einhorn D, Fonseca V, Fulcher GR, Gaciong Z, Gaztambide S, Giles T, Horton E, Ilkova H, Jenssen T, Kahn SE, Krum H, Laakso M, Leiter LA, Levitt NS, Mareev V, Martinez F, Masson C, Mazzone T, Meaney E, Nesto R, Pan C, Prager R, Raptis SA, Rutten GE, Sandstroem H, Schaper F, Scheen A, Schmitz O, Sinay I, Soska V, Stender S, Tamás G, Tognoni G, Tuomilehto J, Villamil AS, Vozár J, Califf RM: Effect of valsartan on the incidence of diabetes and cardiovascular events. N Engl J Med 2010; 362:1477–1490.

18 Astrup A, Rössner S, Van Gaal L, Rissanen A, Niskanen L, Al Hakim M, Madsen J, Rasmussen MF, Lean ME, NN8022–1807 Study Group: Effects of liraglutide in the treatment of obesity: a randomised, double-blind, placebo-controlled study. Lancet 2009; 374:1606–1616.

19 Gaede P, Vedel P, Larsen N, Jensen GV, Parving HH, Pedersen O: Multifactorial intervention and cardiovascular disease in patients with type 2 diabetes. N Engl J Med 2003;348:383–393.

20 Shaw JE, Sicree RA, Zimmet PZ: Global estimates of the prevalence of diabetes for 2010 and 2030. Diabetes Res Clin Pract 2010;87: 4–14.

Robert W. Schrier, MD
Department of Medicine, University of Colorado School of Medicine
Denver, C281
Aurora, CO 80045 (USA)
E-Mail Robert.Schrier@ucdenver.edu

Author Index

Subject Index